Bionanotechnology Towards Green Energy

Bionanotechnology Towards Green Energy explains the role of bionanotechnology in the next generation technologies of green energy from an interdisciplinary and sustainability perspective. Chapters cover different roles of bionanotechnology such as applications of bio-nano enabled materials/coatings, scaling-up of green energy production, design and synthesis of bio-inspired nanomaterials and their applications, bio-nanofluid-based photovoltaic thermal systems, the use of bio-templated and biomimetic materials, and so forth. It focuses on waste-to-energy conversion and fixing intricate environmental issues.

Key features:

- Provides detailed coverage of green energy production through bionano-technological intervention
- Reviews future research needs in bionanotechnology in the green energy sector and scientific challenges in the mitigation of energy crises
- Deals with cutting-edge research on microbial synergism in biohydrogen production and storage
- Discusses the fabrication of bio-nano/hybrid electrode materials for super-capacitors and energy storage devices
- Includes extensive illustrations, case studies, summary tables, and up-to-date references

This book is aimed at researchers and professionals in bionanotechnology, energy sciences, and environmental engineering.

Advances in Bionanotechnology

Series Editors: Ravindra Pratap Singh, *Department of Biotechnology, Indira Gandhi National Tribal University, Anuppur, Madhya Pradesh, India,* **Jay Singh**, *Department of Chemistry, Institute of Science, Banaras Hindu University, Varanasi, Uttar Pradesh, India* and **Charles Oluwaseun Adetunji**, *Department of Microbiology, Edo State University Uzairue, Iyamho, Edo State, Nigeria*

Bionanotechnology is a multi-disciplinary field that shows immense applicability in different domains, namely chemistry, physics, material sciences, biomedical, agriculture, environment, robotics, aeronautics, energy, electronics and so forth. This book series will explore the enormous utility of bionanotechnology for biomedical, agricultural, environmental, food technology, space industry, and many other fields. It aims to highlight all the spheres of bionanotechnological applications and its safety and regulations for using biogenic nanomaterials that are a key focus of the researchers globally.

Bionanotechnology Towards Sustainable Management of Environmental Pollution
Edited by Naveen Dwivedi and Shubha Dwivedi

Natural Products and Nano-formulations in Cancer Chemoprevention
Edited by Shiv Kumar Dubey

Bionanotechnology towards Green Energy
Innovative and Sustainable Approach
Edited by Shubha Dwivedi and Naveen Dwivedi

For more information about this series, please visit: www.routledge.com/Advances-in-Bionanotechnology/book-series/CRCBIONAN

Bionanotechnology Towards Green Energy

Innovative and Sustainable Approach

Edited by
Shubha Dwivedi
and Naveen Dwivedi

CRC Press
Taylor & Francis Group
Boca Raton London New York

CRC Press is an imprint of the
Taylor & Francis Group, an **informa** business

First edition published 2023

by CRC Press
6000 Broken Sound Parkway NW, Suite 300, Boca Raton, FL 33487-2742

and by CRC Press
4 Park Square, Milton Park, Abingdon, Oxon, OX14 4RN

CRC Press is an imprint of Taylor & Francis Group, LLC

ISBN: 9781032327167 (hbk)
ISBN: 9781032327181 (pbk)
ISBN: 9781003316374 (ebk)

DOI: 10.1201/9781003316374

Typeset in Times
by Deanta Global Publishing Services, Chennai, India

Contents

Preface

As we know, high population density and advanced lifestyles increase the energy demand worldwide and lead to the degradation of natural resources and the environment. However, waste generation and energy crises are major challenges in most developing countries. Therefore, there is an exigent demand for the development of sustainable, eco-friendly, cost-effective, and reliable technology for energy production and to meet all environmental challenges. This will only be possible with green technological intervention in the field of science, technology, and public policy innovations; in addition, the world must make fundamental transformations in how energy is used and produced over the coming years.

Over time, we have also discovered the importance of being efficient in our use of energy, reducing our environmental impact, and enhancing our energy security. The major energy challenges are still how to create innovations in energy technologies, regulations, and policies and furthermore how to expand energy sources in ways that are clean, green, reliable, affordable, and sustainable. Waste-to-energy conversion has also become a promising alternative for many countries as an effective and sustainable waste management solution and as an advanced method of waste disposal.

The realization of green energy with zero carbon footprint is a great stride to a more environmentally friendly future and providing significant economic and social benefits. Sustainable green energy is a vital input for the social and economic development of any nation. Now green energy looks set to be part of the future of the world, offering a sustainable, cleaner alternative than many of today's energy sources. Hence, bionanotechnology can play a substantial role in developing clean, green, affordable, and sustainable energy with significant health and environmental benefits. Bionanotechnology takes advantage of the unique properties of biological materials such as plants and their leaves, bacteria, fungi, algae, peptides, and proteins by using their self-assembling nature for the nano-engineering of molecular templates and supra-molecular structures. Bionanotechnology has the ability to enhance the efficiency of energy generation and its consumption in our buildings, transportation systems, industry, and homes. It can serve all these purposes so that, globally, we can reach these societal goals of affordable energy availability, security, and sustainability.

The purpose and aims of this book are to provide a state-of-the-art understanding and application of bionanotechnology in the expansion of next generation technologies of green energy covering current issues from an interdisciplinary and sustainability approach. In this book, an interdisciplinary team of researchers has taken a sustainable approach to bionanotechnology in the field of green energy with the advancement and highlighting of the technical, scientific, regulatory, safety, and societal impacts. The book provides a solid understanding of the subject and current challenges in energy production using bionanotechnology-based interventions. Readers will learn all about the recent and sustainable progress in both theoretical and practical aspects and future potential applications of bionanotechnology in the production of green energy.

We are thankful to the publishing (engineering) team, CRC Press, and all our contributors whose great efforts have made this book a success.

Shubba Dwivedi, Meerut, Uttar Pradesh, India
Naveen Dwivedi, Meerut, Uttar Pradesh, India

Editors

Shubha Dwivedi, a researcher by profession, is Associate Professor and Head of the department in the Department of Biotechnology, School of Life Science and Technology, IIMT University, Meerut, India. She received her MSc degree in Biochemistry from Jiwaji University, Gwalior, and an MBA degree from Punjab Technical University. Dr Dwivedi completed her MSc dissertation work at the Institute of Genomics and Integrative Biology (CSIR-IGIB), Delhi, and PhD research work at the Sustainable Processing and Water Treatment Research Lab, Indian Institute of Technology Roorkee. She has more than 16 years of experience in teaching and research work in the areas of biochemical engineering and environmental biotechnology. Her area of work has been eco-friendly technology development for the removal of hazardous contaminants from wastewater. She has published more than 47 research papers in international journals and conference proceedings. Moreover, she has published more than a dozen book chapters with CRC Press, AAP, Wiley, Elsevier, and Springer, etc. Dr Dwivedi has published one book titled *Introduction to Biotechnology* with University Science Press, New Delhi. She is also engaged in three international book projects with Wiley and CRC Press. Recently, she was appointed session lead at an IPR workshop organized in RAIB-2021 at Precious Cornerstone University, Nigeria. She has chaired sessions at many national and international conferences. She has been invited as a keynote speaker to various institutes of national and international repute. Presently she is engaged in many international projects with various universities such as research exchange programmes, organizing workshops and conferences, etc.

Naveen Dwivedi is currently working as Professor and Head in the Department of Biotechnology at the S.D. College of Engineering and Technology, Muzaffarnagar, Uttar Pradesh, India. He obtained his MTech degree in Biotechnology from the Institute of Engineering and Technology, Lucknow. Dr Dwivedi completed his MTech dissertation work at the Fermentation Division of the Central Drug Research Institute (CSIR-CDRI), Lucknow, and PhD research work at the Bioenergy and Environmental Engineering Research Lab, Indian Institute of Technology Roorkee. Dr Dwivedi is broadly interested in the fields of bioenergy and environmental biotechnology, nanobiotechnology, the biological remediation of wastewater, bioprocess engineering, and fermentation biotechnology. Dr Dwivedi has more than 18 years of teaching experience. He has about 50 research publications in peer reviewed journals and conferences. He has also contributed 20 chapters to books published by Elsevier, CRC Press, AAP, Springer, and Wiley. He is an author of the book *Introduction to Biotechnology* published by University Science Press, New Delhi. Presently he is also engaged in editing various book projects of international repute with Wiley, Elsevier, and CRC Press. Dr Dwivedi has received many research awards and project grants. Dr Dwivedi has delivered several invited talks and guest lectures on various topics in biotechnology, nanobiotechnology, and bioenergy in different places in India and abroad. He is the member of the editorial board and reviewer of several international journals. He is a senior member of the Universal Association of Civil, Structural and Environmental Engineers (IRED), New York, USA, and the Society of Chemical Industry (SCI).

Contributors

Afan Ahmed Department of Biotechnology, Sharda University, Greater Noida, Uttar Pradesh 201310, India

Akshay Raj Research and Development Cell, IIMT University, Meerut, India

Ashootosh Mandpe Department of Civil Engineering, Indian Institute of Technology Indore, Indore 453552, India

Bindu Naik Department of Biotechnology, Graphic Era Deemed to Be University, Uttarakhand 248002, India

Comfort Okoji Department of Plant Biology and Biotechnology, University of Benin, Benin City, Nigeria

Deepa Sharma Department of Chemistry, IIMT University, Meerut, Uttar Pradesh, India

Dinesh K. Sharma Department of Electrical and Electronics Engineering, School of Engineering and Technology, IIMT University Meerut 250002, India

Dioha I.J. Department of Biology and Forensic Science, Admiralty University of Nigeria, Delta State, Nigeria, and Center for Renewable Energy, Admiralty University of Nigeria, Delta State, Nigeria

Disha Tandulkar CSIR-National Environmental Engineering Research Institute (CSIR-NEERI), Nehru Marg, Nagpur 440020, India

Neha Saxena School of Basic Sciences and Technology, IIMT University, Meerut, India

Harrison Ogala Department of Plant Biology and Biotechnology, University of Benin, Benin City, Nigeria

Isitua C.C. Department of Biology and Forensic Science, Admiralty University of Nigeria, Delta State, Nigeria

Jayani J. Wewalwela Department of Agricultural Technology, University of Colombo, Pitipana, CO, 10206, Sri Lanka

Manukonda Suresh Kumar CSIR-National Environmental Engineering Research Institute (CSIR-NEERI), Nehru Marg, Nagpur 440 020, India

Md. Merajul Islam Department of Chemistry, School of Basic Sciences and Technology, IIMT University, Meerut, Uttar Pradesh, India

Musa S.I. Department of Biology and Forensic Science, Admiralty University of Nigeria, Delta State, Nigeria

Nathan Moses Department of Biology and Forensic Science, Admiralty University of Nigeria, Delta State, Nigeria

Nathan Moses Department of Plant Biology and Biotechnology, University of Benin, Benin City, Nigeria

Nathaniel Iboyi Department of Plant Biology and Biotechnology, University of Benin, Benin City, Nigeria

Naveen Dwivedi	Department of Biotechnology, S.D. College of Engineering and Technology, Muzaffarnagar, Uttar Pradesh, India
Pallavi Singh	Department of Biotechnology, Graphic Era Deemed to Be University, Uttarakhand 248002, India
Priya Singh	Department of Biotechnology, S.D. College of Engineering and Technology, Muzaffarnagar, Uttar Pradesh, India
Rachita Sharma	Department of Biotechnology, S.D. College of Engineering and Technology, Muzaffarnagar, Uttar Pradesh, India
Ranjeet Kumar Mishra	Department of Chemical Engineering, MS Ramaiah Institute of Technology, 560054, Bangalore, Karnataka, India
Rashi Srivastava	Chemical and Biochemical Engineering, Indian Institute of Technology, Patna, India
Sachin Sharma	Department of Electrical and Electronics Engineering, School of Engineering and Technology, IIMT University Meerut 250002, India
Salman Khan	School of Life Science and Technology, IIMT University, Meerut, Uttar Pradesh, 250001, India
Sanjeev Maheshwari	Department of Electronics Engineering, IIMT Engineering College, Meerut 250001, Uttar Pradesh, India
Saurabh Kumar Jha	Department of Biotechnology, Sharda University, Greater Noida, Uttar Pradesh 201310, India
Sheila Ojei	Department of Plant Biology and Biotechnology, University of Benin, Benin City, Nigeria
Shubha Dwivedi	Department of Biotechnology, IIMT University, Meerut, India
Sneha Ullhas Naik	Department of Chemical Engineering, MS Ramaiah Institute of Technology, 560054, Bangalore, Karnataka, India
Sonam Paliya	Department of Biological sciences and Bioengineering, Indian Institute of Technology Indore, Indore 453 552, India
Sunil Kumar	CSIR-National Environmental Engineering Research Institute (CSIR-NEERI), Nehru Marg, Nagpur 440020, India
Syeda Minnat Chistie	Department of Chemical Engineering, MS Ramaiah Institute of Technology, 560054, Bangalore, Karnataka, India
Tirtharaj Datta	Department of Biotechnology, Sharda University, Greater Noida, Uttar Pradesh 201310, India
Vaibhav Sharma	Research and Development Cell, IIMT University, Meerut, India
Vipin K. Sharma	Department of Mechanical Engineering, School of Engineering and Technology, IIMT University, Meerut250002, India
Y.M.S.M. Yapa	Department of Agricultural Technology, University of Colombo, Pitipana, CO, 10206, Sri Lanka

1 Recent Advances of Bio-Nanotechnology Potential in the Scaling Up of Sustainable Green Energy Production

Sneha Ullhas Naik, Syeda Minnat Chistie, and Ranjeet Kumar Mishra

CONTENTS

DOI: 10.1201/9781003316374-1

1

1.1   INTRODUCTION

Biomass is now being taken into account for the production of fuels, energy, and chemically synthesised products. Moreover, biofuel has benefits over fossil fuel in terms of emissions. Currently, fossil fuel is the prime source of combustible energy for the whole world (Mishra, 2022). The consumption of the latter releases excessive greenhouse gases (GHG) and suspended materials into the atmosphere, which has innumerable adverse environmental effects. Also, the rapid diminution of non-renewable resources has compelled investigators to advance the latest technologies and strategies. Renewable sources have become the premier alternative to fossil fuels. In 2007, the United States Congress laid down an updated renewable fuel standard (RFS2) where biofuel was used as a substitute to diminish the utilization of petroleum fuel in domestic consumption (Brown, 2015). The plant cell wall mainly consists of hemicellulose, cellulose, and lignin, where pentose and hexose sugar monomers are derived from cellulose and hemicellulose, and polyphenol aromatics from lignin (Jonsson & Martín, 2016). The chemical structure and functionality of lignocellulosic biomass (LB) are as follows: the structure of the cell walls is composed of cellulose, whereas hemicellulose contributes to the interlinkage of non-cellulosic and cellulosic polymeric structures by sharing electron pairs between atoms (Figure 1.1) (Sankaran et al., 2021b). The cellulose-containing polymers are encased in lignin that resembles phenol-formaldehyde resin, which acts as an adhesive to grasp the lignocellulosic structure (LBS) and provides extra resilience to the cell walls for protection against pests and moisture (Sankaran et al., 2021a). They also act as a hydrophobic fence that supports the transference of water and numerous indispensable nutrients through the cell wall.

Presently, substantial research is focused on the LB, especially the conversion of LB into bioenergy that can replace fossil fuel sources without damaging the environment. In a systematic effort to progressively move towards a sustainable economy, several inexhaustible energy sources have been discovered to date (Sankaran et al., 2021a). Thus, energy or fuel extracted from biomass may be a substitute for petroleum-rich fuel for transportation or yield the most refined chemicals. Generally, LB has around 75% carbohydrates due to cellulose and hemicellulose, making it more attractive for energy generation (Sanusi et al., 2021). Moreover, 5-hydroxymethylfurfural (HMF), furfural, and levulinic acid found in LB during bio-refining are encouraging the growth of energy fuels by biochemical amalgamation (Elumalai et al., 2018). The LB is predominantly obtained from farming and forestry sources, having a projected yearly production of $1.7–2.0 \times 10^{11}$ tonnes (Elumalai et al., 2018). In 2011, in the European Union (EU), ~38 million tonnes (MT) of LB were consumed for obtaining biofuel, out of which several crops contributed

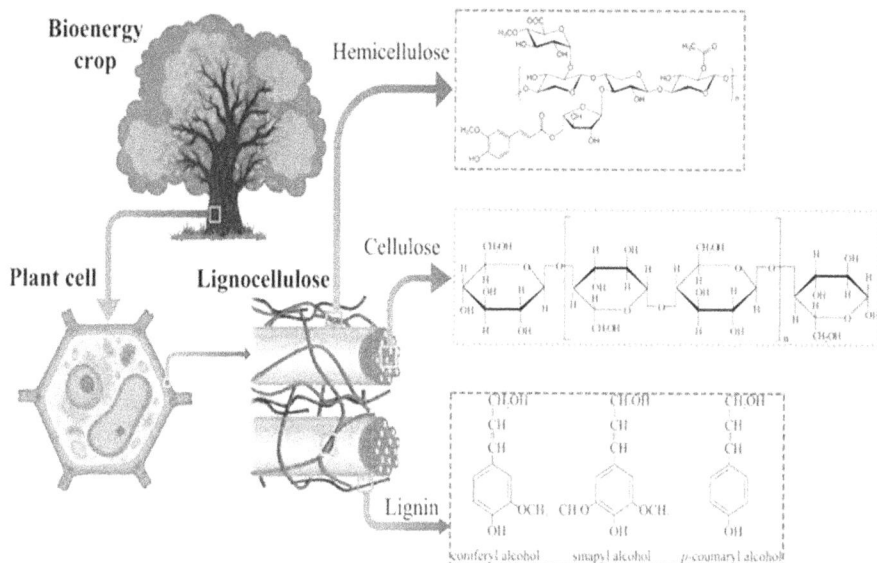

FIGURE 1.1 Schematic presentation of bioenergy crop (lignocellulosic biomass). Adapted from Sankaran et al. (2021b).

an estimate of 1.2 billion tonnes of LB (Scarlat et al., 2015). LB is unmanageable despite these compensations because of the transparent nature of inert lignin. In addition, pretreatment methods are essential to make cellulose responsive to facilitate the cleavage of bonds in molecules with the addition of the elements of water that successively eliminate agitated sugars from the biofuel production. Furthermore, in order to achieve the increased value and yield of bioenergy from LB, there are major needs to overcome technological constraints such greater cost and inadequacy in the existing framework. Keeping in mind these limitations, the present book chapter describes several optimization approaches to pre-treatment methods as well as enzymatic and fermenting techniques to improve the generation of bioenergy in a fuel-efficient and economical way. Recently, the newly discovered fascination in NMs and their unique properties has been extensively studied to increase the biofuel production.

As per the International Standardization Organization (ISO), there is no single definition available that can be accepted internationally. The ISO initiated the first approaches in this direction. However, the interesting field of nanotechnologies does not reveal itself through appropriate definitions but rather through the illustration of fundamental concepts and research approaches that play a decisive role in this connection. In NTs, engineering with elementary units (such as biological and organic nature) is employed as if it was a Lego kit. On the other hand, even structures assessing only one-thousandth of the diameter of one nano-hair can be formed through size reduction. The nanotechnology process is not new but currently demonstrates the development of technologies and analysis techniques. The use of NTs in engineering applications has a positive impact on the modern available analysing techniques or establishing processes. The wide application of nanotechnologies is not limited to nanoscale structures but also adds a good understanding of the principles effective at the molecular level and the technological development of materials and constituents.

1.2 BASICS OF NANOMATERIALS AND THEIR CONTRIBUTION TO BIOENERGY GENERATION

As per the ISO, any materials that have internal, external, or surface structures ranging from 1 to 100 nm are recognized as nanomaterials (Nanotechnologies—Terminology, 2008). Nanoparticles (NPs) are categorized into the following groups: carbonaceous, natural, inert, and compounds. The properties of the latter are subjected to chemical analysis, which enhances their performance in varied conditions. The physicochemical features of NPs vary substantially from their macro equivalents and find application in bioenergy, biochar, applied science, pharmaceuticals, ionized solvents, and retail goods (He et al., 2020; Sankaran et al., 2021b; Shuttleworth et al., 2014). NPs have a more rapid rate of reaction than bulkier substances which is crucial in bioenergy production techniques. The rapid rate of reaction interlinked with minute NPs delivers a large surface area to volume ratio (S/V) that directly increases the quantity of functional spots for several reactions or mechanisms to happen (Misson et al., 2015b). In the context of bioenergy generation, nanomaterials (NPs) are anticipated in various properties for meeting newly expanding worldwide fuel demands because of their extraordinary characteristics: larger superficial area, high stimulating movement, translucent, tensile, stiffness, adsorption proficiency, and effective storage (Hamawand et al., 2020). All the above-mentioned benefits boost the production, decomposition, and firmness of cellulose enzymes for many bioenergy products. Srivastava et al. (2014) testified that an effective cellulose production method is crucial for enzymatic disintegration; the consumption of NPs such as Fe_3O_4 and Fe_3O_4/alginate displayed higher hydrolysis efficiency and improved the yield of cellulose by 35 and 40%, respectively (Srivastava et al., 2014). Further, the same groups of researchers explored the efficacy of nickel cobaltite ($NiCO_2O_4$) NPs and showed that cellulose production was enhanced by around 40% with elevated thermal equilibrium at 80°C for 7 h (Srivastava et al., 2014). The acidity and thermal stability of crude cellulose were also explored using zinc oxide NPs and revealed an enhancement in thermal stability at 65°C for 10 h with the maximum acidity (pH) of 10.5. All of these are mostly used in the production of bioethanol and bio-hydrogen (Srivastava et al., 2016a). Dutta et al. (2014) studied the calcium hydroxyapatite-based NPs to boost the thermic resistance and heighten the manufacture of D-xylose up to 35% (Dutta et al., 2014). According to the author, the introduction of xylanase and cellulase activated by NPs are now popular due to their improved performance at higher temperatures up to 80°C. Additionally, they stated that using cofactor calcium in NPs increased the thermostability and functionality of the cellulase and xylanase enzymes (Dutta et al., 2014).

The other domain of NPs, such as magnetic nanoparticles (MNPs), were found to be very effective compared to other NPs as their magnetic properties permit smooth reclaimability and possess the ability to combine the many specified products, outstanding ecological, less alkylating to biomass cells, as well as comfort of blending (Pena et al., 2012). The surface area to volume ratio (S/V) of MNPs is expended as transporters for the binding of enzymes, which tends to improve the firmness in biofuel production. Xie and Ma (2009) stated that binding the support with magnetic Fe_3O_4 NPs exhibited an outstanding pH endurance and thermal stability with transformation to biodiesel up to 90%, whereas the bounded lipase can be reused four times without affecting its operation (Xie & Ma, 2009). MNPs such as TiO_2 were employed for the binding of enzymes for the decomposition of lignocellulosic biomass for obtaining bioethanol (Ahmad & Sardar, 2014). Further, other MNPs like a composite of TiO_2-ZnO and MnO_2 catalyst are widely employed for the production of biodiesel (Madhuvilakku & Piraman, 2013). The consumption of NPs was demonstrated to have a hydrolyzable impact that matches that detected with chemosynthesized pre-treatment in LB processing. Pena et al. treated wheat straw with perfluoroalkyl sulfonic (PFS) and alkyl sulfonic (AS) acid-functionalized MNPs, and found that after treatment, it yields 46% sugar, which is higher than the control (35%). Further, the introduction of PFS NPs and AS acid leads to balanced and defined hemicellulose decomposition, with similar acidity to H_2SO_4 (Wang et al., 2012). Furthermore, the combined treatment of alkaline with MNPs of Fe_3O4 on rice straw can be used for the generation of biogas (Khalid et al., 2019). The utilization of MNPs boosted the biogas and methane yield by 100 and 129% (Khalid et al., 2019). The generation of sugar can be enhanced by the treatment and utilization of NPs. It was found that the treatment of yellow poplar sawdust with 0.8% (w/v) H_2SO_4 yields 96% (w/w) hemicellulose at 175°C. The introduction of H_2SO_4 with PFS and AS NMs convert hemicellulose by 66 and 61%, respectively, at 50- and 400-fold acid levels (Arora et al., 2020). In addition, some of the studies explored outstanding paramagnetic properties of MNPs that may be used in biogas and CH_4 manufacture. For example, Abdelsalam et al. (2017) testified that MNPs of Fe_3O_4 having a concentration of around 20 mg L^{-1} lower the required time to accomplish the maximum bio-CH_4 production (Abdelsalam et al., 2017). Also, MNPs of Fe_3O_4 sponsor active dissipation of Fe ions in the suspension, which qualified an inexhaustible contribution of Fe ions in the bioreactor, which in turn stimulated the anaerobic metabolism, therefore increasing biogas and CH_4 production. Table 1.1 encapsulates some of the main uses of NPs in producing biofuel from biomasses.

Nanotechnology (NT) has vital applications in chemistry, physics, biotechnology, and engineering and works in the range of 1–100 nm (Fajardo et al., 2022). There are many lengths of basic elements, molecules, pellucid materials, and clumps that are manifested in nanomaterials (NMs) presented in Figure 1.2, including zero-dimension (NPs, nanoclusters, and quantum dots), uni-dimension (carbon nanotubes and multi-walled nanotubes), bi-dimension (example: graphene layers and ultrathin films), and tri-dimension (nanostructured materials) materials (Fajardo et al., 2022). There is a variety of shapes and sizes of NMs available such as dendrimers, nano-capsules, nano-spheres, nanotubes, etc. (Fajardo et al., 2022). Currently, NMs are used in numerous applications such as target dosage tailored to the tissue, and decreasing harmful contamination, as well as augmented bioavailability, drug effectiveness, and

TABLE 1.1

Application of NPs for Bioenergy Production from Biomass Feedstock

Feedstock	Nanoparticles	Products Achieved	Yield (Wt.%)	Reference
Weed (*Carthamus oxyacantha, Asphodelus tenuifolius,* and *Chenopodium album*)	Nickel, cobalt	Biodiesel, biogas	60 and 23.75	Ali et al., 2020
Oleic acid and glyceryl trioleate	Silica-coated Fe/Fe₃O₄	Biodiesel	>95	Wang et al., 2015
Weed (*Parthenium hysterophorus* L.)	Nickel, cobalt	Biodiesel, biogas	44 and 17.66	Tahir et al., 2020
Weed (*Cannabis sativa* L.)	Nickel, cobalt	Biodiesel, biogas	53.33 and 12	Tahir et al., 2020
Spent tea	Cobalt	Biodiesel, biogas	40.79 and 28	Mahmood & Hussain, 2010
Wet microalgal biomass (*Neochloris oleoabundans* UTEX 1185)	Fe₃O₄	Biodiesel, bioethanol	57.49 and 81	Banerjee et al., 2019
Microalgae (*Chlorella vulgaris*)	CaO	Biodiesel	92.03	Pandit & Fulekar, 2019
Pine sawdust and Gulmohar seeds	CaO, CuO, and Al₂O₃	Biofuel	39.39 and 36.68	Mishra & Mohanty, 2019
Microalgae (*Chlorella vulgaris*)	Fe₃O₄, silica	Biodiesel	97.10	Chiang et al., 2015
Waste cooking oil	α-Fe₂O₃	Biodiesel	87–92	Ajala et al., 2020
Potato peels	NiO	Bioethanol	50	Sanusi et al., 2021
Potato peels	Fe₃O₄	Bioethanol	93	Sanusi et al., 2021
Neem oil	Cu/ZnO	Biofuel	97.10	Gurunathan & Ravi, 2015
soybean oil (LIZA)	Fe₃O₄	Biodiesel	96	Santos et al., 2015
Corncob	Fe₃O₄	Bioethanol	53.70	Rekha & Saravanathamizhan, 2021
Sugarcane bagasse	K₂CO₃	Biohydrogen	1.21 mol/mol substrate	Reddy et al., 2017
Neem seeds	Fe₃O₄	Biofuel	49–54	Mishra & Mohanty, 2018
Sugarcane bagasse	Fe₃O₄	Biohydrogen	3427 mL/L cumulative H₂	Srivastava et al., 2021
Algal biomass	Fe₃O₄	Biohydrogen	937 mL/L cumulative H₂	Shanmugam et al., 2020a
Grass biomass (*Lolium perenne* L.)	Zero-valent iron NP (Fe⁰)	Biohydrogen	64.7 mL/g dry grass	Yang & Wang, 2018
Sweet sorghum stover biomass	Fe₃O₄, SiO₂	Biohydrogen	2.8 mol H₂/mol reducing sugar	Shanmugam et al., 2020b
Madhuca longifolia seeds	CuO, and Al₂O₃	Biofuel	51.20	Mishra & Mohanty, 2020b
Waste shells of egg and mussel	CaO-based/Au	Biodiesel	90–97	Bet-Moushoul et al., 2016
Madhuca indica seeds	K₂CO₃ and TiO₂	Biofuel	56.60	Mishra & Mohanty, 2020a
Sunflower oil	Cs/Al/ Fe₃O₄	Biofuel	94.80	Feyzi et al., 2013
Cascabela thevetia seeds	CaO and Al₂O₃	Biofuel	45.26–46.87	Mishra et al., 2020

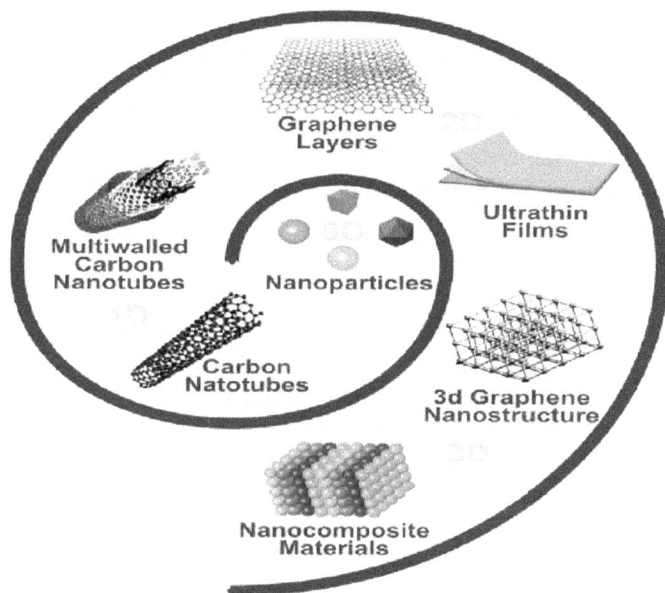

FIGURE 1.2 The diverse dimensions of nanomaterials are presently being industrialized. Adapted from Fajardo et al. (2022).

lessening of adverse side effects (Fajardo et al., 2022). Also, the physicochemical features of NMs result in their extensive utilization in food preservation, water and wastewater treatment, and healthcare (medical sectors), among others. Overall, NMs are used in the bio-sectors to produce higher-grade fuel with improved properties like decreasing viscosity, moisture content, oxygenated content, and increasing heating value, carbon content, etc. (Wei Guo, 2011).

1.3 NANO-CATALYSTS

1.3.1 METAL OXIDE NCS

Nano-catalysts (NCs) have a nanoscale dimension, and a high surface to volume ratio (S/V) activates the catalyst due to greater surface area. Catalysts are categorized into heterogeneous and homogeneous catalysts. Biofuels are divided into bioethanol and biodiesel. A commonly used biofuel, bioethanol has a total estimate of 90% consumption globally (Busic et al., 2018). However, the usage of sugar and starch forms a 6-carbon sugar via enzymatic modification of starchy biomass and zymolysis by the terminal distillation of ethanol-yielding fuel. Bioethanol production involves pre-treatment of lignocellulosic materials to yield cellulose and hemicellulose, which releases hydrolysed 5 and 6 carbon sugars after zymolysis, segregation of solidified remnants and dehydrated cellulose, and at last, distillation to fuel grade (Busic et al., 2018). Biodiesel can be produced through the transesterification of vegetable oils and fats via the inclusion of methanol (other alcohols) and a catalyst, producing glycerol, a co-product. The catalyst intensifies the reaction rate of the ester-based substances, processes them, and helps in the maximization of biodiesel production (Hashmi et al., 2016). Some of the metal oxides are discussed below. Figure 1.3 provides a pictorial presentation of metal oxide NPs.

1.3.2 METAL OXIDE SUPPORTED BY METAL NCS

Calcium oxide (CaO) is most prominently used as a solid base NC due to its inexpensive and highly active nature. Kouzu et al. (2009), in a study, established that increased functioning was due to the leaching of calcium ions into polar phases (Kouzu et al., 2009). These deliquesced Ca ions help in the transesterification reaction. The performance of the catalyst and its lifespan were substantially boosted by the introduction of zinc and lanthanum impurities (Yu et al., 2011). The synthesis of copper-doped zinc oxide NCs by chemical co-precipitation elevated the biodiesel production using neem seed oil. A study done by Baskar et al. (2017) showed that biodiesel can be manufactured from mahua seed oil by introducing manganese-doped ZnO as a catalyst (Baskar et al., 2017). Kaur et al. (2011) showed that with a wet impregnation method in NPs, a good yield of lithium impregnated CaO catalyst

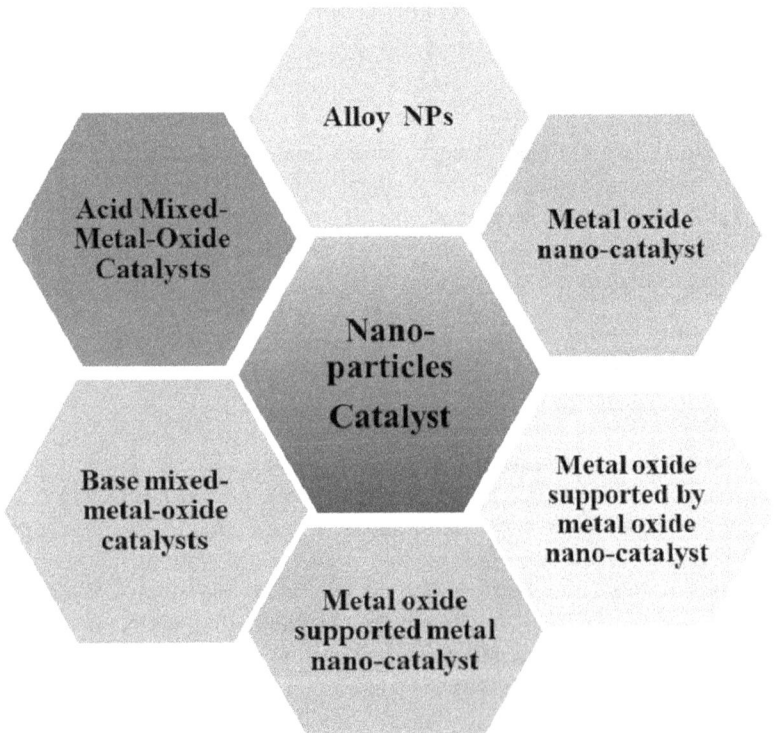

FIGURE 1.3   Types of nano-catalysts used for bioenergy production.

can be prepared. Mishra and Mohanty (2018) showed the production of biofuel from neem seeds by employing K_2CO_3 which improved the yield and properties of fuel (Mishra & Mohanty, 2018). The same authors also showed biodiesel production from the mahua seeds using metal oxide catalysts (Mishra & Mohanty, 2020b). Transesterification of Karanja and Jatropha oil using calcium oxide impregnated with 1.75 wt.% of lithium yields 3.4 and 8.3 wt.% of free fatty acids, within 1 and 2 hours, respectively, at 65°C (Kaur & Ali, 2011). Figure 1.3 provides a pictorial presentation of metal oxide supported by metal NCs.

1.3.3 ALLOY NPs

Carbo-coated Cu-Co bimetallic NPs are developed via the immediate heating treatment of bimetallic oxide forebearers accumulated with polyethylene-glycol (Chen et al., 2017). The continuous formation of carbon layers on the surface of NPs acts as a shield preventing oxidation and deactivation. The chemo-selective hydrogenolysis of 2,5-hydroxymethylfurfural (HMF) showed excellent performance of NCs in producing 5-dimethylfuran (DMF). The yield of DMF increased by around 99.4% and surpassed the outcome of reinforced inert metal catalysts. This method is a new and progressive way to produce cheap and enhanced hydrotreating catalysts for industrial organic processes and the optimization of sustainable biomass (Mahmood & Hussain, 2010). Other sources for biodiesel production include blended NCs encapsulated with enzymes and metallic NPs. Deng et al. (2011) produced hydrotalcite-based particles by a co-precipitation method through microwave-hydrothermal treatment (MHT) using urea and finally calcined at 773 K (Deng et al., 2011). Figure 1.3 provides a pictorial presentation of alloy NPs.

1.3.4 METAL OXIDE SUPPORTED BY METAL OXIDE NCs

Biodiesel is made up of fatty acid ethyl esters and fatty acid methyl esters (FAEE and FAME), where the esterification and transesterification of fats and oils take place using homogeneous or heterogeneous acid/base catalysts (Basumatary, 2013). Heterogeneous catalysts are predominately used as they are easily segregable, environmentally friendly, and are highly efficient compared with homogeneous catalysts (Borges & Díaz, 2012). However, the cons of heterogeneous catalysts are that they have feeble strength, are costly and complicated to produce, have a comparatively minimum surface area and lessened activity, are susceptible to atmospheric CO_2 and polluted water, need extreme temperature and pressure conditions, and have high alcohol-oil ratios (Chang et al., 2014). The introduction of solid mixed-metal-oxide catalysts, consisting of numerous metal oxides in their arrangement, has increased the production of biodiesel. Figure 1.3 provides a pictorial presentation of metal oxide aided by metal oxide NCs.

1.3.4.1 Base Mixed-Metal-Oxide Catalysts

These catalysts are mixed with MgO or CaO impurities, enhancing the catalyst's performance (surface and bulk) (Liu et al., 2021; Ngamcharussrivichai et al.,

2008). For obtaining efficient biodiesel, the transesterification reaction is carried out by these active catalysts efficiently. A definite proportion of free fatty acids in the raw materials (fats or oils) introduced in these catalytic processes is necessary; or else it causes base catalyst poisoning and degrades the standard of the biodiesel. The unique catalytic activity of CaO and MgO-base mixed-metal-oxide catalysts has been a topic of interest for the research community (Liu et al., 2021; Ngamcharussrivichai et al., 2008). However, these catalysts are not easily recoverable and are susceptible to water and acid (Chang et al., 2014). Thus, the transesterification of oil-bearing low acid and water contents is possible using these catalysts (Chang et al., 2014). Figure 1.3 provides a pictorial presentation of base mixed-metal-oxide catalysts.

1.3.4.2 Acid Mixed-Metal-Oxide Catalysts (AMMOC)

For the transformation of high free fatty acids into raw oils, different types of AMMOC have been used. The heterogeneous catalyst (HGC) can be categorized as acids or bases that can be selected as per requirement. Biodiesel manufactured from acid mixed-metal-oxide catalysts is a good replacement for conventional methods because of its straightforwardness and the rapid advancement of esterification and transesterification reactions without soap formation (Vasić et al., 2020). The use of acidic catalysts directly yields biodiesel from lower quality, higher pH, and water-rich oils. Further, the green methodology of biodiesel synthesis inspired the introduction of feasible, solid acid catalysts as substitutes for such liquid acid catalysts such that the application of dangerous materials and the production of harmful wastes are evaded, and the ease of catalyst removal after the reactions. Figure 1.3 and Table 1.2 provide a pictorial presentation of acid mixed-metal-oxide catalysts for the generation of fuel.

1.4 APPLICATION OF NTS FOR BIOMASS TRANSFORMATION

LB has become a feasible solution for the growing ecological issues. It acts as a predominant reserve for extensive energy generation worldwide. A projected study in the US brought to notice that by 2030, 450 million dry tons (MDT) of LB per year could be supplied, which would produce 67 million gallons of ethanol annually (Valdivia et al., 2016). A multistage process of conversion of LB into zymolizable sugars and bioenergy requires precise techniques throughout. The overall transformation stages of LB to outstanding output follow certain processes: pre-treatment of LB to disintegrate its structure to carry out the biocatalytic process and decomposition to generate fermentable sugar and convert into biofuel or other biochemicals. The steps involved in the production of bioenergy from biomass have been listed in Figure 1.4. The LB can be converted into final products such as bioethanol, biodiesel, or biochemical products using a variety of pre-treatment processes. However, these traditional methods are expensive, inappropriate, and have drawbacks. These issues have been resolved with the introduction of a unique, economical, systematic, and eco-friendly method, namely, nanotechnology.

TABLE 1.2
Biodiesel Production from Acid Mixed-Metal-Oxide Catalysts

Catalyst	Oil/Feed	Catalyst Loading	Alcohol/ Oil Ratio	Temp	Time	FAME	Reference
$SO_4^{2-}/$ ZrO_2-TiO_2	Oleic acid	5 wt.%	16:100	65°C	6 h	Esterification: 42.2% Trans-esterification: 88%	Shao et al., 2013
Fe_2O_3-SiO_2	Crude Jatropha oil	15 wt.%	218:1	220°C	3 h	Trans-esterification >99%	Suzuta et al., 2012
WO_3-SnO_2	Soybean oil	5 wt.%	30:1	110°C	5 h	Trans-esterification: 79.2%	Xie & Wang, 2013
SnO_2-SiO_2	Soybean oil	5 wt.%	24:1	180°C	5 h	Trans-esterification: 81.7% Esteri-fication: 94.6%	Xie et al., 2012
$SO_4^{2-}/$ TiO_2-SiO_2	Waste soybean oil	10 wt.%	20:1	120°C	3 h	Trans-esterification: 88%	Shao et al., 2013

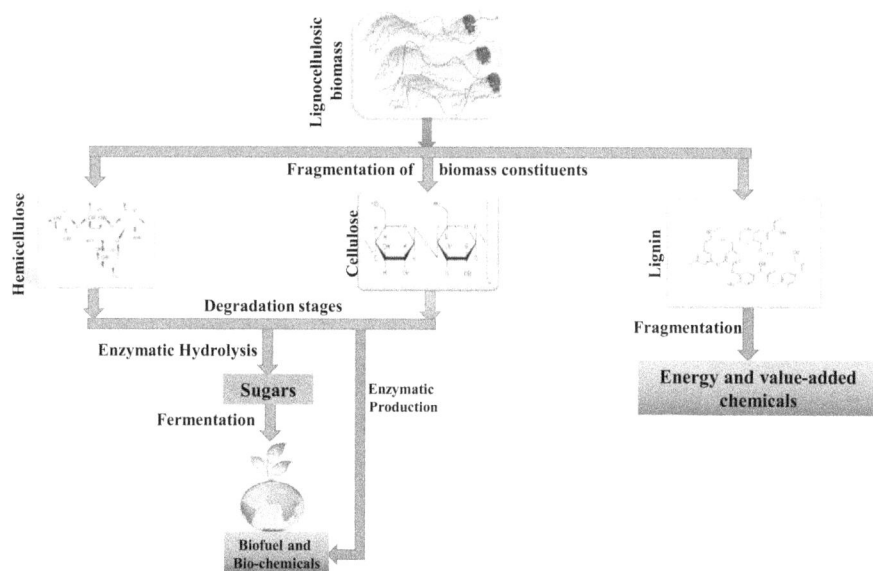

FIGURE 1.4 Schematic presentations of lignocellulosic biomass and major steps involved in bioenergy production.

1.4.1 Pre-Treatment of Lignocellulosic Feedstocks

The hydrolysis of lignocellulosic biomass (LB) releases various reducing sugars, which play a pivotal role in the production of biofuels. Also, LB inhibits biochemical breakdown, which is called biomass recalcitrance. Certain elements account for biomass recalcitrance: the crystalline structure of cellulose, the extent of lignification and structural homogeneity, and the intricacy of the cell wall (Baruah et al., 2018). The primary stage in any bioenergy process is the pre-treatment of LB, where complex structures of LB are broken down to release polymers. Studies have shown the perks of nanotechnology and nanomaterials in the large-scale production of biofuel by applying techniques such as pyrolysis gasification, hydrogenation, and anaerobic digestion. The favourable characteristic of NMs is their smaller size which allows them to readily penetrate through the lignocellulosic biomass cell wall to release monomeric and oligomeric sugars. Moreover, at the molecular level, nanomaterials improve the chemical constituents of biomass (Sankaran et al., 2021b). Further, MNPs strengthen the performance of LB by removing the solid phase and retrieving fermented sugars and are easily recyclable and more economical (Mariño et al., 2021). The use of NPs has a hydrolytic effect that is similar to the chemical pre-treatment of LB processing. A study was done by Pena and his colleague where they treated wheat straw with perfluoroalkyl sulfonic (PFS) and alkyl sulfonic (AS) acid-treated MNPs, which gave 46% more yield than usual (35%) (Pena et al., 2012). The introduction of NPs to PFS and AS acid probably would have caused the counteraction and designated hemicellulose hydrolysis, having an equivalent acid strength of sulphuric acid solutions (Sankaran et al., 2021b). Another study showed an increase in biogas and CH_4 yield by 100% and 129% by incorporating alkaline pre-treatment with magnetite (Fe_3O_4) NPs and rice straw (Sankaran et al., 2021b).

NPs play a dynamic role in sugar production by efficiently increasing mass and heat transfer, enzymatic and cellular metabolic activities, and catalyst properties (Sanusi et al., 2021). An advanced emerging approach to LB pre-treatment is nano-shear hybrid alkaline (NSHA) pre-treatment, which plays a key role in enhancing economic feasibility and the bioethanol production process (Singh et al., 2020). The reaction is carried out in a specific reactor, called nano mixing, where the elevated shearing work axis permits adequate lignin discharge along with cellulose and hemicellulose immediately. Wang developed the NSHA techniques for the pre-treatment of corn stover, where they found that the removal of hemicellulose and lignin from the primary source left behind 82% of cellulose content (Wang et al., 2013). The collaborative performance of cellulose enzymes degenerated biomass recalcitrance and produced a nanoscale polysaccharide cluster, which was easily transformed into simple sugars (Sankaran et al., 2021b). In another study, the addition of poly dialkyl dimethyl ammonium chloride (PDAC) cationic polyelectrolyte reduced the use of and also chemicals needed for the NSHA pre-treatment of corn stover (Singh et al., 2020). Scanning electron microscopy (SEM) and transmission electron microscopy (TEM) are some of the techniques used for the study of morphology and chemical constituents of preheated biomass. Further, lignin produces a spherical aggregate with PDAC polyelectrolyte that eventually modifies the shape of cell walls (Sankaran et al., 2021b).

1.4.2 NANOMATERIALS FORM ENZYME IMMOBILIZATION

Hydrolysis of LB using NPs can be done by physical adsorption on covalent bonding of various enzymes or NMs, engaging functionalized NPs. To expand the enzyme catalytic behaviour of nanostructures, immobilized enzymes are widely used (Sankaran et al., 2021a). The immobilization process can increase the stability and resistance of enzymes under adverse conditions of pH and temperature and allows the re-utilization of these proteins' indefinite reaction cycles. There has been a range of applications of the immobilized biocatalyst in the food, pharmaceutical, and chemical industries. Enzymes are predominantly used in transitional steps for the manufacture of bioactive substances to overcome the issues related to complex chemical synthesis (Cipolatti et al., 2020). The basic types of immobilization methods are mainly adsorption, covalent bonding, encapsulation, and cross-binding, as shown in Figure 1.5 (Mohamad et al., 2015). Each of them has assets and liabilities; for example, the adsorption method is an inexpensive and direct methodology. The inability to interact and the leakage of enzymes are the consequences when pH, temperature, or polarity are modified. Since the covalent bond improves enzyme counteraction and the induced bonds are resistant enough to enzyme discharge, henceforth, they are used for enzyme immobilization. On the other hand, the encapsulation process produces a micro-environment that protects the enzyme from adverse conditions. A distinct immobilization approach that creates enzyme accumulation is known as cross-binding enzyme aggregates (CLEAS). The model used to immobilize enzymes consists of NPs and organic metal frames (Sankaran et al., 2021b). Table 1.3 shows various NMs introduced for enzyme immobilization in handling LB. The use of different types of nanomaterials or catalysts gives a substantially greater yield with improved properties.

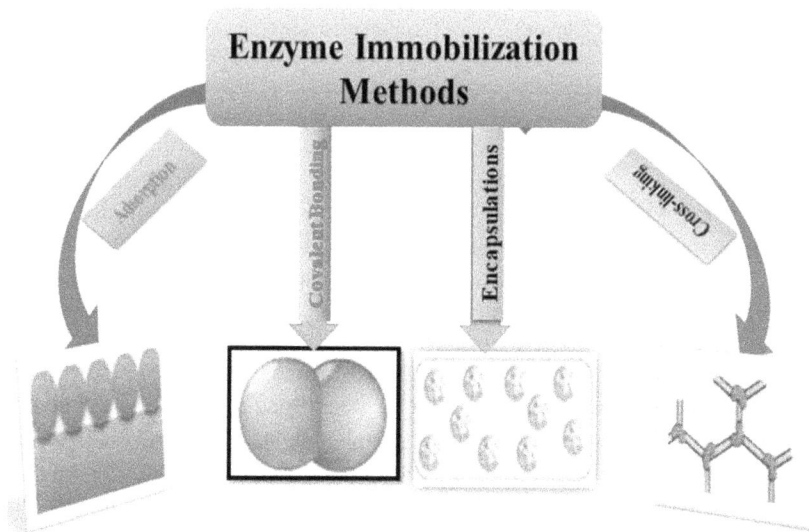

FIGURE 1.5 Different types of strategies used in enzyme immobilization.

TABLE 1.3
Various Nanomaterials Used as Support for Enzyme Immobilization

Nanomaterial Support	Type of Biomasses/ Substrates	Enzyme Used	References
Zinc MNPs	Hemp hurds (natural cellulosic substrates)	Cellulase	Abraham et al., 2014
ZnO NPs	Cellulose	Cellulase (*Aspergillus fumigatus* AA001)	Srivastava et al., 2016b
Silica nanocatalysts	Cellulose	Cellulase	Chang et al., 2011
Magneto-responsive graphene	Cellulose	Cellulase	Gokhale et al., 2013
MNPs encapsulated in polymer nanospheres	Cellulose	Cellulase	Lima et al., 2017
Chitin-functionalized MNPs	P-nitrophenyl glycoside, pNPG	B-glucosidase (*Thermotoga maritima*)	Alnadari et al., 2020
Fe_3O_4-coated chitosan MNPs	-	Xylanase	Liu et al., 2014
Magnetic chitosan	Straw cellulose	B-glucosidase	Zheng et al., 2013
Silica coated crystalline Fe/Fe_3O_4 core/shell MNPs	Hemp hurd	Cellulase	Wang et al., 2015
Multi-walled carbon nanotubes (MWCNTs)	Cellulose	Cellulose (*Aspergillus niger*)	Ahmad & Khare, 2018
Cyclodextrin-based MNPs (B-cyclodextrin-Fe3O4)	Rice straw	Cellulose (*Aspergillus niger*)	Huang et al., 2015
Superparamagnetic NPs	Cellulose	B-glucosidase A(BglA) and cellobiohydrolase D(celD)	Song et al., 2018

Carbon nanotubes (CNTs), which are made of rolled-up graphene sheets and formed into cylindrical tubes, are frequently employed to immobilise enzymes (Eatemadi et al., 2014). A study was done, and it was found that CNT is used in adsorption to bind with glucose oxidase. Another study found that lipase incorporated with CNT has 62% of the original lipase's a-helix composition (Ji et al., 2010). Immobilization restrains several prohibitory effects by stopping inhibitors from binding at the enzyme's active site. The thermal firmness of cellulose enzymes is of huge importance to industrial lignocellulosic operations as increased temperatures increase biomass deterioration. The major disadvantage of cellulose mobilization is that many experimental studies employ man-made cellulose as the substratum, for example, microcrystalline cellulose or carboxymethyl cellulose. A compatible diameter of around 20–200 mm of microcrystalline allows smooth mass transport when compared to original LCM substrates. A study done by Mo and Qiu (2020) found that the formation of scaffolds with a bulkier specific area and open porous structure gave more sites for enzyme immobilization by making it convenient for the substrate to get into contact with the enzyme (Mo & Qiu, 2020).

1.4.3 Production of Bioenergy from Nano-Biocatalysts

By applying nanotechnology, an effective microbial method for assimilating a new metabolic pathway can be implemented to obtain viable biofuels (Singhvi & Kim, 2020). The nano-bio catalyst (NBC) is an advancing innovation that collectively combines advanced NTs and biotechnology for enhancing enzyme activity, stability, competency, and engineering interpretation in bioprocess applications. NBC allows excellent catalytic activity and stability due to configurational changes on immobilization and confined nano environments (Misson et al., 2015a). It has been found that the use of nanomaterials for the fermentation of LB has become more economical. Various enzymes like celluloses, cellobiose, and B-glucosidase were immobilized on several NPs, which illustrated superior features versus free enzymes in the bioethanol fermentation process. A study done recently by Alnadari et al. (2020) determined that covalently binding heat-resistant B-glucosidase extracted from *Aspergillus niger* on MNPs recovered half of the quantity of enzymes by successive regeneration (Alnadari et al., 2020). Immobilization has several benefits, including the capacity to regenerate and recycle enzymes, recovery of unbound enzyme yield, and little interference with the finished product. One of the essentials in bio-hydrogen production using the dark fermentation process is the hydrogenase enzyme (Sankaran et al., 2021b). Compared with conventional hydrogen production processes, bio-hydrogen by dark fermentation is more energy efficient and environmentally favourable (Tang et al., 2022a). This applied nanotechnology exhibited great potential in enhancing ferredoxin-oxidoreductase activities, which raised the electron transfer rates and intensified catalytic activity (Bunker & Smith, 2011). The majority of the testified research addressed the inter-communication of enzymes and NPs in augmenting the electron transmission rate which has the following consequences: the elimination rate of chemical oxygen demand (COD) is relatively lower (about 20%), the agglomeration of liquid metabolites decreases the pH and reduces hydrogen production, and finally, an increase in the partial pressure of H_2 restricts H_2 producing microbial activity, which in turn decreases H_2 production (Tang et al., 2022b).

1.4.4 Hydrogen Fuel or Fuel Cell

The world faces a crucial problem of power crises, which can be overcome by using fuel cells (Pandiyan & Prabaharan, 2020). Fuel cells produce direct current with a low voltage from the catalytic electrochemical reaction and with an oxidizer (Guo, 2012). These fuel cells are zero emissive in nature as they use hydrogen and oxygen fuels which after combustion give just water. Also, they are more efficient for charging electrical equipment and electric vehicles. Indirectly, fuel cells pollute the environment while producing hydrogen, but the implementation of nanotechnology makes the fuel cells non-polluting, as a result restricting global warming (Karthik Pandiyan & Prabaharan, 2020). The major components of a fuel cell include a pump to carry the fuel and to eliminate the water formed in the reaction; a blower or a compressor, to lower the temperature of the power system and to carry oxygen to the fuel cell; and finally, a humidifier, to provide heat and humidity to the approaching hydrogen and oxygen streams. Fuel cells for small and irregular vehicles have complicated structures as they must bear bumps and temperature changes. As a result, using them in vehicles is not so simple (Guo, 2012). Purified hydrogen produced from water hydrolysis in a proton exchange membrane electrolysis cell (PEMEC) acts as a feeder in the fuel cell. In the fuel cell, water is fed into the anodic side of the electrochemical cell, where it gets oxidized, producing oxygen and protons.

$$H_2O \rightarrow \frac{1}{2}O_2 + 2H^+ + 2e^- \tag{1.1}$$

In the gaseous phase, O_2 is given out; however, the electrons flow in the external circuit, and protons pass through the membrane and reach the cathodic site, where they are reduced by the incoming electrons from the external circuit producing hydrogen (Lamy, 2016).

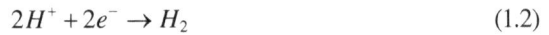

$$2H^+ + 2e^- \rightarrow H_2 \tag{1.2}$$

Other than hydrogen, materials such as carbon act as ionic energy bearers. In a fuel cell, most of the energy stored in the electrochemical reaction is liberated directly as electricity and a smaller portion of it is released as heat (Guo, 2012). Some of the common fuel cells are listed as follows.

1.4.4.1 Alkaline Fuel Cells (AFCs)

AFCs play an important role in providing high power densities and achieving a long lifespan in certain operations. AFC was the first fuel cell technology applied practically and paved the way for feasible electricity generation from hydrogen. AFCs became the basis for the application of fuel cells in vehicles (McLean et al., 2002). The operating temperature of AFC ranges from 333 to 363 K, with the conversion ranging from 50 to 60% (Guo, 2012). AFCs use an aqueous solution of 30–35% KOH as an electrolyte. AFC technology has the capacity for great growth in the research and development sector. The presence of CO_2 impurities in AFCs clogs the electrodes and hence degrades their performance (McLean et al., 2002). Currently, the use of NTs is advancing the use of AFCs.

1.4.4.2 Proton Exchange Membrane Fuel Cells (PEMFCs)

PEMFCs were first introduced in 1993 by the New Generation of Vehicles Program (PNGV) in the United States. The pros of PEMFCs are the low operating temperature, constant operation at high current density, lower weight, rigidity, low cost, and long shelf life (Wee, 2007). The operating temperature ranges from 323 to 353 K, with the conversion ranging from 50 to 60%. The electrolyte is made up of a porous carbon-containing platinum catalyst. PEMFCs use hydrogen-oxygen from the air, along with water which does not cause any corrosion. PEMFCs require a noble metal catalyst (such as platinum) to separate the electrons and protons produced by hydrogen, which also prevents toxic carbon monoxide from reducing its efficiency (Guo, 2012). The basic requirements for commercializing PEMFCs are a steady and inexpensive supply of pure hydrogen, the presence of more efficient power sources than PEMFCs, benefits to health and the environment, the framework of traditional power supply and demand (Wee, 2007). Further, PEMFCs are used to produce energy (electricity) by utilizing anode and cathode ions (Mao et al., 2012). The typical working principle of PEMFCs is shown in Figure 1.6. It can be seen that electrons and protons are produced by ionizing the anode on the cathode. The electrons are passed through an external circuit whereas protons are passed through the membrane. Finally, the homogenization of the electrons and protons on the cathode results in the reduction of oxygen to water.

1.4.4.3 Molten Carbonate Fuel Cells (MCFCs)

MCFCs are fuel cells operating at a high temperature of more than 873 K. They use electrolytes, which contain 62–70 mole% of Li_2CO_3 and 30–38 mole% of H_2CO_3; sometimes NO_2CO_3 and other salts can also be added. These fluids are bound in the opening of a porous ceramic model composed of sedimented MgO or $LiAlO_2$ powders. The electrodes used in MCFCs are gas-diffused metallic plates that are porous in nature (Dicks, 2004). MCFCs are cost-effective as non-precious metal is used as the catalyst for the electrode. There is no need for external reformers in the

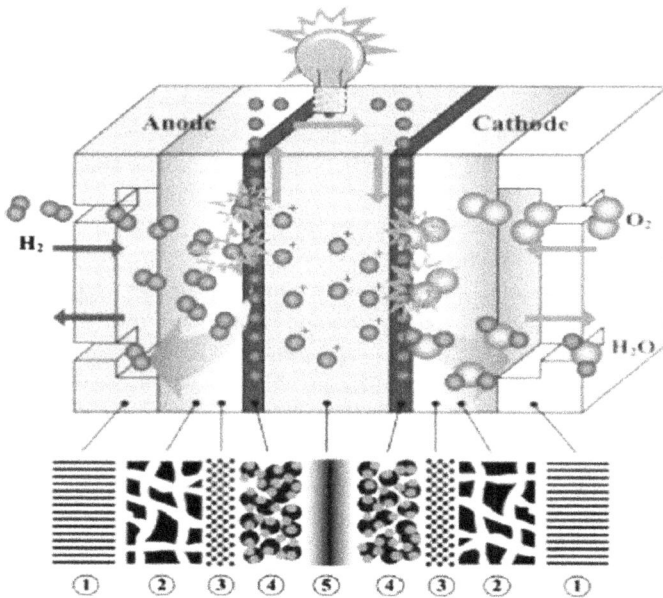

FIGURE 1.6 Schematic presentations of working principle of a PEMFC (direct hydrogen) and the involved porous components. [(1) Bipolar plate; (2) backing layer; (3) micro porous layer; (4) electro catalyst layer; and (5) membrane.] (Adopted from Mao et al., 2012).

case of MCFCs for converting energy-rich fuels to hydrogen (Guo, 2012). The ability of MCFCs to work with different types of fuel is more economical for building stable power plants with high power output. The disadvantage of MCFCs is their long-lasting agitation of molten carbonates, decreasing their performance (Dicks, 2004).

1.4.4.4   Solid Oxide Fuel Cells (SOFCs)

The SOFC consists of a solid ceramic electrolyte, which is a metallic oxide (Ormerod, 2003). The operating temperature of SOFCs ranges from 1073 to 1273 K with a conversion rate of 55–65%. The electrolyte is composed of a mixture of rare earth materials, known as yttria-stabilized zirconium dioxide. SOFCs can use gases produced from coal directly as their cell type is sulphur resistant (Guo, 2012). At the cathode of SOFC, oxygen gets reduced to oxygen ions, which are then transferred through the static electrolyte to the anode. At the anode, a reaction with the fuel takes place and produces water, O_2, electricity, and heat. Due to the high operating temperature of SOFC, it can be directly used on practical hydrocarbon fuels without the use of a fuel reformer (Ormerod, 2003).

Though hydrogen is an ideal raw material for the fuel cells, there are many issues in storing and purifying hydrogen, designing membranes for fuel cells, and the production of commercialized hydrogen from water rather than natural gas to support economic hydrogen production (Guo, 2012). A worldwide formulation of small fuel cells may be created by fuel cells that have an emphasis on downsizing by reducing the centrifugation of small fuel cells. Micro- or nano-electro-mechanical systems are a technological method to miniaturize fuel cells that are powered by liquid gases such as butane and propane (Guo, 2012). In this technology, the fuel reformer is supplied with fuel, air, and water by using the vapour pressure of the gases in order to reduce the power utilization and size of the system (Tanaka et al., 2004). The basic components of this system are a reforming catalytic combustor and polyelectrolyte. Micro-fuel cells are desirable as a power source for mobile electronics due to their elevated energy density potential and uninterrupted operation. The preferred fuel cells of micro-electro-mechanical systems (MEMS) for powering portable electronics are direct methanol fuel cells (DMFC) and polymer electrolyte fuel cells (PEFC). Micro DMFCs use methanol as fuel and are widely used due to the ease of storage of the liquid fuel. The catalyst activity in a DMFC is low at lower temperatures, thus resulting in low power density. MEMS-based DMFC functions are inefficient due to the occurrence of graphite-based electrodes, which show low compatibility (Zhang et al., 2007). Further, miniature solid oxide fuel cells (SOFCs) focus on thin-film technology and can function at a temperature ranging from 773 to 1073 K. But as the temperature decreases, the efficiency also decreases. An extra cooling system is introduced to maximize the capacity of fuel cells (Guo, 2012). SOFCs have the capacity for clean and efficient energy production for amalgamated heat and power for remote applications (Ormerod, 2003). The massive commercialization of fuel cell power plants seems inexorable. Due to PAFC, power plants with a satisfactory lifespan for immobilized applications have been developed. So far, it seems very reasonable that lifespan requirements will be reached by fuel cells that work even at low temperatures (Perry & Fuller, 2002). MEMS is expected to mainly contribute to the larger circulation of and simple access to knowledge, compatible lifestyle with environment, and development in social welfare.

1.4.5 SOLAR ENERGY

The demand for energy has grown so much in recent years, which has paved the way to developing more reliable, economical, and everlasting renewable sources of energy. Solar energy is one of the most favourable and abundant sources of energy that can combat the energy crises (Kannan & Vakeesan, 2016). The earth's surface receives solar radiation annually, which accounts for more than 10,000 times the yearly overall energy usage (Guo, 2012). Photovoltaic technology has been introduced which converts this solar energy directly into electricity. These devices are more economical as they are simple in design, easy to handle, and give more output for smaller input. Photovoltaic (PV) devices are made up of semiconductors like silicon. The principle of this device is that it activates the electrons by supplying additional energy, which creates holes and free electrons in the semiconductor and thus produces electricity (Kannan & Vakeesan, 2016). This photovoltaic technology also has many cons, as it is complex and expensive to manufacture and install. These issues can be resolved by introducing nanotechnology, which helps in increasing its efficiency and also lowers the storage cost (Guo, 2012).

Solar cells based on a single crystalline semiconductor wafer are costly. Solar cells made using inorganic thin-film structures are cheaper but are less efficient (14%) (Abdin et al., 2013). Non-silicon thin films made up of materials like cadmium telluride or copper indium gallium selenide are comparatively inexpensive and have an efficiency of 19%. Researchers are also looking for new thin-film solar cells made up of organic materials which will be optically transparent and more flexible by trapping more solar radiation than existing technology. The use of organic materials in solar cells is more demanding as they create an impermeable vapour barrier and thus reduce their performance (Guo, 2012). Another challenge is the capturing of a huge amount of solar energy. The single p-n junction, which is the heart of the solar cell, has photons that produce the electron-hole pairs, resulting in current flow. One of the sides of the junction is doped with a distinct semiconductor material, one doped with n-type, which collects electrons flowing uni-directionally, and the other collects holes flowing in the opposite direction. The p-n junction absorbs photons at a particular wavelength. Multi-junction narrow-film solar cells consisting of a pile of p-n junctions allow efficient electricity production from a spectrum of wavelengths. Also, there are quantum dot solar cells that absorb radiation of all wavelengths from the sun, thus creating electron-hole pairs from a temperature gradient. Hybrid photovoltaic solar cells are produced by incorporating an inorganic semiconductor possessing outstanding electronic properties and highly flexible organic polymers, which make these PV devices more efficient and economical (Guo, 2012). Although nanotechnology has wide applications in the development of solar cells, it has some disadvantages. The PV cells also have certain disadvantages, as they produce more greenhouse gases, are expensive, and use many toxic materials for their production. Suitable technology should be developed to resolve these challenges so that we can miraculously utilize this abundant source of energy. Nanostructured materials are widely used in the manufacturing of solar cells because of their good energy sustainability (Schoonman et al., 2005). For a study, the potential application of nanostructured material in the manufacturing of solar cells is explored. Figure 1.7 demonstrates the application of nanostructure material in the solar cell. The use of NPs in solar plates enhanced the energy storage capacity of the battery which can be further utilized in different applications.

1.4.6 BIOTECHNOLOGY

The continuous use of fossil fuels and their slow replenishment have driven us to look for alternatives such as biological sources as they are more economical, environmentally friendly, and available in abundance. In the previous decade, scientists have been trying to apply biological processes using DNA in industrial applications by

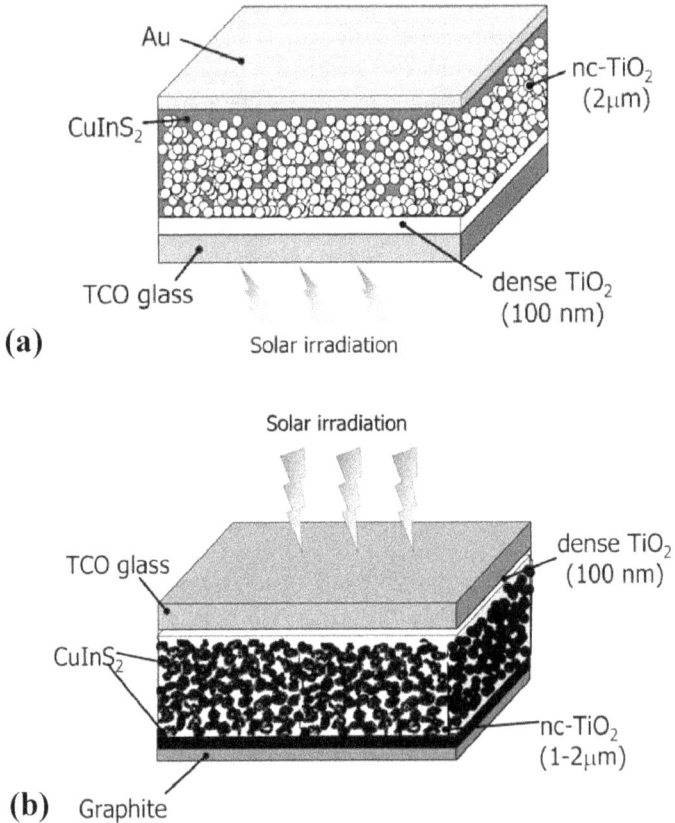

FIGURE 1.7   (a) 3D nanostructured solar cell. $CuInS_2$ applied by atomic layer deposition, and (b) 3D nanostructured solar cell. $CuInS_2$ is applied by aerosol spray pyrolysis (adopted from Schoonman et al., 2005).

introducing a platform called biotechnology (Guo, 2012). Below are some emerging examples of bio-nanotechnologies.

1.4.6.1 Biofuels

Biofuel is a sustainable alternative to fossil fuels; thus, the primary goal of the research and development sector is to produce biologically manufactured fuels from non-food materials such as biomass, algae, etc. To produce biofuels from biomass, more competent biomass conversion methods must be practiced to make cost-competitive biofuels. Alcohol is usually produced from corn, but studies are being done to produce alcohol from cellulosic biomass such as forest residues, crop residues, municipal solid wastes, energy crops, and agricultural residues, which will make it economical (Zabed et al., 2016). The basic steps in converting cellulosic material to alcohol are hydrolysis and fermentation. Biomass can play a crucial role in producing alternatives to fossil fuel as well as electricity and heat, but biomass energy production on a very wide scale can lead to serious environmental issues. The usage of land is also a concerning problem regarding biomass. There will be competition between agricultural land and feedstock production, and then comes the loss of fertility of the lands due to excessive use (Guo, 2012). Bio-oil can be obtained from algae using NPs without damaging the organism (Trindade, 2011). A pictorial presentation is demonstrated in Figure 1.8 for the generation of biofuel from algae using nanotechnology. The NPs used in this study are shown on the left-hand image before the expectant oil algae are added. However, the right-hand image shows the contact between algae and NPs, which results in the extraction of biofuel without damaging the algae (Trindade, 2011). Moreover, keeping the algae alive can effectively lessen the generation costs and the production cycle.

1.4.7 WIND AND OCEAN ENERGY

Inexhaustible sources of energy decrease the usage of fossil fuels. Wind energy is environmentally friendly, and provides worldwide energy reliability in the era when

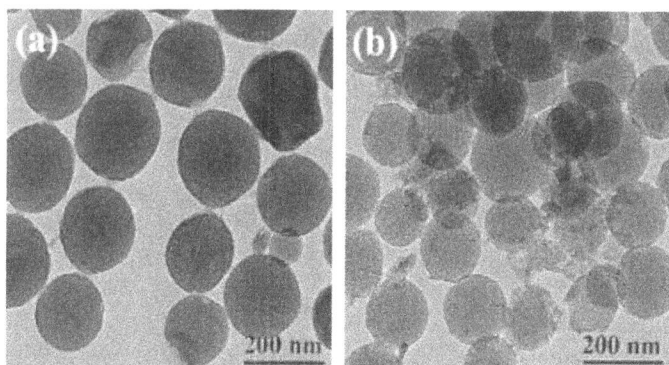

FIGURE 1.8 Harvesting of bio-oil using nano-particles from algae without harming the organism (adopted from Trindade, 2011).

our global fossil fuels are being depleted at a very fast pace (Herbert et al., 2007). When wind energy is transformed into an efficient form of energy, it is known as wind power, like electricity from wind turbines, pumps for the transportation of water, sails for setting ships in motion, and windmills for power generation. In the context of material science, a very pivotal role will be played in evolving coatings and durable materials which are needed for the blades of turbines and towers, to resist corrosion, wind speed, etc. Through NTs, there is a huge opportunity to design implanted sensor long-lasting shielding coatings in order to increase the lifespan of wind energy devices (Guo, 2012). NTs will increase the lifetime of wind turbine blades; nano-based prepregs are very lightweight, so the weight is reduced; and nano lubricants, nanofluids, and nano-sanctioned wires will increase the efficiency (Patel & Mahajan, 2017). To exploit wave energy in order to produce electricity, highly developed ocean technologies are required, like tidal barriers and current turbines (Guo, 2012). Ocean energy has the capability to play a tremendous role in the future because it contributes to lowering carbon release and stimulates economic growth in coastal and isolated areas. There are many challenges when it comes to ocean energy like finance, marketing, management, environmental issues, and the presence of grid connections in isolated areas (Magagna & Uihlein, 2015). These challenges are the reasons for resistance to ocean energy's development and commercialization. It has been found that oceans will contribute an estimated 7% of global electricity production by 2050 (Esteban & Leary, 2012). Employment and economic development will increase in the ocean energy sector by 2030 (Esteban & Leary, 2012). It is also important to make sure the development of technologies in oceans will not pose a threat to the marine environment. Ocean thermal energy conversion (OTEC) systems manufacture electricity from the natural thermal difference of the ocean, using the heat from warm water to produce steam to propel a turbine and pumping cold water to condense the steam; these systems can be built on both onshore and offshore floating platforms. Overall it is estimated that around 10 trillion watts of power, nearly equal to the recent energy demand of the globe, could be manufactured by OTEC without disturbing the thermal structure of the ocean (Pelc & Fujita, 2002). Moreover, triboelectric nano-generators (TENG) technology is a novel technology to produce wave energy. It works on the principle of the coupling of the triboelectrification effect and the electrostatic induction effect (Zhu et al., 2020).

1.4.8 NUCLEAR ENERGY

Nuclear power has appeared to be one of the consistent sources of electricity over the years (Zinkle & Was, 2013). It makes a major contribution to the vast energy supplies of the world. The systematic analysis of nuclear energy using fission and fusion technologies is the practical problem in the field of nuclear energy. In recent times nuclear fission has been put in practice to harness nuclear energy. To increase the efficiency and proper implementation of nuclear fission, there are a few points that need to be acknowledged. The first point is to properly figure out the physico-chemical effects on material fatigue, strain, and corrosion due to radiation. The second point is upgrading the spent fuel processing and improving separation technologies. The third point is thorough analysis of actinide and lanthaside's physical and chemical properties along with the radioactive consequences for polymers, ion exchange substances, and rubbers, and finally, the fourth point covers the adequate storage, elimination, and recycling of nuclear waste (Guo, 2012).

The latest designed nanomaterials in the nuclear energy sector have improved the production and safety of nuclear power. Nuclear nanotechnology (NNT) deals with the usage of NMs for nuclear energy implementation in the future. NMs have been used in nuclear fuel removal and production, fission product encapsulation, radiation sensing and controlling, radioactive waste segregation, and exhausted nuclear fuel recovery (Khanal et al., 2020). On the other hand, nuclear fission still requires a lot of research and development. There is a need for magnetic instruments made of materials with greater performances and resistivity in addition to the ongoing investigations on plasma physics (Guo, 2012). Alpha decay of actinides is the major origin of radiation effects from nuclear waste. Though quantifiable changes in volume, stored energy, and microstructure happen, there is no notable turnaround in volume and stored energy of solid nuclear-waste forms (Weber & Roberts, 1983).

1.5   FUTURE SCOPE OF NANOTECHNOLOGIES

NT is widely used to produce renewable and sustainable biofuel using NMs such as NPs, nanofibers, nanotubes, nano-sheets, and nano-catalysts. NTs are employed in biogas production by using MNPs which have effective paramagnetic properties and resistance to demagnetization. MNPs also immobilize the enzymes which are used in biodiesel production (Rai & Da Silva, 2017). Further, conventional medicines and drugs used for the treatment of cancer have low sensitivity and are linked with low toxicity. Advanced medicines are developed using NPs that can boost the medical industry (Surendiran et al., 2009). NTs have many applications in defence, such as nano-medicines and bandages, providing a more robust and lighter fabric. NTs is also used for the development of many nano-devices and sensors that are utilized for varying applications (Abed & Jawad, 2022). The use of NTs in agriculture sectors provides better yields, water purification, agrochemicals that are resistant to pests and insects, and genetic modifications in plants and animals (Pramanik et al., 2020). They are also used in food packaging and processing and ensure proper food safety. They also act as nutritional food additives and hence increase the shelf life of food, thus increasing its commercialization (Alfadul & Elneshwy, 2010). Textiles manufactured from NTs have unique characteristics like self-sterilizing antimicrobial properties and can act as sensors. As a result, they have an emerging application in the military, fashion, sports, and healthcare (Coyle et al., 2007). Figure 1.9 demonstrates the different possible applications of NTs in the near future.

Although NTs are used widely in different sectors successfully, advanced research is still needed to establish the application of NTs. Soon, NTs could also enable the creation of tools to harvest energy from the environment. New nanomaterials and concepts are currently being developed that show potential for producing energy from movement, light, variations in temperature, glucose, and other sources with high conversion efficiency. Further, NTs can be used in energy-related applications.

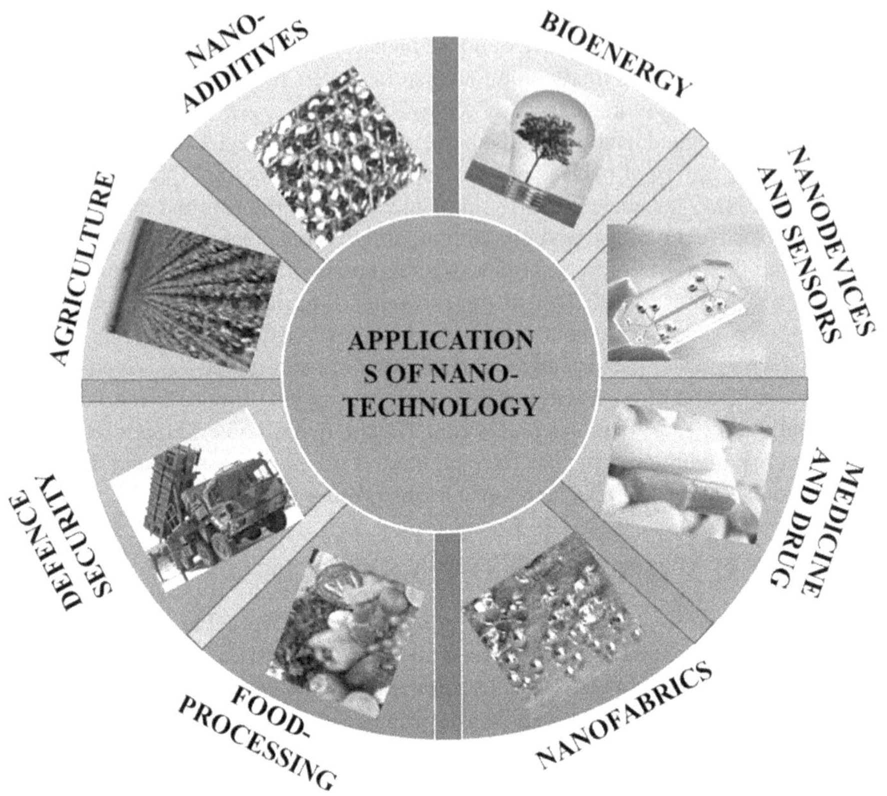

FIGURE 1.9   Application of nanotechnology in different sectors.

For example, super-capacitors or nanotubes made from biochar can be used in different types of industrial applications such as the columns of HPLCs, electrolytes for fuel cells, carbon capture, or other equipment to detect contamination. NTs can also be used in space, where the use of NTs can save lots of space and reduce the weight of the carrier. Moreover, it is expected that the cost will be reduced, but there is a substantial lack of literature that demonstrates the cost analysis of NTs in space technology (Inshakova et al., 2020). Nano-sponges, also known as carbon nano-super-capacitors, are an example of NTs, but still, there is a need for research on this. There is space to make carbon nano-super-capacitors and ultra-capacitors from biochar-based materials that ultimately reduce the cost of the products and can be easily degraded. The PV cells used in solar cells have some drawbacks, as they produce more greenhouse gases, are expensive, and use many toxic materials. Thus, an appropriate technology should be industrialized to face these challenges so that energy can be utilized substantially. The issues associated with LB and its production will be addressed soon using advanced NTs; then bioenergy generation will become a leading source in the growing fuel sector. Across the world, research is focused on the production of renewable energy from waste with minimum cost to replace fossil fuel partially or completely. Apart from the generation of renewable and sustainable energy, the proper handling, manufacture, and systemic use of fossil fuels is also a key area of research. It is true that these renewable sources of energy using nanotechnology will take over globally once these fossil fuels are exhausted.

1.6 CONCLUSIONS

The increasing demand for fossil fuels and their faster rate of degradation than replenishment have paved the way for researchers to find an alternative such as bio-fuel. The biofuel here is produced from lignocellulosic biomass, which is found in abundance, making it more efficient, economical, and environmentally friendly. To meet the challenges of the transformation of biomass, NTs have seen wide application over recent years. One of the major pros of NCs is that they can be reused for several cycles without losing their efficiency. Besides the advantages, there are certain disadvantages; for example, the minute NPs such as metal NCs interrupt the usual activity of human cells by contaminating the bloodstream. These NPs, when used in biofuel, are emitted from the vehicle exhaust, becoming hazardous to human health by depositing into the lungs and causing several respiratory disorders. One method which is used as a pre-treatment technique for obtaining biofuel is the hydrolysis process of biomass which consumes more energy. As a solution to this drawback, microwave reactors are used, which minimizes the length of reaction and energy consumption. It is always said that with advancement come other changes that are also needed; the same applies to NTs with the development in the processing of a considerable variety of biomass.

ACKNOWLEDGEMENTS

The authors would like to thank the Department of Chemical Engineering, M S Ramaiah Institute of Technology, Bangalore, Karnataka, India, for its support in providing the necessary facilities to complete the present book chapter.

REFERENCES

Abdelsalam, E., Samer, M., Attia, Y., Abdel-Hadi, M., Hassan, H., Badr, Y. (2017). Influence of zero-valent iron nanoparticles and magnetic iron oxide nanoparticles on biogas and methane production from anaerobic digestion of manure. *Energy*, 120, 842–853.

Abdin, Z., Alim, M.A., Saidur, R., Islam, M.R., Rashmi, W., Mekhilef, S., Wadi, A. (2013). Solar energy harvesting with the application of nanotechnology. *Renew. Sustain. Energ. Rev.*, 26, 837–852.

Abed, M.S., Jawad, Z.A. (2022). Nanotechnology for defence applications. In: N. M. Mubarak, S. Gopi, P. Balakrishnan (eds.) *Nanotechnology for Electronic Applications*. Springer, pp. 187–205.

Abraham, R.E., Verma, M.L., Barrow, C.J., Puri, M. (2014). Suitability of magnetic nanoparticle immobilised cellulases in enhancing enzymatic saccharification of pretreated hemp biomass. *Biotechnol. Biofuels*, 7(1), 90.

Ahmad, R., Khare, S.K. (2018). Immobilization of *Aspergillus niger* cellulase on multiwall carbon nanotubes for cellulose hydrolysis. *Bioresour. Technol.*, 252, 72–75.

Ahmad, R., Sardar, M. (2014). Immobilization of cellulase on TiO 2 nanoparticles by physical and covalent methods: A comparative study. *J. Biochem. Biophys.*, 51, 314–320.

Ajala, E., Ajala, M., Ayinla, I., Sonusi, A., Fanodun, S. (2020). Nano-synthesis of solid acid catalysts from waste-iron-filling for biodiesel production using high free fatty acid waste cooking oil. *Sci. Rep.*, 10(1), 1–21.

Alfadul, S., Elneshwy, A. (2010). Use of nanotechnology in food processing, packaging and safety–review. *Afr. J. Food Agric. Nutr.*, 10(6), 58068.

Ali, S., Shafique, O., Mahmood, S., Mahmood, T., Khan, B.A., Ahmad, I. (2020). Biofuels production from weed biomass using nanocatalyst technology. *Biomass Bioenergy*, 139, 105595.

Alnadari, F., Xue, Y., Zhou, L., Hamed, Y.S., Taha, M., Foda, M.F. (2020). Immobilization of β-glucosidase from *Thermatoga maritima* on chitin-functionalized magnetic nanoparticle via a novel thermostable chitin-binding domain. *Sci. Rep.*, 10(1), 1663.

Arora, A., Nandal, P., Singh, J., Verma, M.L. (2020). Nanobiotechnological advancements in lignocellulosic biomass pretreatment. *Mater. Sci. Technol.*, 3, 308–318.

Banerjee, S., Rout, S., Banerjee, S., Atta, A., Das, D. (2019). Fe2O3 nanocatalyst aided transesterification for biodiesel production from lipid-intact wet microalgal biomass: A biorefinery approach. *Energy Convers. Manag.*, 195, 844–853.

Baruah, J., Nath, B.K., Sharma, R., Kumar, S., Deka, R.C., Baruah, D.C., Kalita, E. (2018). Recent trends in the pretreatment of lignocellulosic biomass for value-added products. *Front. Energy Res.*, 6, 141.

Baskar, G., Gurugulladevi, A., Nishanthini, T., Aiswarya, R., Tamilarasan, K. (2017). Optimization and kinetics of biodiesel production from Mahua oil using manganese doped zinc oxide nanocatalyst. *Renew. Energy*, 103, 641–646.

Basumatary, S. (2013). Transesterification with heterogeneous catalyst in production of biodiesel: A review. *J. Chem. Pharm. Res.*, 5(1), 1–7.

Bet-Moushoul, E., Farhadi, K., Mansourpanah, Y., Nikbakht, A.M., Molaei, R., Forough, M. (2016). Application of CaO-based/Au nanoparticles as heterogeneous nanocatalysts in biodiesel production. *Fuel*, 164, 119–127.

Borges, M.E., Díaz, L. (2012). Recent developments on heterogeneous catalysts for biodiesel production by oil esterification and transesterification reactions: A review. *Renew. Sustain. Energ. Rev.*, 16(5), 2839–2849.

Brown, T.R. (2015). A techno-economic review of thermochemical cellulosic biofuel pathways. *Bioresour. Technol.*, 178, 166–176.

Bunker, C.E., Smith, M.J. (2011). Nanoparticles for hydrogen generation. *J. Mater. Chem.*, 21(33), 12173–12180.

Bušić, A., Marđetko, N., Kundas, S., Morzak, G., Belskaya, H., Ivančić Šantek, M., Komes, D., Novak, S., Šantek, B. (2018). Bioethanol production from renewable raw materials and its separation and purification: A review. *Food Technol. Biotechnol.*, 56(3), 289–311.

Chang, F., Zhou, Q., Pan, H., Liu, X.F., Zhang, H., Xue, W., Yang, S. (2014). Solid mixed-metal-oxide catalysts for biodiesel production: A review. *Energy Technol.*, 2(11), 865–873.

Chang, R.H.-Y., Jang, J., Wu, K.C.W. (2011). Cellulase immobilized mesoporous silica nanocatalysts for efficient cellulose-to-glucose conversion. *Green Chem.*, 13(10), 2844–2850.

Chen, B., Li, F., Huang, Z., Yuan, G. (2017). Carbon-coated Cu-Co bimetallic nanoparticles as selective and recyclable catalysts for production of biofuel 2, 5-dimethylfuran. *Appl. Catal. B*, 200, 192–199.

Chiang, Y.D., Dutta, S., Chen, C.T., Huang, Y.T., Lin, K.S., Wu, J.C., Suzuki, N., Yamauchi, Y., Wu, K.C.W. (2015). Functionalized Fe3O4@ silica core–shell nanoparticles as microalgae harvester and catalyst for biodiesel production. *ChemSusChem*, 8(5), 789–794.

Cipolatti, E.P., Valério, A., Henriques, R.O., Cerqueira Pinto, M.C., Lorente, G.F., Manoel, E.A., Guisán, J.M., Ninow, J.L., de Oliveira, D., Pessela, B.C. (2020). Production of new nanobiocatalysts via immobilization of lipase B from *C. antarctica* on polyurethane nanosupports for application on food and pharmaceutical industries. *Int. J. Biol. Macromol.*, 165(B), 2957–2963.

Coyle, S., Wu, Y., Lau, K.-T., De Rossi, D., Wallace, G., Diamond, D. (2007). Smart nanotextiles: A review of materials and applications. *MRS Bulletin*, 32(5), 434–442.

Deng, X., Fang, Z., Liu, Y.-h., Yu, C.-L. (2011). Production of biodiesel from Jatropha oil catalyzed by nanosized solid basic catalyst. *Energy*, 36(2), 777–784.

Dicks, A.L. (2004). Molten carbonate fuel cells. *Curr. Opin. Solid State Mater. Sci.*, 8(5), 379–383.

Dutta, N., Mukhopadhyay, A., Dasgupta, A.K., Chakrabarti, K. (2014). Improved production of reducing sugars from rice husk and rice straw using bacterial cellulase and xylanase activated with hydroxyapatite nanoparticles. *Bioresour. Technol.*, 153, 269–277.

Eatemadi, A., Daraee, H., Karimkhanloo, H., Kouhi, M., Zarghami, N., Akbarzadeh, A., Abasi, M., Hanifehpour, Y., Joo, S.W. (2014). Carbon nanotubes: Properties, synthesis, purification, and medical applications. *Nanoscale Res. Lett.*, 9(1), 393–393.

Elumalai, S., Agarwal, B., Runge, T.M., Sangwan, R.S. (2018). Advances in transformation of lignocellulosic biomass to carbohydrate-derived fuel precursors. In: S. Kumar, R. K. Sani (eds.) *Biorefining of Biomass to Biofuels*. Springer, pp. 87–116.

Esteban, M., Leary, D. (2012). Current developments and future prospects of offshore wind and ocean energy. *Appl. Energy*, 90(1), 128–136.

Fajardo, C., Martinez-Rodriguez, G., Blasco, J., Mancera, J.M., Thomas, B., De Donato, M. (2022). Nanotechnology in aquaculture: Applications, perspectives and regulatory challenges. *Aquacult. Fish*, 7(2), 185–200.

Feyzi, M., Hassankhani, A., Rafiee, H.R. (2013). Preparation and characterization of Cs/Al/Fe3O4 nanocatalysts for biodiesel production. *Energy Convers. Manag.*, 71, 62–68.

Gokhale, A.A., Lu, J., Lee, I. (2013). Immobilization of cellulase on magnetoresponsive graphene nano-supports. *J. Mol. Catal.*, 90, 76–86.

Guo, K.W. (2012). Green nanotechnology of trends in future energy: A review. *Int. J. Energy Res.*, 36(1), 1–17.

Gurunathan, B., Ravi, A. (2015). Process optimization and kinetics of biodiesel production from neem oil using copper doped zinc oxide heterogeneous nanocatalyst. *Bioresour. Technol.*, 190, 424–428.

Hamawand, I., Seneweera, S., Kumarasinghe, P., Bundschuh, J. (2020). Nanoparticle technology for separation of cellulose, hemicellulose and lignin nanoparticles from lignocellulose biomass: A short review. *Nano Struct. Nano Objects*, 24, 100601.

Hashmi, S., Gohar, S., Mahmood, T., Nawaz, U., Farooqi, H. (2016). Biodiesel production by using CaO-Al2O3 nano catalyst. *Int. J. Eng. Res.*, 2(3), 43–49.

He, Z., Zhang, Z., Bi, S. (2020). Nanoparticles for organic electronics applications. *Mater. Res. Express*, 7(1), 012004.

Herbert, G.J., Iniyan, S., Sreevalsan, E., Rajapandian, S. (2007). A review of wind energy technologies. *Renew. Sustain. Energ. Rev.*, 11(6), 1117–1145.

Huang, P.-J., Chang, K.-L., Hsieh, J.-F., Chen, S.-T. (2015). Catalysis of rice straw hydrolysis by the combination of immobilized cellulase from *Aspergillus niger* on β-cyclodextrin-Fe3O4 nanoparticles and ionic liquid. *BioMed Res. Int.*, 2015. 409103.

Inshakova, E., Inshakova, A., Goncharov, A. (2020). Engineered nanomaterials for energy sector: Market trends, modern applications and future prospects. In: *IOP Conference Series: Materials Science and Engineering*. IOP Publishing, pp. 032031.

Ji, P., Tan, H., Xu, X., Feng, W. (2010). Lipase covalently attached to multiwalled carbon nanotubes as an efficient catalyst in organic solvent. *AIChE J.*, 56(11), 3005–3011.

Jönsson, L.J., Martín, C. (2016). Pretreatment of lignocellulose: Formation of inhibitory by-products and strategies for minimizing their effects. *Bioresour. Technol.*, 199, 103–112.

Kannan, N., Vakeesan, D. (2016). Solar energy for future world:-A review. *Renew. Sustain. Energ. Rev.*, 62, 1092–1105.

Karthik Pandiyan, G., Prabaharan, T. (2020). Implementation of nanotechnology in fuel cells. *Mater. Today Proc.*, 33, 2681–2685.

Kaur, M., Ali, A. (2011). Lithium ion impregnated calcium oxide as nano catalyst for the biodiesel production from karanja and jatropha oils. *Renew. Energy*, 36(11), 2866–2871.

Khalid, M.J., Waqas, A., Nawaz, I. (2019). Synergistic effect of alkaline pretreatment and magnetite nanoparticle application on biogas production from rice straw. *Bioresour. Technol.*, 275, 288–296.

Khanal, L.R., Sundararajan, J.A., Qiang, Y. (2020). Advanced nanomaterials for nuclear energy and nanotechnology. *Energy Technol.*, 8(3), 1901070.

Kouzu, M., Yamanaka, S.-y., Hidaka, J.-s., Tsunomori, M. (2009). Heterogeneous catalysis of calcium oxide used for transesterification of soybean oil with refluxing methanol. *Appl. Catal. A*, 355(1–2), 94–99.

Lamy, C. (2016). From hydrogen production by water electrolysis to its utilization in a PEM fuel cell or in a SO fuel cell: Some considerations on the energy efficiencies. *Int. J. Hydrog. Energy*, 41(34), 15415–15425.

Lima, J.S., Araújo, P.H.H., Sayer, C., Souza, A.A.U., Viegas, A.C., de Oliveira, D. (2017). Cellulase immobilization on magnetic nanoparticles encapsulated in polymer nanospheres. *Bioprocess Biosyst. Eng.*, 40(4), 511–518.

Liu, H., Wang, K., Cao, X., Su, J., Gu, Z. (2021). A new highly active La 2 O 3–CuO–MgO catalyst for the synthesis of cumyl peroxide by catalytic oxidation. *RSC Adv.*, 11(21), 12532–12542.

Liu, M.-q., Dai, X.-j., Guan, R.-f., Xu, X. (2014). Immobilization of *Aspergillus niger* xylanase A on Fe3O4-coated chitosan magnetic nanoparticles for xylooligosaccharide preparation. *Catal. Commun.*, 55, 6–10.

Madhuvilakku, R., Piraman, S. (2013). Biodiesel synthesis by TiO2–ZnO mixed oxide nanocatalyst catalyzed palm oil transesterification process. *Bioresour. Technol.*, 150, 55–59.

Magagna, D., Uihlein, A. (2015). Ocean energy development in Europe: Current status and future perspectives. *Int. J. Mar. Energy*, 11, 84–104.

Mahmood, T., Hussain, S.T. (2010). Nanobiotechnology for the production of biofuels from spent tea. *Afr. J. Biotechnol.*, 9(6), 858–868.

Mao, S.S., Shen, S., Guo, L. (2012). Nanomaterials for renewable hydrogen production, storage and utilization. *Prog. Nat. Sci. Mater. Int.*, 22(6), 522–534.

Mariño, M.A., Fulaz, S., Tasic, L. (2021). Magnetic nanomaterials as biocatalyst carriers for biomass processing: Immobilization strategies, reusability, and applications. *Magnetochemistry*, 7(10), 133.

McLean, G.F., Niet, T., Prince-Richard, S., Djilali, N. (2002). An assessment of alkaline fuel cell technology. *Int. J. Hydrog. Energy*, 27(5), 507–526.

Mishra, R.K. (2022). Pyrolysis of low-value waste switchgrass: Physicochemical characterization, kinetic investigation, and online characterization of hot pyrolysis vapours. *Bioresour. Technol.*, 347(2022), 126720.

Mishra, R.K., Mohanty, K. (2018). Thermocatalytic conversion of non-edible neem seeds towards clean fuel and chemicals. *J. Anal. Appl. Pyrol.*, 134, 83–92.

Mishra, R.K., Mohanty, K. (2019). Thermal and catalytic pyrolysis of pine sawdust (*Pinus ponderosa*) and gulmohar seed (*Delonix regia*) towards production of fuel and chemicals. *Mater. Sci. Technol.*, 2(2), 139–149.

Mishra, R.K., Mohanty, K. (2020a). Effect of low-cost catalysts on yield and properties of fuel from waste biomass for hydrocarbon-rich oil production. *Mater. Sci. Technol.*, 3, 526–535.

Mishra, R.K., Mohanty, K. (2020b). Pyrolysis characteristics, fuel properties, and compositional study of *Madhuca longifolia* seeds over metal oxide catalysts. *Biomass Convers. Biorefin.*, 10(3), 621–637.

Mishra, R.K., Muraraka, A., Mohanty, K. (2020). Optimization of process parameters and catalytic pyrolysis of *Cascabela thevetia* seeds over low-cost catalysts towards renewable fuel production. *J. Energy Inst.*, 93(5), 2033–2043.

Misson, M., Zhang, H., Jin, B. (2015a). Nanobiocatalyst advancements and bioprocessing applications. *J. R. Soc. Interface*, 12(102), 20140891.

Misson, M., Zhang, H., Jin, B. (2015b). Nanobiocatalyst advancements and bioprocessing applications. *J. R. Soc. Interface*, 12(102), 20140891.

Mo, H., Qiu, J. (2020). Preparation of chitosan/magnetic porous biochar as support for cellulase immobilization by using glutaraldehyde. *Polymers*, 12(11), 2672.

Mohamad, N.R., Marzuki, N.H.C., Buang, N.A., Huyop, F., Wahab, R.A. (2015). An overview of technologies for immobilization of enzymes and surface analysis techniques for immobilized enzymes. *Biotechnol. Biotechnol. Equip.*, 29(2), 205–220.

Nanotechnologies—Terminology, I. (2008). *Definitions for Nano-objects—Nanoparticle, Nanofibre and Nanoplate*. International Organization for Standardization.

Ngamcharussrivichai, C., Totarat, P., Bunyakiat, K. (2008). Ca and Zn mixed oxide as a heterogeneous base catalyst for transesterification of palm kernel oil. *Appl. Catal. A*, 341(1–2), 77–85.

Ormerod, R.M. (2003). Solid oxide fuel cells. *Chem. Soc. Rev.*, 32(1), 17–28.

Pandit, P.R., Fulekar, M. (2019). Biodiesel production from microalgal biomass using CaO catalyst synthesized from natural waste material. *Renew. Energy*, 136, 837–845.

Pandiyan, G.K., Prabaharan, T. (2020). Implementation of nanotechnology in fuel cells. *Mater. Today Proc.*, 33, 2681–2685.

Patel, V., Mahajan, Y. (2017). Techno-commercial opportunities of nanotechnology in wind energy. In: B. Raj, M. Van de Voorde and Y. Mahajan (eds.) *Nanotechnolgy for Energy Sustainability*, pp. 1079–1106. https://doi.org/10.1002/9783527696109.

Pelc, R., Fujita, R.M. (2002). Renewable energy from the ocean. *Mar. Policy*, 26(6), 471–479.

Pena, L., Angle, B., Burton, B., Charrow, J. (2012). Follow-up of patients with short-chain acyl-CoA dehydrogenase and isobutyryl-CoA dehydrogenase deficiencies identified through newborn screening: One center's experience. *Genet. Med.*, 14(3), 342–347.

Perry, M.L., Fuller, T.F. (2002). A historical perspective of fuel cell technology in the 20th century. *J. Electrochem. Soc.*, 149(7), S59.

Pramanik, P., Krishnan, P., Maity, A., Mridha, N., Mukherjee, A., Rai, V. (2020). Application of nanotechnology in agriculture. In: *Environmental Nanotechnology*, Volume 4. Springer, pp. 317–348.

Rai, M., Da Silva, S.S. (2017). *Nanotechnology for Bioenergy and Biofuel Production*. Springer.

Reddy, K., Nasr, M., Kumari, S., Kumar, S., Gupta, S.K., Enitan, A.M., Bux, F. (2017). Biohydrogen production from sugarcane bagasse hydrolysate: Effects of pH, S/X, Fe2+, and magnetite nanoparticles. *Environ. Sci. Pollut. Res. Int.*, 24(9), 8790–8804.

Rekha, B., Saravanathamizhan, R. (2021). Catalytic conversion of corncob biomass into bioethanol. *Int. J. Energy Res.*, 45(3), 4508–4518.

Sankaran, R., Markandan, K., Khoo, K.S., Cheng, C., Leroy, E., Show, P.L. (2021a). The expansion of lignocellulose biomass conversion into bioenergy via nanobiotechnology. *Front. Nanotechnol.*, 96.

Sankaran, R., Markandan, K., Khoo, K.S., Cheng, C.K., Ashokkumar, V., Deepanraj, B., Show, P.L. (2021b). The expansion of lignocellulose biomass conversion Into bioenergy via nanobiotechnology. *Front. Nanotechnol.*, 3(96), 793528.

Santos, E.C., dos Santos, T.C., Guimarães, R.B., Ishida, L., Freitas, R.S., Ronconi, C.M. (2015). Guanidine-functionalized Fe_3O_4 magnetic nanoparticles as basic recyclable catalysts for biodiesel production. *RSC Adv.*, 5(59), 48031–48038.

Sanusi, I.A., Suinyuy, T.N., Kana, G.E. (2021). Impact of nanoparticle inclusion on bioethanol production process kinetic and inhibitor profile. *Biotechnol. Rep. (Amst)*, 29, e00585.

Scarlat, N., Dallemand, J.-F., Monforti-Ferrario, F., Nita, V. (2015). The role of biomass and bioenergy in a future bioeconomy: Policies and facts. *Environ. Dev.*, 15, 3–34.

Schoonman, J., Perniu, D., Duta, A. (2005). Deviations from stoichiometry and molecularity in the solar cell material CuInS2. In: *International Materials Forum 2005–Frontiers in Materials Science & Technology*.

Shanmugam, S., Hari, A., Pandey, A., Mathimani, T., Felix, L., Pugazhendhi, A. (2020a). Comprehensive review on the application of inorganic and organic nanoparticles for enhancing biohydrogen production. *Fuel*, 270, 117453.

Shanmugam, S., Krishnaswamy, S., Chandrababu, R., Veerabagu, U., Pugazhendhi, A., Mathimani, T. (2020b). Optimal immobilization of *Trichoderma asperellum* laccase on polymer coated Fe3O4@ SiO2 nanoparticles for enhanced biohydrogen production from delignified lignocellulosic biomass. *Fuel*, 273, 117777.

Shao, G.N., Sheikh, R., Hilonga, A., Lee, J.E., Park, Y.-H., Kim, H.T. (2013). Biodiesel production by sulfated mesoporous titania–silica catalysts synthesized by the sol–gel process from less expensive precursors. *Chem. Eng. Sci.*, 215, 600–607.

Shuttleworth, P.S., Parker, H., Hunt, A., Budarin, V., Matharu, A., Clark, J. (2014). Applications of nanoparticles in biomass conversion to chemicals and fuels. *Green Chem.*, 16(2), 573–584.

Singh, N., Dhanya, B.S., Verma, M.L. (2020). Nano-immobilized biocatalysts and their potential biotechnological applications in bioenergy production. *Mater. Sci. Technol.*, 3, 808–824.

Singhvi, M., Kim, B.S. (2020). Current developments in lignocellulosic biomass conversion into biofuels using nanobiotechology approach. *Energies*, 13(20), 5300.

Song, J., Lei, T., Yang, Y., Wu, N., Su, P., Yang, Y. (2018). Attachment of enzymes to hydrophilic magnetic nanoparticles through DNA-directed immobilization with enhanced stability and catalytic activity. *New J. Chem.*, 42(11), 8458–8468.

Srivastava, N., Alhazmi, A., Mohammad, A., Haque, S., Srivastava, M., Pal, D.B., Singh, R., Mishra, P., Dai Viet, N.V., Yoon, T. (2021). Biohydrogen production via integrated sequential fermentation using magnetite nanoparticles treated crude enzyme to hydrolyze sugarcane bagasse. *Int. J. Hydrog. Energy*, 47(72), 30861–30871

Srivastava, N., Rawat, R., Sharma, R., Oberoi, H.S., Srivastava, M., Singh, J. (2014). Effect of nickel–cobaltite nanoparticles on production and thermostability of cellulases from newly isolated thermotolerant *Aspergillus fumigatus* NS (Class: Eurotiomycetes). *Appl. Biochem. Biotechnol.*, 174(3), 1092–1103.

Srivastava, N., Srivastava, M., Mishra, P., Ramteke, P.W. (2016). Application of ZnO nanoparticles for improving the thermal and pH stability of crude cellulase obtained from *Aspergillus fumigatus* AA001. *Front. Microbiol.*, 7, 514.

Surendiran, A., Sandhiya, S., Pradhan, S., Adithan, C. (2009). Novel applications of nanotechnology in medicine. *Indian J. Med. Res.*, 130(6), 689–701.

Suzuta, T., Toba, M., Abe, Y., Yoshimura, Y. (2012). Iron oxide catalysts supported on porous silica for the production of biodiesel from crude Jatropha oil. *J. Am. Oil Chem. Soc.*, 89(11), 1981–1989.

Tahir, N., Tahir, M.N., Alam, M., Yi, W., Zhang, Q. (2020). Exploring the prospective of weeds (*Cannabis sativa* L., *Parthenium hysterophorus* L.) for biofuel production through nanocatalytic (Co, Ni) gasification. *Biotechnol. Biofuels*, 13(1), 1–10.

Tanaka, S., Chang, K.-S., Min, K.-B., Satoh, D., Yoshida, K., Esashi, M. (2004). MEMS-based components of a miniature fuel cell/fuel reformer system. *Chem. Eng. Sci.*, 101(1–3), 143–149.

Tang, T., Chen, Y., Liu, M., Du, Y., Tan, Y. (2022). Effect of pH on the performance of hydrogen production by dark fermentation coupled denitrification. *Environ. Res.*, 208, 112663.

Trindade, S.C. (2011). Nanotech biofuels and fuel additives. In: *Biofuel's Engineering Process Technology*. Intech, pp. 103–114.

Valdivia, M., Galan, J.L., Laffarga, J., Ramos, J.L. (2016). Biofuels 2020: Biorefineries based on lignocellulosic materials. *Microb. Biotechnol.*, 9(5), 585–594.

Vasić, K., Hojnik Podrepšek, G., Knez, Ž., Leitgeb, M. (2020). Biodiesel production using solid acid catalysts based on metal oxides. *Catalysts*, 10(2), 237.

Wang, D., Ikenberry, M., Pe, L., Hohn, K. (2012). Acid-functionalized nanoparticles for pretreatment of wheat straw. *Journal of Biomaterials and Nanobiotechnology,* 3(3), 342–352.

Wang, H., Covarrubias, J., Prock, H., Wu, X., Wang, D., Bossmann, S.H. (2015). Acid-functionalized magnetic nanoparticle as heterogeneous catalysts for biodiesel synthesis. *J. Phys. Chem. C*, 119(46), 26020–26028.

Wang, W., Ji, S., Lee, I. (2013). Fast and efficient nanoshear hybrid alkaline pretreatment of corn stover for biofuel and materials production. *Biomass Bioenergy*, 51, 35–42.

Weber, W.J., Roberts, F.P. (1983). A review of radiation effects in solid nuclear waste forms. *Nucl. Technol.*, 60(2), 178–198.

Wee, J.-H. (2007). Applications of proton exchange membrane fuel cell systems. *Renew. Sustain. Energ. Rev.*, 11(8), 1720–1738.

Wei Guo, K. (2011). Green nanotechnology of trends in future energy. *Recent Pat. Nanotechnol.*, 5(2), 76–88.

Xie, W., Ma, N. (2009). Immobilized lipase on Fe3O4 nanoparticles as biocatalyst for biodiesel production. *Energy Fuels*, 23(3), 1347–1353.

Xie, W., Wang, H., Li, H. (2012). Silica-supported tin oxides as heterogeneous acid catalysts for transesterification of soybean oil with methanol. *Ind. Eng. Chem. Res.*, 51(1), 225–231.

Xie, W., Wang, T. (2013). Biodiesel production from soybean oil transesterification using tin oxide-supported WO3 catalysts. *Fuel Process. Technol.*, 109, 150–155.

Yang, G., Wang, J. (2018). Improving mechanisms of biohydrogen production from grass using zero-valent iron nanoparticles. *Bioresour. Technol.*, 266, 413–420.

Yu, X., Wen, Z., Li, H., Tu, S.-T., Yan, J. (2011). Transesterification of *Pistacia chinensis* oil for biodiesel catalyzed by CaO−CeO2 mixed oxides. *Fuel*, 90(5), 1868–1874.

Zabed, H., Sahu, J.N., Boyce, A.N., Faruq, G. (2016). Fuel ethanol production from lignocellulosic biomass: An overview on feedstocks and technological approaches. *Renew. Sustain. Energ. Rev.*, 66, 751–774.

Zhang, Y., Lu, J., Shimano, S., Zhou, H., Maeda, R. (2007). Development of MEMS-based direct methanol fuel cell with high power density using nanoimprint technology. *Electrochem. Commun.*, 9(6), 1365–1368.

Zheng, P., Wang, J., Lu, C., Xu, Y., Sun, Z. (2013). Immobilized β-glucosidase on magnetic chitosan microspheres for hydrolysis of straw cellulose. *Process Biochem.*, 48(4), 683–687.

Zhu, J., Zhu, M., Shi, Q., Wen, F., Liu, L., Dong, B., Haroun, A., Yang, Y., Vachon, P., Guo, X., He, T., Lee, C. (2020). Progress in TENG technology—A journey from energy harvesting to nanoenergy and nanosystem. *Ecomat.*, 2(4), e12058.

Zinkle, S.J., Was, G. (2013). Materials challenges in nuclear energy. *Acta Mater.*, 61(3), 735–758.

2 Bio-Inspired Nanomaterials

Applications in Artificial Photosynthesis and Energy Generation

Naveen Dwivedi and Shubha Dwivedi

CONTENTS

2.1 INTRODUCTION

Progress in bionanotechnology has led to the development of well-designed and sustainable materials in recent years, which have found a wide range of applications in areas ranging from medical to environmental engineering and high-energy storage as well as energy harvested interests in vital sciences which are more viable (Arico et al., 2005; Davis., 1997; Wiesner and Bottero, 2007; Walcarius et al., 2013; Zhang et al., 2013; Du et al., 2012, p. 1–6). Any physical activity in this world is carried out by the input of energy. However, the law of energy conservation clearly states that energy can neither be created nor destroyed, it can only be transformed from one form to another. All activities are driven by the flow of energy. The word 'energy' is

DOI: 10.1201/9781003316374-2

derived from the Greek word 'en-ergon', which means 'in-work'. Energy is the most important requirement for the economic and industrial development of any country.

Due to the limited availability of fossil fuel-based resources for energy generation, the switch towards renewable resources has started. These resources are abundantly available in nature these includes solar, wind, biomass, ocean, geothermal, hydro etc. All these resources are categorized as green and eco-safe resources, though there are some minor disadvantages in terms of cost-effectiveness and the fact that the energy flow depends on various natural phenomena beyond human control. In today's world, many researchers are working on the preparation of bio-inspired nanomaterial for energy production. Bio-inspired nanomaterials have been used in rechargeable lithium batteries as electrode materials. However, the lack of an effectual technology for producing solar fuels that can compete with fossil fuel utilization makes it exemplary.

The elementary idea is the transformation of light energy into electrical energy using photosynthetic microorganisms. The microbes will use their photosynthetic apparatus and the incoming light to break down the water molecule. The generated protons and electrons are collected using a bio-electrochemical system. Bio-photovoltaic systems are a green and eco-safe energy generating technology. BPV systems are sometimes also described as live solar panels. According to the available report of the International Energy Agency (IEA), between the years 2008 and 2035, there will be an increase of 36% of the world's primary energy demand and 2:2% electricity requirement (IEA) (Manaktala and Singh, 2016). Therefore renewable energy resources play a crucial role in bringing the world towards a more reliable, secure and sustainable energy path considering global climate change and CO_2 emissions.

As an estimate, 1h of sunlight received by our planet is equivalent to all the energy consumed by humans in a whole year. Solar energy could be a foremost primary energy source, and it must be stored and despatched on demand to the end user. Natural photosynthesis is an exclusively sustainable approach to storing solar energy in the form of chemical bonds. There is a huge requirement for such technology which has a year-round average conversion efficiency significantly higher than currently available by natural photosynthesis; as a result it could reduce land-area requirements and be independent of food production. Consequently, various research groups are working on the construction of an 'artificial leaf' which will be capable of capturing, converting and storing solar energy in the form of chemical bonds of a high-energy density fuel such as hydrogen and also producing oxygen from water.

2.2   NATURAL PHOTOSYNTHESIS AND ARTIFICIAL PHOTOSYNTHESIS

Following is the detail about natural and artificial photosynthesis.

2.2.1   NATURAL PHOTOSYNTHESIS

Photosynthesis is the most vital and ample source of solar energy conversion system on the planet and the most well tested over time; 86% of the energy in the form of fossil fuels being used today is provided by photosynthetic plants and bacteria. Photosynthetic pigments harvest photons to generate the assimilatory power of the reduction of carbon into energy-rich carbohydrates.

The harvesting of solar energy is easy compared to its conversion into chemical energy, which is the real herculean task. Researchers are searching for different ways of harnessing and converting solar energy into more easily usable forms to solve the issues related to geo-political problems and climate change. The awareness of the possibilities of utilizing bionanotechnology in artificial photosynthesis for innovating the design of solar panels has been reaching new heights in this field (Lewis and Nocera, 2006; Lal, 2018). Three apparently distinct domains of use of bio-inspired nanomaterial research, biofuel production, biological modulation and bio-photovoltaics, have been developed over the past several decades. Because of their shape- and size-based optical properties and excellent synthetic control the bio-inspired nanomaterials have unique technological advantages as light absorbers or energy transducers. Bio-inspired nanomaterials have also been incorporated into biological systems including proteins, biomolecules, bacteria and eukaryotic cells in a large collection of major studies and applications.

Nature has developed well-organized and proficient systems; mimicking natural processes in all their numerous forms will open a *Pandora box* of opportunities. Utilizing nanoparticles that can be prepared by a top-down or bottom-up approach contributes exceptional properties to the other known materials. Nanoparticles can be utilized to mimic nature's energy-efficient process of photosynthesis.

In the past few decades, a rapid explosion of bionanomaterials research and bionanotechnology has recently paved the way to develop strategies to modify existing pathways or to explore incipient pathways in biological systems (Yang et al., 2011; Ricco et al., 2018; Liang et al., 2016; Rengifo et al., 2014).

Photosynthesis is the technique of transforming solar energy into chemical power which is then stored as carbohydrates, the biomolecule shown in Figure 2.1. In the developmental stage of an ecosystem, green plants, algae and photosynthetic microorganisms perform this phenomenon without any significant change. The

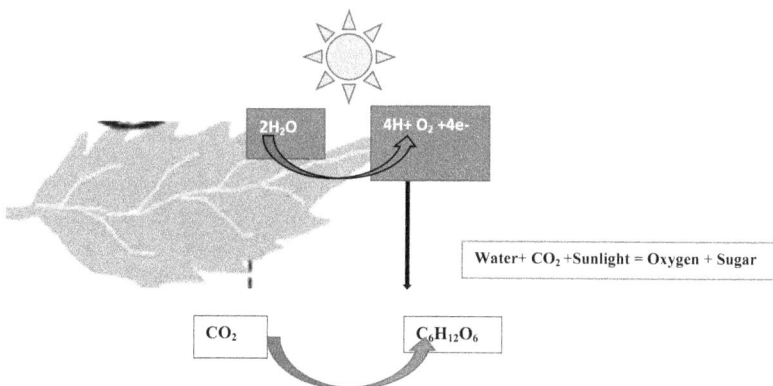

FIGURE 2.1 Illustration of natural photosynthesis.

photosynthetic products are rich energy compounds. The potential chemical energy of these compounds is the resultant of the light energy. Chlorophylls play a vital role in the absorption of light energy. Chlorophyll a and b are the two most abundant chlorophylls. Chlorophyll a is found in all autotrophic plants whereas chlorophyll b is absent in blue-green, brown and red algae. The other chlorophylls c, d and e, are found only in algae. Chlorophylls are located in grana thylakoids. The portion of the electromagnetic spectrum which participates in photosynthesis is from 300 to 900 nm. The visible spectrum is effective in the green plants for photosynthesis which ranges from 400 to 750 nm. Photosynthetic purple bacteria absorbs wavelength of range 300 to 900 nm, while green bacteria absorbs the range of visible spectrum between 375 and 800 nm. The efficiency of the different regions of the visible spectrum was measured on the basis of oxygen released during the process of photosynthesis. It was also earlier stated that the amount of oxygen released was found to be highest in blue and red absorption bands of the spectrum.

This energy-driven biological process takes place in the cellular membranes present in both prokaryotes and eukaryotes. These biological membranes are mainly made up of lipid bilayers, forming an operational framework that supports nearly all biological phenomena and functions for the survival of life. The lipid bilayer of the membranes is structurally composed of polymers with hydrophilic heads and hydrophobic tails.

The cell membrane also consists of proteins in addition to lipids, which are either surrounded or flank both the sides of the lipids making it topologically diverse (internal and external) and creating a mosaic pattern in the membrane. In order to perform highly specific functions like communication, locomotion, intracellular and extracellular transportation across membranes and the production of energy, the divergent pattern and other mechanism of membrane makes highly specific to perform each function.

In the years between 1915 and 1997 various researchers carried out their work on photosynthesis. In the history of photosynthesis, this field won the Nobel Prize nine times: in 1915, 1930, 1937, 1938, 1961, 1965, 1978, 1988 and 1997 as shown in Table 2.1.

Three distinct protein complexes are assimilated in the membrane of chloroplasts in higher plants, viz. two light-harvesting complexes (LHC) or antennae and cytochrome b6f complexes which are connected by different electron carriers allowing the photosynthetic electron transport chain to perform the light reaction of photosynthesis. The light-harvesting systems of photosystem II (LHCII) and photosystem I (LHCI) transfer photons in the form of resonance energy to the reaction centre. Both photosystems have a number of light absorbing pigments making an antenna complex to collect photons and allocate them to the reaction centre (RC), which mainly consists of chlorophyll a. This creates a charge difference across the membrane. On the donor side of Photosystem II (PSII), a strong oxidant complex is present which splits water into molecular oxygen, protons and electrons; this is commonly referred to as the oxygen-evolving complex (OEC).

Like various constant organic molecules, the electrons of the chlorophyll molecule are even in number and present in the orbitals of lowest energy. Each of these

TABLE 2.1

Work Carried out by Various Researchers on Photosynthesis Which Was the Basis of a Nobel Prize

S. No	Nobel Laureate	Basis	Year
1	Richard Martin Willstatter	Plant pigments, especially chlorophyll	1915
2	Hans Fischer	The constitution of haemin and chlorophyll and especially for his synthesis of haemin	1930
3	Norman Haworth	On carbohydrates and vitamin C	1937
	Paul Karrer	For his investigations on carotenoids, flavins and vitamins A and B2	
4	Richard Kuhn	On carotenoids and vitamins	1938
5	Melvin Calvin	Carbon dioxide assimilation in plants	1961
6	Robert Burns Woodward	For his outstanding achievements in the art of organic synthesis	1965
7	Peter D. Mitchell	The understanding of biological energy transfer through the formulation of the chemiosmotic theory	1978
8	Johann Deisenhofer, Robert Huber and Hartmut Miche	For the determination of the three-dimensional structure of a photosynthetic reaction centre	1988
9	Paul D. Boyer and John E. Walker	For their elucidation of the enzymatic mechanism underlying the synthesis of adenosine triphosphate (ATP)	1997
	Jens C. Skou	For the first innovation of an ion-transporting enzyme, Na+, K+-ATPase	1997

molecular orbitals has two electrons spinning in opposite directions. A molecule in which all spins are paired has no magnetic effect and has maximum stability; it is considered that the molecule is in its ground state. Two types of spin are found in electrons, viz. right-handed spin and left-handed spin. Both the sides, electrons are paired in such a way that they cancel each other during spin. The moment a photon of light energy is received by the chlorophyll molecule, one electron from the electron pair is raised to a higher orbit. Absorption of light occurs in a femtosecond (10^{-15} sec). Due to the complex structure of chlorophyll, it has many electrons, slightly specific to different energy requirements, which in turn are taken up by absorption at different wavelengths. Since the lifespan of the first singlet excited state is very short, i.e., 10^{-13} sec, the electron returns to the ground state rapidly dissipating the energy by heat or by energy transfer.

2.2.2 Artificial Photosynthesis

The first scheme of artificial photosynthesis was described in 1912 by an Italian chemist, Giacomo Ciamician, who recognized the unsustainability of fossil fuels (Ciamician, 1912). He proposed the idea of replicating nature's way of creating and storing energy as a substitute. The Swedish Consortium for Artificial Photosynthesis was established in 1994, the very first of its kind, as an association between groups of three different universities, Stockholm, Uppsala and Lund. The Swedish Consortium for Artificial Photosynthesis, 1994, was built with multidisciplinary objectives and approaches to focus on learning from natural photosynthesis and applying this knowledge in biomimetic systems to create artificial photosynthesis, a new approach. Research on artificial photosynthesis is experiencing a boom at the start of the 21st century.

Eventually the light energy captured in the form of photons is converted into chemical energy in two energy-rich molecules, viz. NADPH and ATP. The energy contained in these two molecules is used for carbon dioxide assimilation, and fixing atmospheric carbon. These essential components of photosynthesis for the collection and conversion of solar energy work at nanoscale, and it provides a basic road map for green, renewable, recycled and clean energy strategies that capture sunlight. The complete process of photosynthesis is extremely complicated, and to exactly mimic or copy the structures and components is a tough task. Even though artificial systems try to incorporate elementary principles of the natural process of photosynthesis, that certainly does not result in the exact reproduction of the natural phenomenon. Artificial photosynthesis is actually the bio-mimicking of natural photosynthesis with the use of bionanotechnology. The bio-mimicking of photosynthesis aims to capture solar energy for the production of electricity or fuel with enhanced efficiency for sustainable development and efficient energy management. Part of the idea of artificial photosynthesis is that efficient bio-inspired nanomaterial should absorb light in the near-infrared as well as visible range of the solar spectrum.

Generally, the science of solar energy focuses, first, on the absorption and transformation of solar energy and, second, on the storage or extraction of electricity, heat or fuel produced by solar energy. A deep understanding of photosynthesis is essential for the advanced study of artificial reaction centres which will serve a dual objective, viz. initially, it will help in an enhanced understanding of photosynthesis as well as developing bio-inspired nanomaterial systems to convert solar energy into fuel with simpler nanostructures. Artificial photosynthesis is a field related to the advanced bionanotechnological approach.

Artificial photosynthesis (AP) requires the designing of a system that can able to proficiently absorb light, transport and separate electrical charges and finally catalyse a reaction to produce energy. To achieve highly efficient energy, the complete process must operate synchronously. By providing a means for scalable energy storage, the direct production of fuels from sunlight could provide a technological solution to challenges that accompany the large-scale integration of intermittent energy sources, such as solar, into electrical grids. When sunlight is available, artificial photosynthetic systems would produce a fuel (stored energy) that would then be available for subsequent on-demand combustion or for use as a feedstock for a fuel cell. Artificial photosynthesis could also provide an accessible technology for the carbon-neutral production of transportation fuels for the 40% of global transportation (ships, aircraft, heavy-duty trucks) which requires a high-energy-density fuel for viable operation (Lewis, 2016; Wasielewski, 2006; Wasielewski, 2009).

Various studies reveal that the vital procedures in both artificial and natural photosynthesis are essentially the same; however the nanostructure used in artificial photosynthesis is modest and simple in comparison to the complex molecular frame-work in biological systems. The progressions in bionanotechnology, in the field of imaging and employment of nanomaterials, have been used to fabricate these bio-inspired nanoparticles.

Success has been attained in some belongings with the coupling of synthetic materials with existing biological systems; this has also been proven to better adapt to adverse and lethal environmental conditions, extended life cycles and the transformation of non-photosynthetic species into photosynthetic species. There are many biological functions that have been acquired by biological species over time with evolution; however, scientists are expecting that the linking of new functional synthetic materials with pre-existing biological systems will generate bio-inspired nanoparticles which will be more active and would be an effective substitute for natural evolution as well as genetic engineering. In-depth study has been undertaken in the last decade to design bio-inspired nanoparticles systems including composite materials derived from both synthetic materials as well as living organisms. The biological components used to create bio-inspired nanopar-ticles system are purified biomolecules like DNA and RNA, proteins and complex living systems like living cells, tissues and organisms; the synthetic components used include inorganic materials (e.g., silicon dioxide, calcium carbonate, gold, silver, quartz, carbon materials, or iron oxide), organic materials like diverse types of lipids and other organic polymers and hybrid materials like metal-phenolic net-works (MPNs) and metal-organic frameworks (MOFs) (Liang et al., 2016; Rengifo et al., 2014; Kwak et al., 2017; Su et al., 2016; Richardson and Liang, 2018; Lauth et al., 2017; Setyawati et al., 2015; Niu et al., 2017; Lynge et al., 2011; Lykourrinou et al., 2011; Kodis et al., 2002).

The bio-functionality of the biological system and synthetic system affected the selection procedure for the same. Figure 2.2 clearly shows the integrity approach of biological system with synthetic system.

2.2.3 ARTIFICIAL PHOTOSYNTHESIS MECHANICS

There are three key steps in artificial photosynthesis as explained in Table 2.2, which are analogous to natural photosynthesis, as shown in Figure 2.3:

I. The absorption of light to reach an excited state
II. Charge generation and separation
III. Chemical conversion for the production of fuel

There are other biological methods available for biofuel/energy production with the help of engineered photosynthesis, using living organisms such as algae and bacte-ria (Gibson, 2010). Sometimes discussed as 'artificial life', these nano-engineered microorganisms are noted by a tailored and laboratory-synthesized genome, but still require living cytoplasmic input for genome 'boot-up' and reproduction (Gibson, 2010). One recombinant organism of this type of cyanobacteria has been proposed by the J Craig Venter Institute and used to produce hydrogen gas and other fuels.

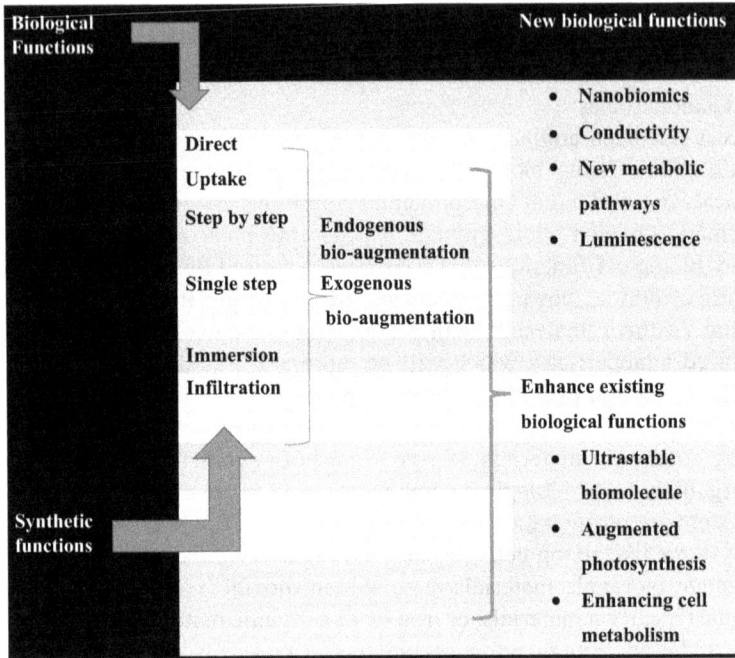

FIGURE 2.2   Integration of biological system with synthetic system.

2.3   ROLE OF BIO-INSPIRED NANOMATERIAL IN ARTIFICIAL PHOTOSYNTHESIS

The increasing demand for green, inexpensive and environmentally sustainable materials has stimulated researchers in different fields to draw motivation from nature in developing materials with unique properties such as miniaturization, ordered organization and malleability. Together with the extraordinary properties of nanomaterials, over the past decades, the field of bio-inspired nanomaterials has taken an interest among researchers. Bionanomaterials (BNMs) have unique properties like high surface area and improved optical properties. For a biological reaction, the rate of reaction depends on the surface area of the reactants, and due to the large surface area, nanomaterial-mediated reactions operate at a high rate. Nanomaterials are of different types, viz. natural, anthropogenic and engineered nanomaterials as explained in Table 2.3.

Different photosensitizers like semiconductors, molecular dyes and quantum dots (QD) are used as light absorbers and generate the charge separation. Artificial photosynthesis comprises the transfer of multiple electrons and protons; as a result, these reactions are kinetically very inactive and henceforth high potential is required to complete these reactions. QDs are excellent light absorbers, competent in electron excitation and have charge separation properties. In addition, they also possess some inherent properties that make them beneficial in exploiting artificial photosynthesis for large-scale solar conversion to energy.

TABLE 2.2
Key Steps in Artificial Photosynthesis

Light Absorption	Water Splitting	CO$_2$ Reduction
The first step in artificial photosynthesis is the absorption of light photons as a source of energy to drive the system. This region of study focuses on the role of photosensitizers that make optimal use of photon exposure, and are capable of combining light energy.	Water splitting involves the disintegration of water into oxygen and hydrogen by the means of a chemical redox reaction. Upon irradiation, semiconductor nanowires absorb light and the oxidization of water occurs, producing oxygen, as well as electrons and protons.	The reduction, or fixation, of carbon dioxide is another dynamic process allied with artificial photosynthesis. This method involves the generating of other hydrocarbon fuels by chemically reducing CO$_2$ with the utilization of hydrogen, in addition to the production of oxygen and hydrogen derived from water.
One of the limiting characteristics of natural photosynthesis is the fact that most pigment molecules in photosynthesizing organisms can only absorb light at wavelengths within the range of about 400–700 nm.	The redox equations involved in water splitting can be seen as: Oxidation: $2H_2O \rightarrow 4e^- + 4H^+ + O_2$ Reduction: $4H^+ + 4e^- \rightarrow 2H_2$ Combined Reaction: $2H_2O \rightarrow 2H_2 + O_2$ A noteworthy attribute of water splitting is that it involves the oxidation of water to produce oxygen as well as the reduction of water to produce hydrogen. Molecular water-oxidation catalysts are specialized for the evolution of oxygen. Molecular water-reduction catalysts perform the production of hydrogen.	Although the exact process has not yet been established, scientists have patched together an idea of the process that takes place during the reduction of CO$_2$ by light, which is similar to that of water splitting. The physical and chemical environment like temperature, pressure, applied energy, etc., plays a huge role in the production of fuel with this process.

In green plants, chloroplasts are the location of synthesis for chemical energy, that is, carbon-based fuels, where photosynthesis takes place. In the presence of light energy, the captured atmospheric CO$_2$ is changed into different forms of sugars. The photosynthetic apparatus which includes the chloroplast system utilizes less than 10% of the sunlight, which is very low; as an approximation, 90% solar light remains uncaptured, and there is a huge research area open for introducing the bio-inspired nanomaterial-based pigment to plant cells to help in capturing a large solar spectrum. With unique properties and higher stability, the bio-inspired nanomaterials can form chloroplast-based photocatalytic complexes with heightened and better-quality functional properties under ex vivo and in vivo conditions (Weise et al., 2004; Zhu et al., 2010; Blankenship et al., 2011). It is very clear that neither the

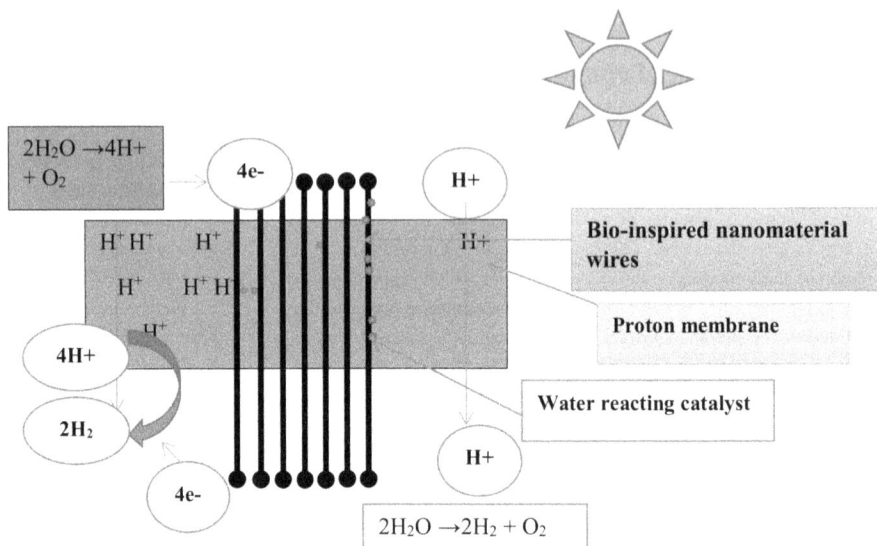

FIGURE 2.3   Diagrammatic representation of artificial photosynthesis.

TABLE 2.3
Types of Nanomaterials

Natural nanomaterials	These are found naturally; there is no human involvement in creating these types of nanomaterials, e.g., soil dust, sea spray, sea salt, etc.
Anthropogenic nanomaterials	These are created with the help of human involvement
Engineered nanomaterials	These are manufactured or designed on the basis of human interest or applications

---

chloroplasts capture all the solar radiation nor are all the absorbed photons involved in the electron flow. With this in mind, it is clear that the focus should be on artificial photosynthesis in which the incorporated bio-inspired nanoparticles will be useful in absorbing the maximum solar light.

The engineered and enhanced photosynthetic reactions might be accredited to the electronic band gap of semiconducting, the single walled carbon nanotubes which converts the absorbed solar light into photosynthetic excitons. The plants have been designed in such a way that they can harvest more light energy by delivering carbon nanotubes into the chloroplast. These carbon nanotubes serve as potential artificial probes permitting the chloroplast to capture wavelengths of light outside the normal range, that is, ultraviolet, green and near-infrared (Cossins, 2014; Giraldo et al., 2014).

Plant photosystems consist of RCs and the antenna chlorophylls, which receive solar light; they are bound in the membrane by weak intermolecular forces. The antenna chlorophyll absorbs photons and transfers them to the reaction centres and then electrons are transferred to the next electron acceptor. Naturally, photosynthetic machinery absorbs light within certain wavelength intervals as described in section of this chapter. It has been seen that if bionanoparticles conjugate with these active centres and antenna chlorophyll, there is an excite enhancement effect created due to the plasmon, as metal nanoparticles have an oscillating free electron; this effect is also called the plasmon enhancement effect. This excited electron can be used for photocurrents or chemical reactions. The association of metal nanoparticles with photosynthetic systems has been reported to enhance the efficiency of photosystems (Nabiev et al., 2010).

Biological artificial photosynthesis offers a transitional technology for combustion-based energy needs while offering an avenue towards carbon-neutrality. This will be most important in the next ten years as the mechanical work carried out by combustion engines is transitioned to electrically driven induction motors, which offer superior energy conversion to mechanical work. Biological artificial photosynthesis will continue after this transition as a cost-effective means of capturing carbon dioxide from other industrial processes and mitigating greenhouse gas accumulation.

Artificial bionanostructures are made up of various metal nanoparticles and a photosynthetic system, which also display strong augmentations of photosynthetic efficiency, and this causes the equivalent growth in light absorption by chlorophylls and energy transfer from chlorophylls to bionanoparticles (Govorov and Carmeli, 2007; Nadtochenko et al., 2008; Mingyu et al., 2007). Different components, methods and routes are given in Figure 2.4.

The design of bio-inspired nanomaterials requires further technological improvement; and more focused discovery is required to assemble materials with the help of top-down and bottom-up approaches after proper understanding of their architectural design. The integration of the approaches, viz. top-down and bottom-up, now quite distinctly representing different operating principles, is the key to bionanomaterials synthesis, as shown in Figure 2.5.

Bionanotechnology is an emergent field of biological engineering and has enormous potential to alter or augment plant function by employing nanomaterial. Bionanotechnology has great prospects and strength to develop new gears for the assimilation of bionanoparticles into plants to augment the existing functions (Ghorbanpour et al., 2017; Giraldo et al., 2014; DeRosa et al., 2010; Nair et al., 2010; Lahiani et al., 2013; Siddiqui, 2014; Cossins, 2014).

With bionanotechnological advancement, plant species can perform various functions:

    I. Plants are capable of self-powering themselves as light sources
    II. Plants are capable of self-powered groundwater sensors
    III. Plants are capable of imaging objects in their environment
    IV. Plants are capable of developing infrared communication devices
    V. Plants are capable of solar energy harnessing and biochemical sensing
    VI. Nano-bionic plants were developed for enhanced photosynthesis and biochemical sensing

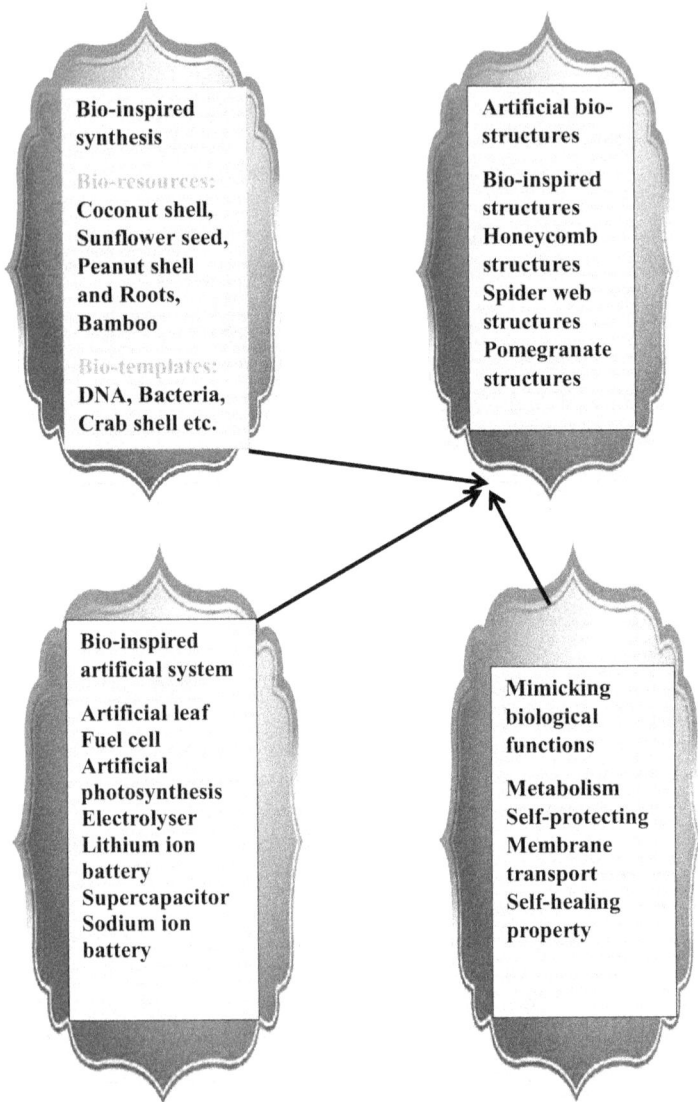

FIGURE 2.4   Different components, methods and routes for artificial photosynthesis.

2.3.1   BRIEF REVIEW OF WORK DONE BY VARIOUS RESEARCHERS IN THE FIELD OF BIO-INSPIRED NANOMATERIAL

In 2013, Olejnik et al. stated that with the use of silver nanowire conjugate up to an average of a ten-fold increase in chlorophyll fluorescence was recorded, and one study by Mackowski and his research group, in 2008, indicates a higher rate of generation of excitations in the chlorophylls.

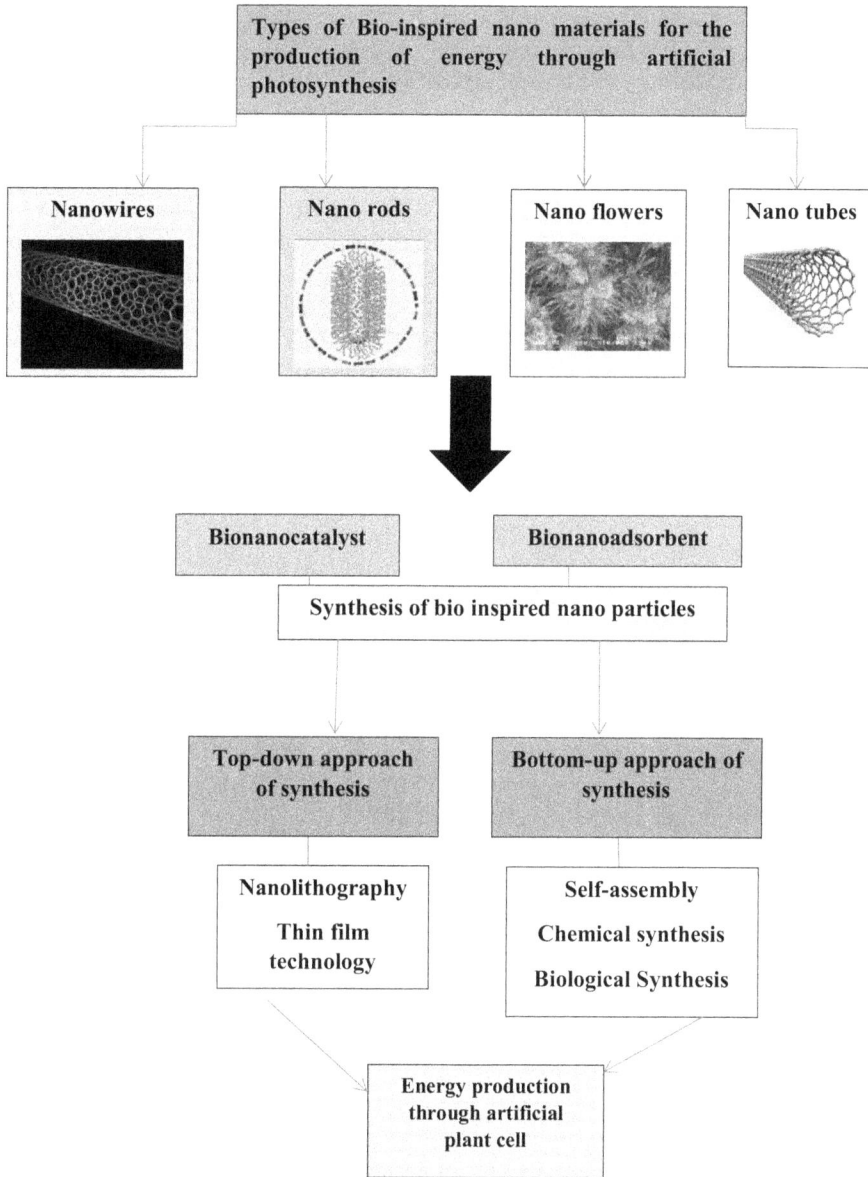

FIGURE 2.5   Types of bio-based biogenic nano structures for purposes in energy production through artificial photosynthesis.

Giraldo et al., in 2014, reported a 49% increase in electron transfer rate in *ex vivo* conditions (in chloroplast extracted from baby spinach leaves) after treatment with single walled nanotubes. Single walled nanotubes also enhanced the light reaction *in vivo* in leaves of *A. thaliana*. Calkins et al., in 2013, observed a similar effect; they observed improved photo-electrochemical activity in spinach thylakoid conjugated with carbon nanotubes under illumination.

Gao et al., in 2013, on the contrary, observed various functions like reduced PSII quantum yield, photochemical quenching, electron transfer rate, chlorophyll fluorescence and higher non-photochemical quenching and water loss as a result of the exogenous application of TiO2-anatase NPs.

Joshi et al., in 2017, Zheng et al., in 2005, and Mingyu et al., in 2007, reported titanium oxide nanoparticles used to seed germination, plant growth, water absorption, photosynthesis and nitrogen metabolism. Titanium oxides nanoparticles also alleviate heat stress through regulating stomatal opening.

Qi et al., in 2013, reported an increased net photosynthetic rate, water conductance and transpiration rate with exogenous application of TiO_2.

Pradhan et al., in 2013, found that manganese nanoparticles induced an increase in the hill reaction rate in mung bean (*Vigna radiata*).

Noji et al., in 2011, reported that the photosystem II (PSII) of *T. vulcanus* conjugated with nanomesoporous silica compound (SBA) maintained the high and stable oxygen-evolving ability of PSII in species.

Falco et al., in 2011, observed the effect of gold nanoparticles (5–20 nm) on PSII chlorophyll a fluorescence quenching in soybean leaves; in this study this effect was shown to be concentration-dependent. Falco et al. detected a swing in fluorescence pattern towards a higher wavelength in gold nanoparticle-treated soybean leaves. The same enhanced PSII quantum efficiency effect was also reported in Indian mustard by Sharma et al., in 2012.

Ze et al. in 2011 worked on the conjugate of *A. thaliana* with TiO_2 nanoparticles. They reported an increased expression of LHCII b and contents of LHCII in the thylakoid membrane of *A. thaliana* after the application of TiO_2 nanoparticles. Further they explained that titanium oxide nanoparticles stimulate the light absorption by chloroplast and also help in the regulation of the distribution of light energy from PSI to PSII by increasing LHCII content, which in turn accelerates the transformation from light energy to electronic energy, water photolysis and oxygen evolution.

Nabiev et al., in 2010, reported that *Rhodobacter sphaeroides* species showed an efficient transfer of electron to reaction centres, and this species was conjugated artificially with the quantum dots (artificial antennae which absorb light efficiently in a wide range of photon energies from the solar spectrum). The competent energy transfer from quantum dots to the bacterial reaction centre clearly offers an opening for the utilization of bionanocrystals to enhance photosynthetic biological functioning (Nabiev et al., 2010).

Nadtochenko et al., in 2008, detected that alumina nanoparticles are used for the isolation of photosynthetic reaction centres which shows enhanced electron transfer efficiency.

Linglan et al., in 2008, observed that Rubisco carboxylation is responsible for a higher photosynthetic carbon reaction as a result of nano-anatase-induced marker genes for Rubisco activase mRNA, enhanced protein levels and activities of Rubisco activase.

Govorov and Carmeli, in 2007, worked on metal nanoparticles which have the ability to influence the energy conversion efficiency in photosynthetic systems. They further stated that the binding to gold and silver nanoparticles with chlorophyll molecule results in a novel hybrid system, which could produce ten times more excited electrons due to plasmon resonance and fast electron–hole separation.

Lei et al., in 2007, reported that nano-anatase was used to promote electron transport chain reaction, the photoreduction activity of PSII, the evolution of O_2 and the photophosphorylation of chlorophyll under both visible and ultraviolet light.

Gao et al., in 2006, reported that the nano-anatase TiO_2 improved light absorbance and the conversion of light energy to electron energy and ultimately to chemical energy, and this promotes carbon dioxide (CO_2) assimilation. Treatment of nano-anatase TiO_2 improved Rubisco-carboxylase activity 2.67 times in spinach as compared to control, which consecutively activates Rubisco carboxylation and eventually the rate of photosynthesis increase.

Hong et al., in 2005, and Yang et al., in 2006, reported that TiO_2 nanoparticles are capable of protecting chloroplasts from ageing during long illumination regimes, promote chlorophyll formation and stimulate Rubisco activity, which in turn results in increased photosynthesis or enhanced photosynthetic carbon assimilation. The same observation was also reported by Sharma et al. in 2012.

Hong et al., in 2005, found that nano-TiO_2 (rutile) influences the photochemical reaction in spinach chloroplasts. The spinach treated with 0.25% nano-TiO_2 showed improved up-hill reaction and oxygen evolution. The noncyclic photophosphorylation activity was found to be higher than cyclic photophosphorylation in chloroplasts. This increase in photosynthesis with nano-TiO_2 might be associated with the activation of a photochemical reaction in spinach chloroplasts.

Zheng et al., in 2005, reported a similar study; they found that when aged spinach was treated with 2.5% nano-TiO_2 rutile, it showed an increase in dry weight, chlorophyll formation, the ribulose bisphosphate carboxylase/oxygenase activity and the photosynthetic rate.

Hong et al., in 2005, observed that bread wheat (*Triticum aestivum* L.) showed an increase in grain number, biomass, stomatal density, xylem-phloem size, epidermal cells and water uptake after seed priming with multi-walled carbon nanotubes.

Bujak et al., in 2009, reported dinoflagellate *Amphidinium carterae* showed solid enhancement in fluorescence intensity of protein-bound chlorophyll molecules conjugated with silver nanowire (Bujak et al., 2009).

Various studies like Barazzouk et al., 2005, Nieder et al., 2010, and Beyer et al., 2011, reported a detailed mechanism of electron transfer from excited fluorophore to Au or Ag nanoparticles.

Further, in 2003, Kuang stated that TiO_2 nanoparticles are responsible for the transfer of charges between light-harvesting complex II (LHCII) and TiO_2 NPs because of their photocatalytic properties which induced a reduction–oxidation reaction (Kuang, 2003).

As per the report of BETO publications, the recent goal, defined up to 2030, of the Bioenergy Technologies Office Advanced Algal Systems programme within the Department of Energy is $2.5–3 gallon of gas equivalent for renewable algal biofuels, while current gasoline prices remain relatively low at \$2.18 per gallon (BETO Publications, 2020). Furthermore, due to the speedily rising world population, food production must increase by more than 50% in the coming decades to overcome food shortages with a more limited amount of productive land and in a changing climate (Hatfield et al., 2011; Xu et al., 2013; Masson-Delmotte et al., 2018; Lowry et al., 2019).

2.3.2 DIFFERENT PRODUCTS OF ARTIFICIAL PHOTOSYNTHESIS

Broadly two types of solar fuels can be foreseen as products of artificial photosynthesis.

Hydrogen: When the substrate in the reaction is water then the resultant fuel is hydrogen. Hydrogen has been recognized as a striking zero-carbon energy carrier that could play a key role in future renewable energy technology. To produce hydrogen, the protons that result from the splitting of water need to be reduced as shown in reaction below:

$$2H_2O \rightarrow 4H^+ + O_2$$

Carbon-based fuels: These are complex carbohydrates consisting of carbon–carbon bonds. However, combining protons with carbon dioxide out of the air to produce carbon-based organic fuels is much more challenging in comparison to producing hydrogen. This procedure involves tough multi-electron chemistry, even for simple energy carriers such as methane, methanol and syngas (a mixture of carbon mono-oxide and hydrogen). Artificial molecular nano-based catalysts and semi-synthetic microorganisms have been developed for accelerating the conversion process. Hydrogen and carbon mono-oxide can be utilized as forerunners for other fuels like various Fischer–Tropsch-based fuels, alcohol and methane, that may be fused into our current energy set-up.

As per the study, in an artificial photosynthesis reaction, two photons are required to transfer one electron. As per the literature, supposing 40 moles of photons per metre squared per day, and then 20 tons of H_2 produced per km^2 per day, formic acid produced 460 per km^2 per day with respect to 440 tons of CO_2 converted per km^2 per day, CO produced 280 per km^2 per day with respect to 440 tons of CO_2 converted per km^2 per day, methanol produced 107 per km^2 per day with respect to 147 tons of CO_2 converted per km^2 per day and biogas produced 40 per km^2 per day with respect to 110 tons of CO_2 converted per km^2 per day.

Reactions of conversions are as follows:

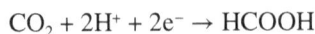

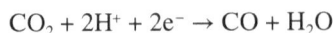

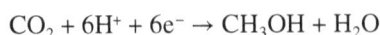

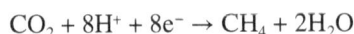

$$2H^+ + 2e^- \rightarrow H_2$$

$$CO_2 + 2H^+ + 2e^- \rightarrow HCOOH$$

$$CO_2 + 2H^+ + 2e^- \rightarrow CO + H_2O$$

$$CO_2 + 6H^+ + 6e^- \rightarrow CH_3OH + H_2O$$

$$CO_2 + 8H^+ + 8e^- \rightarrow CH_4 + 2H_2O$$

2.4 DEVELOPMENT OF CHLOROPLAST AND NANOPARTICLE INTERFACE: A ROUTE FOR THE INSERTION OF NANOPARTICLES INTO CHLOROPLASTS

The goal of the 'Green Revolution' has two objectives, firstly to prevent food shortages and secondly to foster green energy production with the help of bionanotechnology, plant physiology and molecular biology (Long et al., 2015).

The new route for the development of artificial photosynthesis is to strengthen the chloroplast of the plant cell which is the primary site of photosynthesis. Many researchers work on chloroplast bionanotechnology in which a route has been designed for injecting the nanoparticles into the chloroplast; this may lead to the development of a new era of synthetic biology, i.e., bionanotechnology. In enhancing chloroplast performance, bionanotechnology ensures the targeted delivery of chemicals and genetic elements to chloroplasts, nanosensors for chloroplast biomolecules and nano-therapeutics. Chloroplasts can produce renewable fuel that is environmentally sustainable and cost effective.

When the research on chloroplast bionanotechnology was started, various questions came to the minds of researchers like how possible is the conjugation of nanoparticles with plant chloroplasts as nanoparticles are very small in size and shape, charge, hydrophobicity etc., and plant cells also have barriers, like the plant cuticle, cell wall and matrix, which may hinder the interactions. Still, the impact of the physical and chemical properties of nanoparticles and plant membranes on the uptake into chloroplasts is not well understood. However, fresh studies have a better understanding of the translocation of nanoparticles through chloroplast galactolipid-based membranes.

Organelle membrane surfaces of plants create obstacles for delivering nanoparticles with their cargo into chloroplasts. There are various methods of nanoparticle delivery to chloroplast (Economou et al., 2014):

   I. Particle bombardment
  II. Rely on pressure and force to deliver microcarriers
 III. Spontaneous penetration of lipid membranes via diffusion *in vitro*
  IV. Leaf infiltration using a needleless syringe
   V. Topical foliar delivery mediated by surfactants

Obstacles created by the plant cell during the transport of nanoparticles to improve solar energy capturing efficiency is a challenge for researchers, and to overcome these obstacles, some new transport procedures are sometimes included by them. The major challenges are created by the plant cell wall, the plant cell membrane, the cytosol and the chloroplast double membrane; in algae, there can also be an outer epilithic algal matrix. Each of these plant bio-surfaces represents various physical and chemical barriers that can limit nanoparticle uptake by size, charge, hydrophobicity and other properties (Kramer et al., 2014; Giraldo et al., 2014; Wong et al., 2016; Lew et al., 2018).

The plant cell wall contains cellulose microfibrils, pectin and cross-linking glycan (Barros et al., 2015), and it acts as the first substantial fence for nanomaterial to enter into the plant cell. Hu et al., in 2020, also reported the same findings as Karmer presented in 2014. His research showed the role of plant cell wall pore size, charge and hydrophobicity which limits the entry of nanoparticle into cells but is not well understood; some more advanced research is required in this field. In continuation, he states that some specific plant cells permit the entry of nanoparticles, and these are size-specific, e.g., nanoparticles up to 18 nm were capable of permeating cotton leaf cells, while nanoparticles larger than 8 nm could not permeate the maize leaf cells (Hu et al., 2020). Avellan et al., in 2019, worked on hydrophilic nanoparticles; this study involves advanced technology, grounded on high spatial and temporal resolution confocal fluorescence microscopy, and recommends that at a size less than 20 or 10 nm, hydrophilic nanoparticles with a positive charge depending on plant type and leaf anatomy are more efficiently delivered into chloroplasts and plant cells. Conversely, other studies have perceived amphiphilic nanoparticles up to a size of 40 nm to translocate across leaf cells and into other plant organs (Avellan et al., 2019). In addition, Palocci et al., in 2017, reported that poly- and mono-dispersed poly(lactic-co-glycolic) acid nanoparticles inhibit the cell wall in grapevine cells over 50 nm in size whereas the plasma membrane is permeable from 500 to 600 nm with the same nanoparticles (Palocci et al., 2017).

The transport of nanoparticles into the plant cell was also prohibited by the lipid bilayer composed of phospholipids, carbohydrates and proteins. Some researchers reported that high charged nanoparticles can cross the chloroplasts as well as the plasma membrane of the plant cell wall.

In 2018, Lew and his research work suggested the lipid exchange envelope penetration (LEEP) model which recommends a disruption of the lipid bilayer by the ionic cloud surrounding nanoparticles; passive penetration rather than energy-dependent endocytosis is hypothesized as the mechanism for nanoparticle uptake (Lew et al., 2018). Modelling studies of nanoparticle uptake by chloroplasts highlight the importance of nanoparticle charge.

After entering into the plant cell these nanoparticles cross the cell wall and membrane, and nanoparticles must then pass through the cytosol, containing a variety of different biomolecules, including proteins. Nanoparticles passing through the cytosol are expected to be coated with biomolecule coronas, but this is poorly understood within plants. As already suggested by the previous researchers, high charged nanoparticles can easily cross the biomembrane; on the basis of this, in 2020, Prakash and Deswal established that gold nanoparticles interfaced with plant extracts from *Brassica juncea* formed protein coronas increasing the nanoparticle surface charge by approximately 30% after 36 h of interaction. Mass spectrometry showed that 27% of the hard corona formation around the gold nanoparticle comes from the plant energy-yielding pathways including glycolysis, photosynthesis and ATP synthesis (Prakash and Deshwal, 2020).

The last hindrances to reaching the chloroplast are its double lipid bilayers, referred to as the chloroplast membranes. The chloroplast membranes are formed by galactolipids and are highly dynamic (Block et al., 2007). The chemical interactions of nanomaterials with the phospholipid-based membranes of eukaryotic cells have been thoroughly studied (Sanchez ct al., 2012; Wu et al., 2013; Wang et al., 2016;

Lew et al., 2018). However, there are no studies of nanomaterial interactions with the galactolipid-based membranes that form the majority of the chloroplast envelopes. Highly positively or negatively charged nanoparticles interact with the exposed lipids, allowing diffusion and eventual kinetic trapping into isolated chloroplasts without mechanical aid (Wong et al., 2016). These nanoparticles can be larger than chloroplast porin's diameter of 2.5–3 nm, and channel proteins, including mechanosensitive channels, have the largest diameter in chloroplast membranes (Ganesan et al., 2018). High and low aspect ratio nanomaterials, such as carbon nanotubes and carbon dots, respectively, are capable of penetrating plant cells and chloroplasts with high efficiency (Giraldo et al., 2014; Wong et al., 2016; Hu et al., 2020; Santana et al., 2020). However, the role of nanomaterial aspect ratio on entry into cells and chloroplasts has not been systematically explored with nanomaterials with precise control of aspect ratios.

2.5   FUTURE RECOMMENDATIONS

Photosynthesis is such a complicated natural process that to mimic it in full form is a difficult task. In general, the photosystems of chlorophylls existing in nature to capture light energy during photosynthesis are very proficient and the basis of molecular photovoltaic nano-machines which bring about electrical charge separation, creating potential difference across the membrane of high dielectric strength. Copying the natural photosynthetic machinery constitutes a major challenge in the development of a carbon-neutral energy economy. The principles of artificial photosynthesis are the same as in natural photosynthesis, but the nanostructures used are not as complex and self-sustaining as in nature. Bio-inspired artificial nanostructures are the most worthwhile discovery in recent years for which the credit goes to the growth of highly advanced devices in the field of nano biotechnology. The fundamental steps involved in the photosynthesis process are capturing light, separating charge and finally reducing carbon.

Research in the field of chloroplast biotechnology through bionanotechnology approaches may take inspiration from previous research breakthroughs in the bioenergy field, which form a strong base for new discoveries and innovation through improved synthetic biology tools, and enable ground-breaking innovative forms of human-plant interactions, all while managing environmental impacts for applications in the green energy field.

Several recent developments in the field of artificial photosynthesis have provided a noteworthy boon to the future of solar energy conversion to energy and biofuels. Biological artificial photosynthesis offers a transitional technology for combustion-based energy needs while offering an avenue towards carbon-neutrality. This will be of the utmost significance in the next ten years as the mechanical work carried out by combustion engines is transitioned to electrically driven induction motors, which offer superior energy conversion to mechanical work.

Future work is concentrating on generating efficient bio-inspired nanostructures as antennae to capture light and reaction centres to transfer energy to separate charge and finally bringing about the reduction of carbon into a usable form of fuel in a sustainable manner which is the key route of green sustainability. All promising synthetic materials have been simulated and tested, and to some extent efficiency in execution functions has been confirmed for some of them, but there are some disadvantages, for example a lack of self-sustenance over a period of time and the fact that the research work is limited to the lab level only. Pilot scale work in the field of artificial photosynthesis is required in which bio-inspired nanomaterials are used at large scale, and surely in the next century we will be more efficient in the production of energy through bionanotechnological routes. Here, one more recommendation that as earlier discussed the goal of chloroplast bionanotechnology has fulfilled the dual objectives, such as of producing more food and also generating energy by utilizing the solar potential through artificial photosynthesis; in the same way, easy routes should be identified for the synthesis of bio-inspired nanoparticles utilizing waste material. This would certainly be a boon for the coming generation where, again, the dual objectives are to manufacture bio-inspired nanoparticles from the waste generated and solid waste management. This step is a move towards sustainability, zero hunger, zero carbon and zero waste.

2.6 CONCLUSION

Bionanotechnology offers capable and promising new methods, principles and approaches for some of the hardest challenges in the construction of bio-inspired nanoparticles for the generation of energy through artificial photosystem routes and chloroplast biotechnology research.

Competent, vigorous and cost-effective artificial photosynthesis end products could transform and revolutionize global energy. Artificial photosynthesis is an effective solution to the global warming problem since it's a renewable and clean energy source with the objective of zero carbon dioxide emission. Bionanotechnology has mammoth potential to produce novel and improved well-designed properties in photosynthetic organelles and organisms for the enhancement of solar energy harnessing. It is also clear in this chapter that the entry of nanoparticles into the plant cell is responsible for various alterations in the plant cell like metabolic changes that lead to an increase in biomass, fruit/grain yield and so on; hence, further action can be explicated to evaluate the possibility of their uses. As it is also clear that natural photosynthetic apparatus utilizes less than 10% of sunlight, and most of the solar spectrum is unutilized, so there is potential to improve the solar energy conversion efficiency in photosynthetic organisms by creating an engineered photosynthetic apparatus with the help of bio-inspired nanomaterials which have the more potential to harness solar energy. The enhancement in photosynthetic efficiency requires broadening the range of solar light absorption.

REFERENCES

Arico, A.S., Bruce, P., Scrosati, B., Tarascon, J.M., Van Schalkwijk, W. (2005). Nanostructured materials for advanced energy conversion and storage devices. *Nat. Mater.* 4(5), 366–377.

Avellan, A., Yun, J., Zhang, Y., Spielman-Sun, E., Unrine, J.M., Thieme, J., Li, J., Lombi, E., Bland, G., Lowry, G.V. (2019). Nanoparticle size and coating chemistry control foliar uptake pathways, translocation, and leaf-to-rhizosphere transport in wheat. *ACS Nano* 13(5), 5291–5305.

Barazzouk, S., Kamat, P.V., Hotchandani, S. (2005). Photoinduced electron transfer between chlorophyll a and gold nanoparticles. *J. Phys. Chem. B.* 109, 716.

Barros, J., Serk, H., Granlund, I., Pesquet, E. (2015). The cell biology of lignification in higher plants. *Ann. Bot.* 115(7), 1053–1074.

BETO Publications (2020). *Integrated Strategies to Enable Lower-Cost Biofuels*. Bioenergy Technologies Office. Available at: https://www.energy.gov/eere/bioenergy/beto-publications (Accessed April 2, 2021).

Beyer, S.R., Ullrich, S., Kudera, S., Gardiner, A.T., Cogdell, R.J., Kohler, J. (2011). Hybrid nanostructures for enhanced light-harvesting: Plasmon induced increase in fluorescence from individual photosynthetic pigment-protein complexes. *Nano Lett.* 11(11), 4897.

Blankenship, R.E., Tiede, D.M., Barber, J., Brudvig, G.W., Fleming, G., Ghirardi, M., Gunner, M.R., Junge, W., Kramer, D.M., Melis, A., Moore, T.A. (2011). Comparing photosynthetic and photovoltaic efficiencies and recognizing the potential for improvement. *Science* 332(6031), 805–809.

Block, M.A., Douce, R., Joyard, J., Rolland, N. (2007). Chloroplast envelope membranes: A dynamic interface between plastids and the cytosol. *Photosynth. Res.* 92(2), 225–244.

Bujak, L., Piatkowski, D., Mackowski, S., Wormke, S., Jung, C., Brauchle, C., Agarwal, A., Kotov, N.A., Schulte, T., Hofmann, E., Brotosudarmo, T.H.P. (2009). Plasmon enhancement of fluorescence in single light-harvesting complexes from *Amphidinium carterae*. *Acta Phys. Pol. A* 116, S22–S25. https://doi.org/10.12693/APhysPolA.116.S-22.

Calkins, J.O., Umasankar, Y., O'Neill, H., Ramasamy, R.P. (2013). High photo-electrochemical activity of thylakoid–carbon nanotube composites for photosynthetic energy conversion. *Energy Environ. Sci.* 6(6), 1891–1900.

Ciamician, G. (1912). The photochemistry of the future. *Science* 36(926), 385–394.

Cossins, D. (2014). Next generation: Nanoparticles augment plant functions. The incorporation of synthetic nanoparticles into plants can enhance photosynthesis and transform leaves into biochemical sensors. *The Scientist*, News & Opinion. March 16. http://www.Thescientist.com/?articles.view/articleNo/39440/title/Next-Generation-Nanoparticles-Augment-Plant-Functions/

Davis, S. (1997). Biomedical applications of nanotechnology: Implications for drug targeting and gene therapy. *Trends Biotechnol.* 15(6), 217–224.

DeRosa, M.C., Monreal, C., Schnitzer, M., Walsh, R., Sultan, Y. (2010). Nanotechnology in fertilizers. *Nat. Nanotechnol.* 5(2), 91–91.

Du, D., Yang, Y., Lin, Y. (2012). Graphene-based materials for biosensing and bioimaging. *MRS Bull.* 37(12), 1290–1296.

Economou, C., Wannathong, T., Szaub, J., Purton, S. (2014). A simple, low-cost method for chloroplast transformation of the green alga *Chlamydomonas reinhardtii*. In: *Chloroplast Biotechnology: Methods and Protocols*. Ed. P. Maliga. Totowa, NJ: Humana Press, pp. 401–411.

Falco, W.F., Botero, E.R., Falcão, E.A., Santiago, E.F., Bagnato, V.S., ARL, C. (2011). In vivo observation of chlorophyll fluorescence quenching induced by gold nanoparticles. *J. Photochem. Photobiol. A* 225(1), 65–71.

Ganesan, I., Shi, L.X., Labs, M., Theg, S.M. (2018). Evaluating the functional pore size of chloroplast TOC and TIC protein translocons: Import of folded proteins. *Plant Cell* 30(9), 2161–2173.

Gao, F., Hong, F., Liu, C., Zheng, L., Su, M., Wu, X., Yang, F., Wu, C., Yang, P. (2006). Mechanism of nano-anatase TiO_2 on promoting photosynthetic carbon reaction of spinach. *Biol. Trace Elem. Res.* 111(1–3), 239–253.

Gao, J., Xu, G., Qian, H., Liu, P., Zhao, P., Hu, Y. (2013). Effects of nano-TiO_2 on photosynthetic characteristics of *Ulmus elongata* seedlings. *Environ. Poll.* 176, 63.

GerdenisKodis, P.A., Liddell, L., de la Garza, P.C., Clausen, J.S., Lindsey, A.L., Moore, T.A., Moore, D.G., Gust, D. (2002). Efficient energy transfer and electron transfer in an artificial photosynthetic antenna-reaction center complex. *J. Phys. Chem. A* 106(10), 2036–2048.

Ghorbanpour, M., Fahimirad, S. (2017). Plant nanobionics a novel approach to overcome the environmental challenges. In: *Medicinal Plants and Environmental Challenges.* Cham: Springer, pp. 247–257. https://doi.org/10.1007/978-3-319-68717-9_14.

Gibson, D.G., Glass, J.I., Lartigue, C. et al. (2010). Creation of a bacterial cell controlled by a chemically synthesized genome. *Science* 329(5987), 1–6.

Giraldo, J.P., Landry, M.P., Faltermeier, S.M., McNicholas, T.P., Iverson, N.M., Boghossian, A.A., Reuel, N.F., Hilmer, A.J., Sen, F., Brew, J.A., Strano, M.S. (2014). Plant nanobionics approach to augment photosynthesis and biochemical sensing. *Nat. Mater.* 13(4), 400–408.

Govorov, A.O., Carmeli, I. (2007). Hybrid structures composed of photosynthetic system and metal nanoparticles: Plasmon enhancement effect. *Nano Lett.* 7(3), 620–625.

Hatfield, J.L., Boote, K.J., Kimball, B.A., Ziska, L.H., Izaurralde, R.C., Ort, D., Thomson, A.M., Wolfe, D. (2011). Climate impacts on agriculture: Implications for crop production. *Agron. J.* 103(2), 351–370.

Hong, F., Yang, F., Liu, C., Gao, Q., Wan, Z., Gu, F., Wu, C., Ma, Z., Zhou, J., Yang, P. (2005). Influences of nano-TiO_2 on the chloroplast aging of spinach under light. *Biol. Trace Elem. Res.* 4(3), 249–260.

Hong, F., Zhou, J., Liu, C., Yang, F., Wu, C., Zheng, L., Yang, P. (2005). Effect of nano-TiO_2 on photo-chemical reaction of chloroplasts of spinach. *Biol. Trace Elem. Res.* 105(1–3), 269–279.

Hong, F.S., Liu, C., Zheng, L., Wang, X.F., Wu, K., Song, W.P., Lv, S.P., Tao, Y., Zhao, G.W. (2005). Formation of complexes of RuBisCO–Rubiscoactivase from La3+, Ce3+ treatment spinach. *Sci. China Seri B Chem.* 48(1), 67–74.

Hu, P., An, J., Faulkner, M.M., Wu, H., Li, Z., Tian, X., Giraldo, J.P. (2020). Nanoparticle charge and size control foliar delivery efficiency to plant cells and organelles. *ACS Nano* 14(7), 7970–7986.

Joshi, A., Kaur, S., Dharamvir, K., Nayyar, H., Verma, G. (2017). Multi-walled carbon nanotubes applied through seed-priming influence early germination, root hair, growth and yield of bread wheat (*Triticum aestivum* L.). *J. Sci. Food Agri.* 98(8), 3148–3160.

Kramer, M.J., Bellwood, D.R., Bellwood, O. (2014). Large-scale spatial variation in epilithic algal matrix cryptofaunal assemblages on the Great Barrier Reef. *Mar. Biol.* 161(9), 2183–2190.

Kuang, T.Y. (2003). *Mechanism and Regulation of Primary Energy Conversion Process in Photosynthesis.* Nanjing: Science and Technology Press of Jiangsu, pp. 22–68.

Kwak, S.Y., Giraldo, J.P., Wong, M.H., Koman, V.B., Lew, T.T.S., Ell, J., Weidman, M.C., Sinclair, R.M., Landry, M.P., Tisdale, W.A., Strano, M.S. (2017). A nanobionic light-emitting plant. *Nano Lett.* 17(12), 7951–7961.

Lahiani, M.H., Dervishi, E., Chen, J., Nima, Z., Gaume, A., Biris, A.S., Khodakovskaya, M.V. (2013). Impact of carbon nanotube exposure to seeds of valuable crops. *ACS Appl. Mater. Interfac.* 5(16), 7965–7973.

Lal, M. (2018). Concepts in Metabolism. In: *Plant Physiology, Development and Metabolism.* Eds. S.C. Bhatla, M.A. Lal. Singapore: Springer, pp. 159–226.

Lauth, V., Maas, M., Rezwan, K. (2017). An evaluation of colloidal and crystalline properties of CaCO₃ nanoparticles for biological applications. *Mater. Sci. Eng. C Mater. Biol.
Appl.* 78, 305–314.

Lei, Z., Mingyu, S., Chao, L., Liang, C., Hao, H., Xiao, W., Xiaoqing, L., Fan, Y., Fengqing,
G., Fashui, H. (2007). Effects of nanoanatase TiO_2 on photosynthesis of spinach chloroplasts under different light illumination. *Biol. Trace Elem. Res.* 119(1), 68.

Lew, T.T.S., Wong, M.H., Kwak, S.Y., Sinclair, R., Koman, V.B., Strano, M.S. (2018). Rational
design principles for the transport and subcellular distribution of nanomaterials into
plant protoplasts. *Small* 14(44), e1802086.

Lewis, N.S. (2016). Research opportunities to advance solar energy utilization. *Science*
351(6271), 6271. https://doi.org/10.1126/science.aad1920.

Lewis, N.S., Nocera, D.G. (2006). Powering the planet: Chemical challenges in solar energy
utilization. *PNAS* 103(43), 15729–15735.

Liang, K., Richardson, J.J., Cui, J., Caruso, F., Doonan, C.J., Falcaro, P. (2016). Metal-organic
framework coatings as cytoprotective exoskeletons for living cells. *Adv. Mater.* 28(36),
7910–7914.

Linglan, M., Chao, L., Chunxiang, Q., Sitao, Y., Jie, L., Fengqing, G., Fashui, H. (2008).
RuBisCO activase mRNA expression in spinach: Modulation by nanoanatase treatment. *Biol. Trace Elem. Res.* 122(2), 168–178.

Long, S.P., Marshall-Colon, A., Zhu, X.G. (2015). Meeting the global food demand of the
future by engineering crop photosynthesis and yield potential. *Cell* 161(1), 56–66.

Lowry, G.V., Avellan, A., Gilbertson, L.M. (2019). Opportunities and challenges for nanotechnology in the agri-tech revolution. *Nat. Nanotechnol.* 14(6), 517–522.

Lykourinou, V., Chen, Y., Wang, X.S., Meng, L.E., Hoang, T., Ming, L.J., Musselman, R.L.,
Ma, S. (2011). Immobilization of MP-11 into a mesoporous metal–organic framework, MP-11@ mesoMOF: A new platform for enzymatic catalysis. *J. Am. Chem. Soc.*
133(27), 10382–10385.

Lynge, M.E., Vander Westen, R., Postma, A., Städler, B. (2011). Polydopamine-a nature
inspired polymer coating for biomedical science. *Nanoscale* 3(12), 4916–4928.

Mackowski, S., Wörmke, S., Maier, A.J., Brotosudarmo, T.H., Harutyunyan, H., Hartschuh,
A., Govorov, A.O., Scheer, H., Bräuchle, C. (2008). Metal-enhanced fluorescence of
chlorophylls in single light-harvesting complexes. *Nano Lett.* 13(2), 558–564.

Manaktala, S.S., Singh, K.M. (2016). Nanotechnology for energy applications. *ISST J. Electr.
Electron. Eng.* 7(1), 63–69.

Masson-Delmotte, V., Zhai, P., Pörtner, H.O., Roberts, D., Skea, J., Shukla, P.R. (eds.). (2018).
*IPCC, 2018: Global Warming of 1.5°C. An IPCC Special Report on the Impacts of
Global Warming of 1.5°C above Pre-industrial Levels and Related Global Greenhouse
Gas Emission Pathways, in the Context of Strengthening the Global Response to
the Threat of Climate Change, Sustainable Development, and Efforts to Eradicate
Poverty.* Intergovernmental Panel on Climate Change. Available at: https://www.ipcc
.ch/sr15/ (Accessed December 5, 2018).

Mingyu, S., Hong, F., Liu, C., Wu, X., Liu, X., Chen, L., Gao, F., Yang, F., Li, Z. (2007).
Effects of nano-anatase TiO_2 on absorption, distribution of light and photoreduction
activities of chloroplast membrane of spinach. *Biol. Trace Elem. Res.* 118(2), 120.

Mingyu, S., Wu, X., Liu, C., Qu, C., Liu, X., Chen, L., Huang, H., Hong, F. (2007). Promotion
of energy transfer and oxygen evolution in spinach photosystem II by nano-anatase
TiO_2. *Biol. Trace Elem. Res.* 119(2), 183.

Nabiev, I., Rakovich, A., Sukhanova, A., Lukashev, E., Zagidullin, V., Pachenko, V., Rakovich,
Y.P., Donegan, J.F., Rubin, A.B., Govorov, A.O. (2010). Fluorescent quantum dots as
artificial antennas for enhanced light harvesting and energy transfer to photosynthetic

reaction centers. *Angew. Chem. Int. Ed.* 49(40), 7217–7221. https://doi.org/10.1002/anie.201003067.

Nadtochenko, V.A., Nikandrov, V.V., Gorenberg, A.A., Karlova, M.G., Lukashev, E.P., Semenov, A.Y., Bukharina, N.S., Kostrov, A.N., Permenova, E.P., Sarkisov, O.M. (2008). Nanophotobiocatalysts based on mesoporous titanium dioxide films conjugated with enzymes and photosynthetic reaction centers of bacteria. *High Energy Chem.* 42(7), 591–593.

Nair, R., Varghese, S.H., Nair, B.G., Maekawa, T., Kumar, Y.Y.D.S. (2010). Nanoparticulate material delivery to plants. *Plant Sci.* 179(3), 154–163.

Nieder, J.B., Bittl, R., Brecht, M. (2010). Fluorescence studies into the effect of plasmonic interactions on protein function. *Angew. Chem. Int. Ed.* 49(52), 10217–10220. https://doi.org/10.1002/anie.201002172.

Niu, J., Lunn, D.J., Pusuluri, A., Yoo, J.I., O'Malley, M.A., Mitragotri, S., Soh, H.T., Hawker, C.J. (2017). Engineering live cell surfaces with functional polymers via cytocompatible controlled radical polymerization. *Nat. Chem.* 9(6), 537–545.

Noji, T., Kamidaki, C., Kawakami, K., Shen, J.R., Kajino, T., Fukushima, Y., Sekitoh, T., Itoh, S. (2011). Photosynthetic oxygen evolution in mesoporous silica material: Adsorption of photosystem II reaction center complex into 23 nm nanopores in SBA. *Langmuir* 27(2), 705–713.

Olejnik, M., Krajnik, B., Kowalska, D., Twardowska, M., Czechowski, M., Hofmann, E., Mackowski, S. (2013). Imaging of fluorescence enhancement in photosynthetic complexes coupled to silver nanowires. *Appl. Phys. Lett.* 102(8), 083703.

Palocci, C., Valletta, A., Chronopoulou, L., Donati, L., Bramosanti, M., Brasili, E., Baldan, B., Pasqua, G. (2017). Endocytic pathways involved in PLGA nanoparticle uptake by grapevine cells and role of cell wall and membrane in size selection. *Plant Cell Rep.* 36(12), 1917–1928.

Pradhan, S., Patra, P., Das, S., Chandra, S., Mitra, S., Dey, K.K., Akbar, S., Pali, P., Goswami, A. (2013). Photochemical modulation of biosafe manganese nanoparticles on Vignaradiata: A detailed molecular, biochemical, and biophysical study. *Environ. Sci. Technol.* 47(22), 13122–13131.

Prakash, S., Deswal, R. (2020). Analysis of temporally evolved nanoparticle-protein corona highlighted the potential ability of gold nanoparticles to stably interact with proteins and influence the major biochemical pathways in *Brassica juncea*. *Plant Physiol. Biochem.* 146, 143–156.

Qi, M., Liu, Y., Li, T. (2013). Nano-TiO$_2$ improve the photosynthesis of tomato leaves under mild heat stress. *Biol. Trace Elem. Res.* 156(1–3), 323.

Rengifo, H.R., Giraldo, J.A., Labrada, I., Stabler, C.L. (2014). Long-term survival of allograft murine islets coated via covalently stabilized polymers. *Adv. Healthc. Mater.* 3(7), 1061–1070.

Ricco, R.W., Liang, S., Li, J.J., Gassensmith, F., Caruso, C., Doonan, Falcaro P. (2018). Metal organic frameworks for cell and virus biology: A perspective. *ACS Nano* 12(1), 13–23.

Richardson, J.J., Liang, K. (2018). Nano-biohybrids: In vivo synthesis of metal-organic frameworks inside living plants. *Small* 14(3), 1702958.

Sanchez, V.C., Jachak, A., Hurt, R.H., Kane, A.B. (2012). Biological interactions of graphene-family nanomaterials: An interdisciplinary review. *Chem. Res. Toxicol.* 25(1), 15–34.

Santana, I., Wu, H., Hu, P., Giraldo, J.P. (2020). Targeted delivery of nanomaterials with chemical cargoes in plants enabled by a biorecognition motif. *Nat. Commun.* 11(1), 2045.

Setyawati, M.I., Tay, C.Y., Leong, D.T. (2015). Mechanistic investigation of the biological effects of SiO$_2$, TiO$_2$, and ZnO nanoparticles on intestinal cells. *Small* 11(28), 3458–3468.

Sharma, P., Bhatt, D., Zaidi, M.G., Saradhi, P.P., Khanna, P.K., Arora, S. (2012). Silver nanoparticle-mediated enhancement in growth and antioxidant status of *Brassica juncea*. *Appl. Biochem. Biotechnol.* 167(8), 2225.

Siddiqui, M.H., Al-Whaibi, M.H. (2014). Role of nano-SiO_2 in germination of tomato (*Lycopersicum esculentum* seeds Mill.). *Saudi J. Biol. Sci.* 21(1), 13–17.

Su, D., Liu, X., Wang, L., Ma, C., Xie, H., Zhang, H., Meng, X., Huang, Y., Huang, X. (2016). Bioinspired engineering proteinosomes with a cell-wall-like protective shell by self-assembly of a metal-chelated complex. *Chem. Commun.* 52(95), 13803–13806.

Walcarius, A., Minteer, S.D., Wang, J., Lin, Y., Merkoci, A. (2013). Nanomaterials for biofunctionalized electrodes: Recent trends. *J. Mater. Chem. B* 1(38), 4878–4908.

Wang, Z., Zhu, W., Qiu, Y., Yi, X., von dem Bussche, A., Kane, A., Gao, H., Koski, K., Hurt, R. (2016). Biological and environmental interactions of emerging two-dimensional nanomaterials. *Chem. Soc. Rev.* 45(6), 1750–1780.

Wasielewski, M.R. (2006). Energy, charge, and spin transport in molecules and self-assembled nanostructures inspired by photosynthesis. *J. Org. Chem.* 71(14), 5051–5066.

Wasielewski, M.R. (2009). Self-assembly strategies for integrating light harvesting and charge separation in artificial photosynthetic systems. *Acc. Chem. Res.* 42(12), 1910–1921.

Weise, S.E., Weber, A.P., Sharkey, T.D. (2004). Maltose is the major form of carbon exported from the chloroplast at night. *Planta* 218(3), 474–482.

Wiesner, M.R., Bottero, J.Y. (2007). *Environmental Nanotechnology: Applications and Impacts of Nanomaterials.* New York: McGraw-Hill.

Wong, M.H., Misra, R.P., Giraldo, J.P., Kwak, S.Y., Son, Y., Landry, M.P., Swan, J.W., Blankschtein, D., Strano, M.S. (2016). Lipid exchange envelope penetration (LEEP) of nanoparticles for plant engineering: A universal localization mechanism. *Nano Lett.* 16(2), 1161–1172.

Wu, Y.L., Putcha, N., Ng, K.W., Leong, D.T., Lim, C.T., Loo, S.C.J., Chen, X. (2013). Biophysical responses upon the interaction of nanomaterials with cellular interfaces. *Acc. Chem. Res.* 46(3), 782–791.

Xu, Z., Shimizu, H., Yagasaki, Y., Ito, S., Zheng, Y., Zhou, G. (2013). Interactive effects of elevated CO_2, drought, and warming on plants. *J. Plant Growth Regul.* 32(4), 692–707.

Yang, F., Hong, F., You, W., Liu, C., Gao, F., Wu, C., Yang, P. (2006). Influence of nano-anatase TiO_2 on the nitrogen metabolism of growing spinach. *Biol. Trace Elem. Res.* 110(2), 179.

Yang, S.H., Kang, S.M., Lee, K.B., Chung, T.D., Lee, H., Choi, I.S. (2011). Mussel-inspired encapsulation and functionalization of individual yeast cells. *J. Am. Chem. Soc.* 133(9), 2795–2797.

Ze, Y., Liu, C., Wang, L., Hong, M., Hong, F. (2011). The regulation of TiO2 nanoparticles on the expression of light-harvesting complex II and photosynthesis of chloroplasts of *Arabidopsis thaliana*. *Biol. Trace Elem. Res.* 143(2), 1131–1114.

Zhang, S., Shao, Y., Yin, G., Lin, Y. (2013). Recent progress in nanostructured electrocatalysts for PEM fuel cells. *J. Mater. Chem. A* 1(15), 4631–4641.

Zheng, L., Hong, F., Lu, S., Liu, C. (2005). Effect of nano-TiO_2 on spinach of naturally aged seeds and growth of spinach. *Biol. Trace Elem. Res.* 104(1), 83–91.

Zhu, X.G., Long, S.P., Ort, D.R. (2010). Improving photosynthetic efficiency for greater yield. *Annu. Rev. Plant Biol.* 61, 235–261.

3 Bio-Nanofluids-Based Photovoltaic Thermal System

Deepa Sharma

CONTENTS

3.1   INTRODUCTION

The conventional photovoltaic thermal system works by the conversion of solar energy into electrical energy along with the generation of waste energy in the form of heat. The performance of photovoltaic panels is largely affected by temperature rise of PV cells thereby decreasing the efficiency of PV panels. In a conventional solar photovoltaic power system, the electrical energy is generated from solar energy through a sequence of interconnected components working together.

The heat generated is to be removed by the use of various PV/T control systems used with heat removal agents, previously air and water, and recently bio-nano/nanofluids have been explored for this purpose.

3.2   PHOTOVOLTAIC AND THERMAL SYSTEMS

The PV/T systems are classified on the basis of heat transfer medium as the conventional PV/T systems using bi-fluids, water and air as heat transfer medium and recent

DOI: 10.1201/9781003316374-3

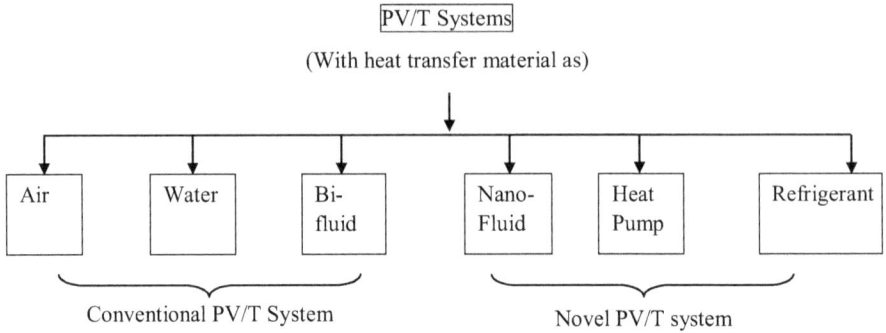

FIGURE 3.1   Classification of PV/T systems on the basis of heat transfer medium.

systems using nanofluids with head pump, refrigeration phase change materials as shown in Figure 3.1.

Further, the PV/T systems are also classified on the basis of the phase of the heat transfer medium.

3.2.1   LIQUID-BASED PV/T SYSTEMS

These PV/T systems use liquid as a heat transfer medium. The process of cooling involves the pumping of fluid through the channels made on the plate for collecting heat. The collecting plate is mounted on the PV/T system; it transfers the generated heat from PV cells to the fluid for absorption. The process of heat transfer from PV cells is required to increase the efficiency of PV cells. Previously, water was used as a liquid cooling system, but recently researchers have included bio-nanofluids, bi-fluid and refrigerant as PV/T system cooling agents.

The first reported study on flat plate PV/T systems is supported by the results of Bargene et al. (1995) showed an enhancement in PV/T system efficiency by more than 60% by using heat transfer fluid (water/bio-nanofluid), as shown in Figure 3.2 (a and b).

The structure of bio-nanofluids PV/T systems is similar to that of conventional flat plate solar collectors. For heating purposes, in the absorber, a tube is attached to the PV panel. The effectiveness of bio-nanofluids for the purpose of the cooling of PV/T cells is greater than air-based PV/T systems.

Chow et al. (2010) revealed the fact that the efficiency of PV/T systems depends upon fin efficiency and bonding between the collector and the sheet underneath the module (Figure 3.2b). Further, as per Zondag et al. (2003), tube system and covered sheet as the recognized PV/T concept for the heating of tap water with an annual average efficiency of 61.3% approximately.

3.2.2   REFRIGERANT COOLING PV/T SYSTEMS

PV/T system efficiency is also found to be enhanced many-fold if refrigerant material is used as a coolant fluid, but its ozone depletion effect has restricted its use. As

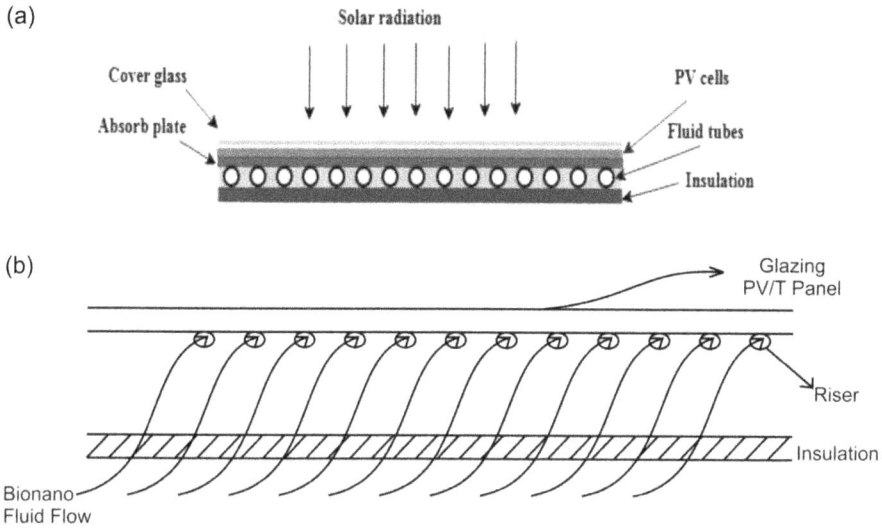

FIGURE 3.2 (a) PV/T system (conventional). (b) Bio-nanofluid based.

a result, it is suggested to use environmentally friendly refrigerants such as carbon dioxide or ammonia. The refrigerant cooling-based PV/T system was experimentally studied by Tsai et al. (2015) under real conditions, and they found enhanced water heating reaching up to 50°C.

3.2.3 BIO-NANOFLUID COOLING PV/T SYSTEMS: AN EFFECTIVE REPLACEMENT OF CONVENTIONAL COOLING IN PV/T SYSTEMS

To prepare bio-nanofluids, nanoparticles with particle size less than 100 mm are dispersed into the base fluid. Bio-nanofluids have noticeable thermal property. Experimental studies show increased heat transfer by bio-nanofluids with small nanoparticle size and increased volume fraction in a model of a PV/T system in which bio-nanofluid flows in laminar flow with a square cross-section. It has been found in experiments that the high thermal conductivity and high heat transfer with bio-nanofluids is possible in compact PV/T systems. The stability of bio-nanofluids also plays an important role in enhancing the thermal conductivity (Faizal et al., 2013a, 2013b, 2014a, 2015; Lari & Sahin, 2017; Khanjari et at., 2016), thereby enhancing PV/T system efficiency up to 55%. The bio-nanofluids are made up of metal, oxides, carbides or carbon nanotubes suspended in base fluids. The use of nanofluids in PV/T is an advanced methodology to enhance the efficiency of the system many-fold by enhanced thermal management and overall energy output as shown in Figure 3.3 (Rejeb et al., 2016).

Further, Sardabadi et al. in 2014 studied experimentally the effects of a bio-nanofluid (silica/H_2O) on the thermal efficiency of PV/T systems (Sardabadi et al., 2014; Sardabadi et al., 2017; Duangthongsuk et al., 2009). The following is a summary

```
                              ┌──────────────────────────┐
                              │   Photo Voltaic Module    │
                              └──────────────────────────┘
┌────────────────────────┐                    │
│  Electric Control Box   │                    ▼
└────────────────────────┘        ┌──────────────────────────┐      ┌──────────────────────┐
                                   │  Nano fluid pre storage   │      │  PV/T Thermal Collector│
                                   │  tank                     │      └──────────────────────┘
                                   └──────────────────────────┘
```

⎧
⎪ * Provide high heat extraction
⎨ (Using AbO₃ & Cu nanoparticles at variable concentration with basic fluids glycol &
⎪ water)
⎩

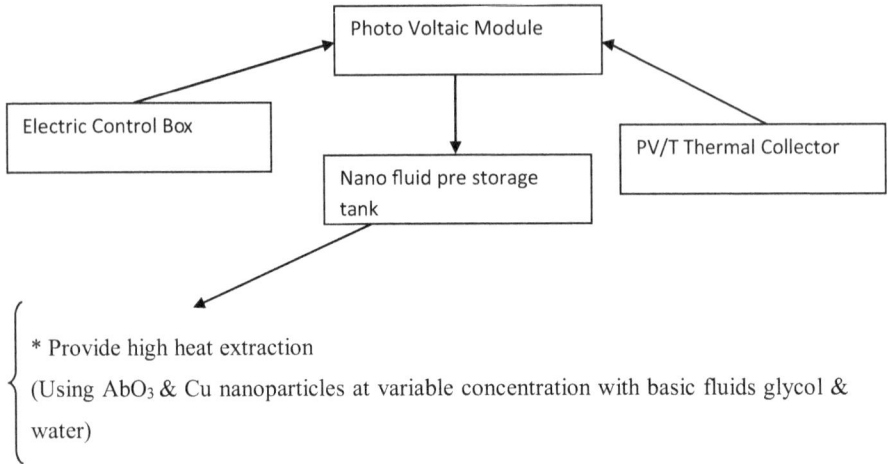

FIGURE 3.3   PV/T system using bio-nanofluid.

of a review of the literature in which the performance of bio-nanofluid-based PV/T systems is studied and compared with respect to particle size, mass (% wt) fraction and thermal efficiency.

Jing et al. used SiO_2 water (bio-nanofluid) of variable particle sizes, 5–50 mm, and mass fraction 2% V, the interpretation of this paper is that the bio-nanofluids with nanoparticle size of 5 mm and 2V% have showed the transmittance of the value of 97%.

Yun Cui et al. (2012) worked with MgO water as a nanofluid of nanoparticle size 10 mm; with mass fraction % wt of range 0.6–0.1, an efficiency of 60% can be achieved as the transmittance of nanofluids is inversely related to mass fraction and film thickness, while Sardarabadi et al. (2014) worked with nanofluid SiO_2, particle size range 11–14 mm; 1–3% wt found an enhancement of efficiency by 3.6–7.9%.

From the various experimental findings, it is interpreted that in PV/T systems, bio-nanofluids as heat transfer media are very effective in enhancing the efficiency of the system; moreover, the presence of a heat collector in a PV/T system also enhances the performance of the PV/T system. The mechanism involved in the system includes the use of lower concentrations of nanofluids as a thin layer on the PV cells as well as below the PV cells; the layer of fluid above the PV panel filters the IR, rays of incident light, and the lower fluid layer removes the heat generated during the working of PV cells; this reduces the operational temperature of PV cells thereby enhancing the efficiency of the whole system. Detailed categorizations of nanoparticles are given in Figure 3.4.

It has been found that the suspension of nanoparticles in fluids enhances the efficiency of conducting heat by heat transfer fluid in PV/T systems. It was Choi et al. in 2020 who found the efficient heat transfer fluid in PV/T system as bio-nanofluids with suspended nanoparticles during their experiments at Argonne National Laboratory, USA. The thermal conductivity of fluids used for heat transfer in PV/T

Nanoparticles Categorization

Metallic Nanoparticleses	Carbon Nanoparticles	Nano Composites

Metal Oxide e.g. MgO ZnO CuO CuO2 Al₂O₃ Fe₂O₃ Fe₃O₄ SiO₂ TiO₂	Metals e.g. Al, Ag, Fe, Cu, Zn, Au	Graphite based	Fullerenes	Carbon Nanotubes	Polymer matrix	Metal Matric

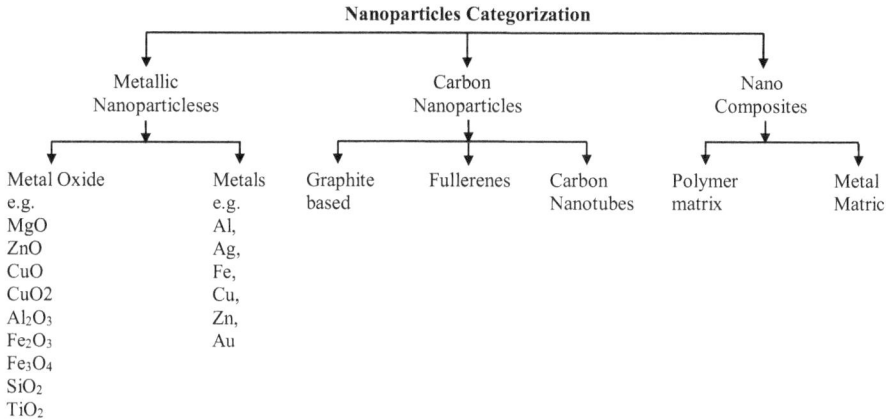

FIGURE 3.4 Thermal conductivity/thermo physical properties of nanofluids and their usefulness in hybrid PV/T systems.

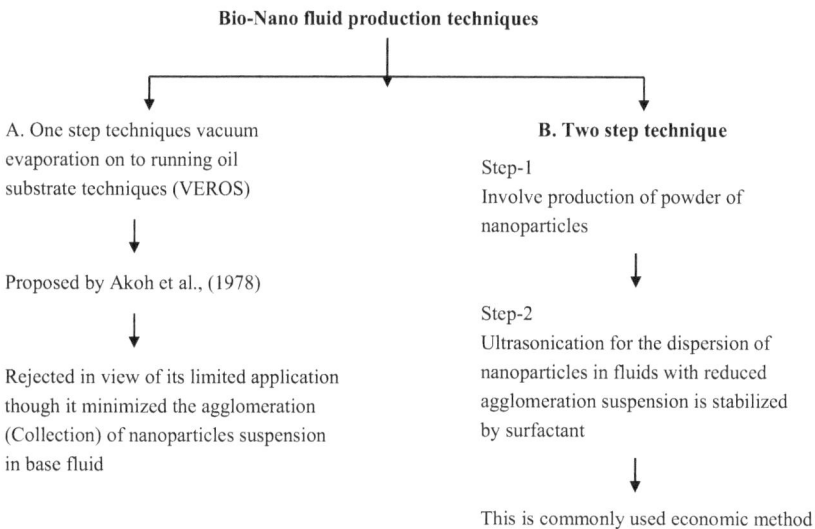

Bio-Nano fluid production techniques

A. One step techniques vacuum evaporation on to running oil substrate techniques (VEROS)

Proposed by Akoh et al., (1978)

Rejected in view of its limited application though it minimized the agglomeration (Collection) of nanoparticles suspension in base fluid

B. Two step technique

Step-1
Involve production of powder of nanoparticles

Step-2
Ultrasonication for the dispersion of nanoparticles in fluids with reduced agglomeration suspension is stabilized by surfactant

This is commonly used economic method

FIGURE 3.5 Bio-nanofluid production techniques.

systems is the important parameter to determine the efficiency of the system; in turn, the thermal efficiency of nanofluids is dependent on nanoparticle size, temperature, stability and dispersion and these characteristics are considered for preparing of the suspension (Choi et al., 2020; Das et al., 2021; Diwania et al., 2020; El-Samie et al., 2020; Jha et al., 2019; Kosan et al., 2020; Ma et al., 2020; Riaz et al., 2020; Zhou et al., 2020a). The preparation of nanofluids includes the suspension of solid particles of nanometer size in base bio-liquid (water, oil, ethylene glycol); the techniques to produce bio-nanofluids (excluding the aggregation and inhomogeneous colloidal suspensions) are as shown in Figure 3.5.

3.2.4 Air-Based PV/T Systems

In view-based PV/T systems, the heat transfer medium is air; for the process, single or double pass with active or passive mode is used. The efficiency of air-based PV/T systems made by using numerical models and simulations in fluid-based PV/T systems. However, seven years ago, Tiwari and his co-researchers in 2016 demonstrated the system in an experiment in which a PV/T system was integrated with a greenhouse in the temperature and humidity conditions of New Delhi in the month of summer season (Tiwari et al., 2016) as shown in Figure 3.6.

Further, in the experiment Tiwari and his research group conducted, the temperature of the greenhouse was found to reach 47°C, the thermal efficiency was found to reach 35% and electrical efficiency was found to be enhanced by 14% (Tiwari et al., 2016). Recently, in the year 2020, Rubbi and his team formulated a bio-nanofluid based on soyabean oil and Maxine (Ti3 C_2) nanoparticles and found better results (Rubbi et al., 2020). In the experiment to study heat transfer and airflow in PV/T systems (Gholampour et al., 2016), a turbulent CFD model was used.

They studied the energy and energy analysis of PV/T flat transpired collectors experimentally. The experimental setup used by them is according to Figure 3.7.

As per Kong et al. (2020), in one study proposed a solar, hot air as transfer medium for agriculture. A comparison was made between two PV/T systems, amorphous silicon thin film and polycrystalline silicon, for their efficiency and performance.

The compositions of both PV/T systems were the same with the same elements, insulation and absorber, cover (glass) and air channels with the common provision of air inlet. The working of the setup is shown in Figure 3.8 and Figure 3.9 (Kong et al., 2020).

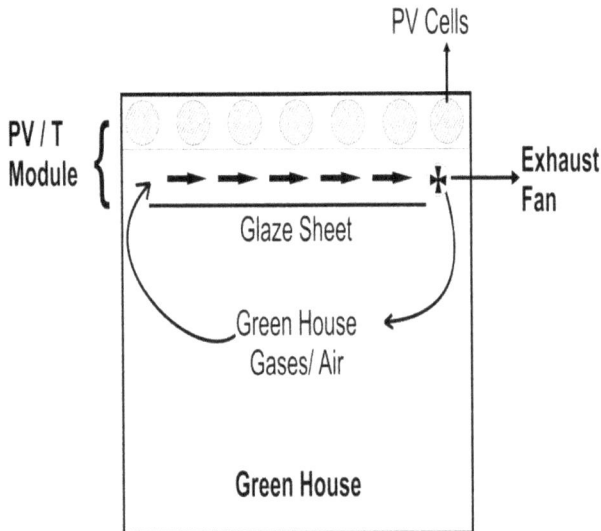

FIGURE 3.6 PV/T module installed in greenhouse to study air-based PV/T system efficiency.

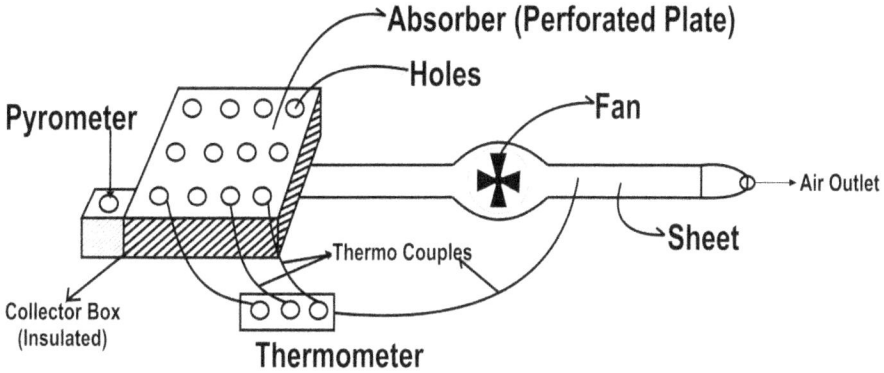

FIGURE 3.7 PV/T experimental turbulent model CFD.

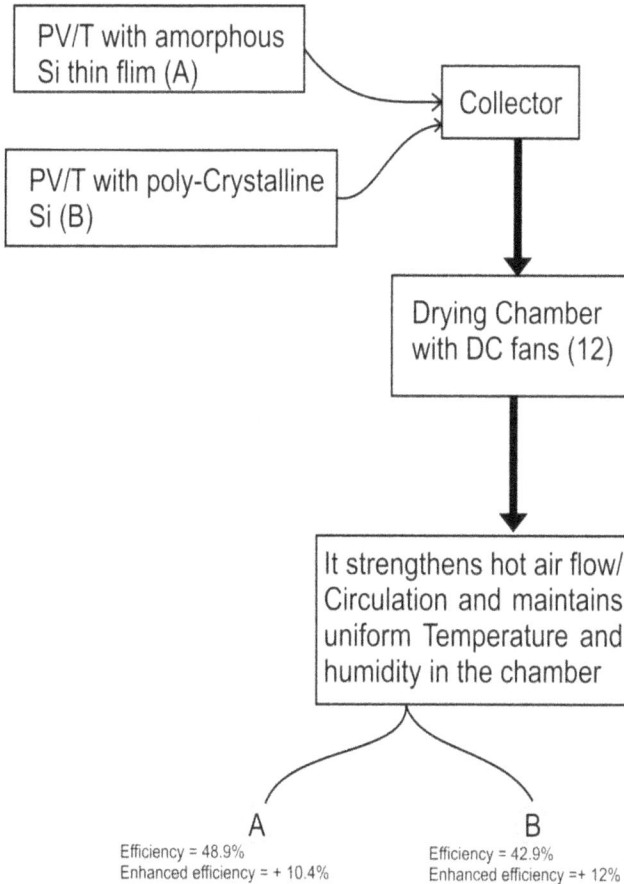

FIGURE 3.8 Flowchart of working of drying system (Kong et al., 2020).

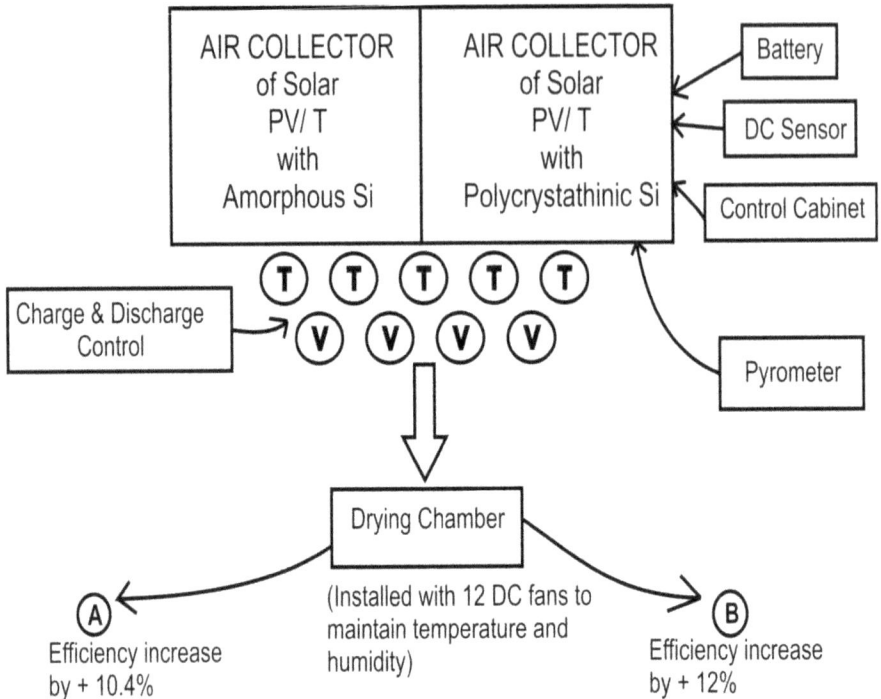

FIGURE 3.9 Flowchart and block diagram of drying system setup proposed.

3.2.5 Bi-Fluid Systems

There are fewer examples of bi-fluid-based PV/T systems; air and water, both, are the potential heat transfer medium. Such systems provide hot water as well as hot air together, at the same time, with air and water exchanges for heat as heat transfer media.

Such bi-fluid systems are economic but with reduced efficiency for thermal to electrical conversion (Sathe and Dhoble, 2017). A highly efficient PV/T system was fabricated by Othman et al. (2016) in which two PV transparent systems pass parallelly connected with a double-pass flat plate air collector, copper water tube and storage tank. Practically, this system shows an increase in electrical efficiency by 17% and thermal efficiency by 76%.

3.2.6 PV/T Integrated Systems

An integrated PV/T system involves the integration of a heat producing PV/T system and heat pump evaporation. The process involved is depicted in the flow diagram in Figure 3.10a and b.

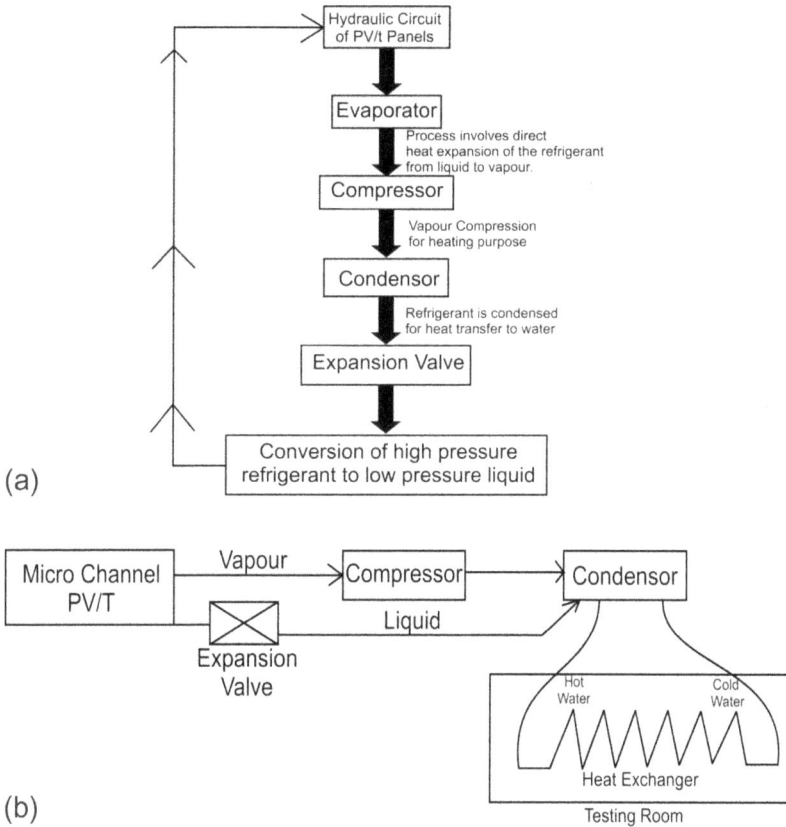

FIGURE 3.10 (a) Flow diagram of working and setup of integrated PV/T system. (b) Block diagram of integrated PV/T model.

3.3 APPLICATIONS OF NANOFLUIDS IN THE HYBRID PV/T SYSTEMS

Recently, performance of the system has been improved by applying nanofluids in hybrid PV/T systems; this has attracted many researchers. The PV/T systems using conventional fluids had lower efficiency; this has encouraged nanofluid use as a heat transfer medium to improve heat transfer properties. Adding nanoparticles to conventional fluids improves their thermal properties, which leads to enhanced efficiency (Bajestan et al., 2016; Sokhansefat et al., 2014). Using nanofluids contributes to the extraction of the PV system's excess heat and its use in other applications. Thus cooling the PV system with the nanofluid will increase the electricity generation of the PV system (Manikandan and Rajan 2016). Numerous researchers have used nanofluids with PV/T systems in two ways, for cooling the PV/T system and as a spectral filter, or combining them, as shown in Figure 3.11 (Zafar et al., 2018).

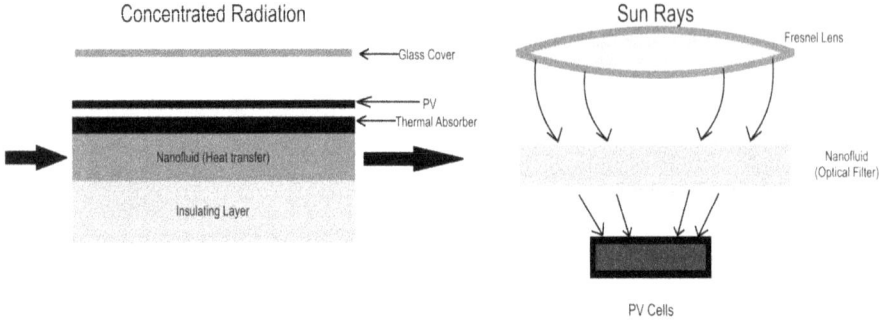

FIGURE 3.11   Schematic figure of the PV/T system with nanofluid as a coolant and spectral.

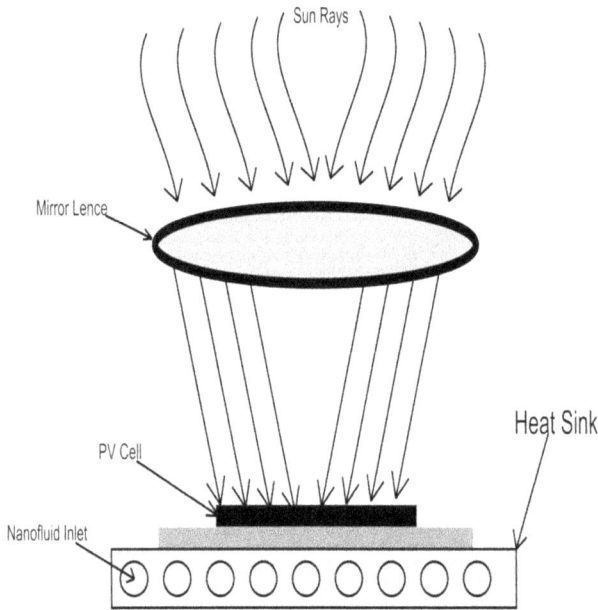

FIGURE 3.12   Schematic diagram of the nanofluid-based PV/T system.

An experiment conducted by Ali et al. (2017) showed that the use of a nanofluid with a PV/T system with a certain percentage of nanoparticles leads to an increase in the electrical efficiency compared to the PV system alone. Thermal efficiency has increased compared with cooling water, thus increasing the system's overall efficiency. A new technique has used mirrors to concentrate the sunlight on the PV system, as shown in Figure 3.12, using nanofluids to increase the electrical and thermal efficiencies of the concentrated PV/T system (CPV/T).

Both Hassani et al. (2016) and Rahbar et al. (2019) indicate that using nanofluid as a cooling fluid and optical filter had a noticeable effect on the CPV/T system's overall efficiency. Many researchers have concentrated on the uses of different sorts of nanoparticles, such as metal, metal oxide, etc., to improve the PV/T system. Various studies have been conducted on improving the PV/T performance using different types of nanofluids, base fluids with different designs of PV/T systems, as shown in Table 3.1, to increase the system's electrical and thermal efficiency.

3.4 DISCUSSION

The effects of conventional fluids used in PV/T systems are depicted in Table 3.2. Using Ag/water nanofluid for cooling and optical filtering, PV/T has achieved higher thermal and overall efficiency as compared to Au-water and Cu-water systems (Saroha et al., 2015). Using Ag-water nanofluid as a coolant in separate channels in PV/T was better than the double-pass channel. If the volume fraction is increased, it leads to a rise in the overall efficiency of the PV/T system (Hassani et al., 2016). A comparison was conducted by Chandrashekhar et al. (2010) between Cu and Al nanofluids to examine the thermal conductivity enhancement at 1% volume fraction, with nanoparticle size 80 nm. The results revealed Cu nanofluid has higher thermal conductivity as compared to Al nanofluid (Chandrashekhar et al., 2013). To improve nanofluids' thermal conductivity, ethylene glycol-based Fe nanofluids were prepared by using sonication at high powered pulses. The Fe nanofluids' conductivity is improved as compared to Cu nanofluids because Fe nanofluids are efficient thermal transporters. Hong et al. (2005) used an active cooling technique to enhance the performance of the PV panels. Zn-water nanofluid used as a work fluid passing during small heat exchanger placed in the back of the PV panels, with different concentration ratios Zn have used. The PV panels' temperature is reduced by increasing the electrical efficiency with a concentration ratio of 0.3% as observed. A summary of prior art since 2017 on the topic for quick reference is presented in Table 3.2.

3.5 CONCLUSION

In view of the studies in the field, it appears that using nanofluids as a working fluid with the PV/T systems instead of the conventional liquids improves the hybrid PV/T system's efficiency. Several studies and researchers have used different nanofluids with different PV/T system designs to improve performance. Most of the studies that used nanofluids have achieved encouraging results by enhancing the PV/T systems' performance. Therefore, using nanofluids as a cooling fluid will contribute to an increase in the electrical and thermal energy of the PV/T system by removing excess heat of the PV cells by cooling them, which contributes to stabilize its work and increases its productivity. Furthermore, nanofluids' thermo physical properties play an intrinsic role by enhancing nanofluids' performance, thus positively reflecting on the PV/T system performance. Other conclusions can be drawn from the review as follows:

- The effect of temperature rise on the thermo physical properties of nanofluids: Rising temperatures have a significant effect on nanofluids' thermo physical properties, where both the density and viscosity of nanofluids are

TABLE 3.1

Some Studies Related to the Usage of Nanofluids with PV/T Applications

Type of Nanoparticles	Base Fluid	Application	Outcomes	Reference
Metal oxide-based nanoparticles	Water, ethylene glycol-water	PV/T	The use of nanofluids with hybrid PV/T systems applications improves the system's performance, whether with the laminar or turbulent regime. Nanoparticles with a larger diameter positively influence overall energy efficiency in the laminar regime, while opposite that with the turbulent regime. Moreover, it is observed that aluminum oxide/water has higher system performance than titanium oxide. Nanofluid that adopts water as a base fluid contributes to increased energy and exergy efficiency compared to ethylene glycol used as a base fluid.	—
Metal-based, oxide-based nanoparticles	Water, ethylene glycol	PV/T, CPV/T	Use of nanofluid as an alternative for the conventional liquids as work fluid or optical filtration as a coolant for the PV/T system leads to enhancement in the system's performance. Also, using some modifications in the system's design, like using honeycomb in the back of the PV panel instead of the channel's cross-section, contributes to enhancing the thermal load distribution.	Ahmed et al., 2018
Metal-based, oxide-based nanoparticles and carbon-based nanoparticles	Different type of base fluids	PV/T	The use of nanofluids helps enhance the PV/T system's performance. It raises the total solar energy yield and increases the system's thermal energy compared with conventional fluids. However, nanofluids have contributed to enhanced heat transfer as a base fluid in the PV/T system, thus increasing electrical and thermal efficiencies.	Sachit et al., 2018
Metal-based, oxide-based nanoparticles and carbon-based nanoparticles	Water, ethylene glycol	PV/T, CPV/T	According to various types of PV/T systems that used nanofluids or other fluids, the studies concluded that the nanofluids used have improved thermal conductivity, which positively impacts the performance. Thus, the possibility for using nanofluid as an optical filter or for cooling in the PV/T systems is confirmed because nanofluids possessed high thermal properties, which showed higher efficiency in the PV/T system than other fluids.	ZaifarSaida et al., 2018
Oxide-based nanoparticles	Water	PV	A study was conducted on the PV module's performance using cotton wick structures that were fixed at the back of the PV module using nanofluids as cooling fluid. The results refer to increasing the PV module temperature up to 65°C without cooling. Using water with the cotton wick as a coolant contributes to lowering PV's temperature to 45°C. Using both CuO/water and Al_2O_3/water as nanofluids reduces the temperature of the PV module about 59°C and 54°C, that is, lower than using water as a coolant with cotton wick structures, and this is because of the adherence of nanoparticles by the wick fibers, which natively influence in wick structures.	Chandra Shekhar et al., 2013

TABLE 3.2
Summary of Prior Art since 2017 on the Topic for Quick Reference

Type of PV/T System Studied	Electrical Efficiency Enhancement by the Type of Heat Transfer Medium	Thermal Efficiency Enhancement by the Type of Heat Transfer Medium	Reference
Water-based	11.71	59.6	—
Water-based	15.5	65.7	Ma et al., 2020
Nanofluid	13.8	79.13	Samylingam et al., 2020
Nanofluid	14.20	84.25	Rubbi et al., 2020
Air-based	13.6	34	Jha et al., 2019
Air-based	13.98	49.5	Arslan et al., 2020
Air-based	12	48.9	Kong et al., 2020
Bi-fluid	13.17	50	Lebbi et al., 2020
Heat pump	13.1	56.6	Zhou et al., 2020
Heat pump	14.6 to 15.8	39.3 to 42.2	Zhou et al., 2020
Heat pump	12.27	53.66	Koşan et al., 2020
Heat pump	–	–	Kong et al., 2020
Heat pump	16.61	30.28	Choi et al., 2020
PCM	16.3	26.87	Gaur et al., 2017
PCM	13.1	73.5	—
CPV/T	7.1	48.8	Samie et al., 2020

reduced with increasing temperature. The addition of surfactant also has an enhancive effect on the properties.

On the other hand, the specific heat of nanofluids increases with a rise in temperature, while the volume fraction causes a decrease in their specific heat. Adding a quantity of surfactant slightly reduces the thermal conductivity of nanofluids. The thermal conductivity of nanofluids is also increased by a rise in temperature as well as a concentration of nanofluids.

• Most of the applications have used nanofluids in the PV/T systems in two ways, either for cooling the PV/T system or as a spectral filter or a combination of both. Using a certain proportion of nanoparticles leads to an increase in the electrical efficiency compared to the PV system alone.

• Using metal nanoparticles with base fluids as a nanofluid leads to enhancing the PV/T system's performance in varying proportions. Some of the nanoparticles improve the PV/T system's thermal efficiency, with adverse effects on electrical efficiency which is decreased, while the use of another type of metal nanoparticles has improved both thermal and electrical efficiency.

Some researchers have used metal oxide nanoparticles and found the PV/T system performed better when parameters such as volume fraction, the concentration of nanoparticles, dispersion of nanoparticles, size and shape of nanoparticles, effectiveness of addition of surfactants, etc., were controlled. Nanofluids produced by using combined metal and metal oxide nanoparticles mixed with base fluid have good features; they have an effect of enhanced thermal conductivity and reduced PV cell temperatures.

As a final conclusion, the use of nanoparticle-based fluids in the PV/T system enhances the thermal efficiency thereby improving the overall efficiency and energy of the PV/T system. The use of carbon-based nanoparticles mixed with base fluid increases the thermal and electrical efficiency of the PV/T system.

This arrangement absorbs the PV's high temperature thereby improving the PV/T system's thermal storage capacity.

REFERENCES

Ahmed, S.A., Sulaiman, F., Saidur, R., Mansour, R.B. (2018). A review on recent development for the design and packaging of hybrid photovoltaic/thermal (PV/T) solar systems. *Renewable and Sustainable Energy Reviews*, 95, 110–129.

Arslan, E., Aktaş, M., Can, Ö.F. (2020). Experimental and numerical in vestigation of a novel photo voltaic thermal (PV/T) collect or with the energy and exergy analysis. *Journal of Cleaner Production*, 276, 123255. https://doi.org/10.1016/j.jclepro.2020.123255

Bajestan, E.E., Moghadam, M.C., Daungthongsuk, H.N.W., Wongwises, S. (2016). Experimental and numerical investigation of nanofluids heat transfer characteristics for application in solar heat exchangers. *International Journal of Heat and Mass Transfer*, 92, 1041–1052.

Bargene, T., Lovvik, O.M. (1995). Model calculation on a flat plate solar heat collector with integrated solar cell. *Solar Energy*, 55(6), 453–462.

Chandrasekar, M., Suresh, S., Bose, A.C. (2010). Experimental investigations and theoretical determination of thermal conductivity and viscosity of Al2O3/water nanofluid. *Experimental Thermal and Fluid Science*, 34(2), 210–216.

Chandrasekar, M., Suresh, S., Senthilkumar, T., Ganesh Karthikeyan, M. (2013). Passive cooling of standalone flat PV module with cotton wick structures. *Energy Conversion and Management*, 71, 43–50.

Choi, H.U., BokKim, Y., HyoSon, C., InYoon, J., HwanChoi, K. (2020). Experimental study on the performance of heat pump water heating system coupled with air type PV/T collector. *Applied Thermal Engineering*, 178, 115427. https://doi.org/10.1016/j.applthermaleng.2020.115427

Chow, T.T. (2010). A review on photovoltaic thermal hybrid solar technology. *Applied Energy*, 87(2), 365–379.

Das, D., Bordoloi, U., Kamble, A.D., Muigai, H.H., Pai, R.K., Kalita, P. (2021). Performance investigation of a rectangular spiral flow PV/T collector with a novel form-stable composite material. *Applied Thermal Engineering*, 182, 116035. https://doi.org/10.1016/j.applthermaleng.2020.116035

Diwania, S., Agrawal, S., Siddiqui, A.S., Singh, S. (2020). Photovoltaic–thermal (PV/T) technology: A comprehensive review on applications and its advancement. *International Journal of Energy and Environmental Engineering*, 11(1), 33–54. https://doi.org/10.1007/s40095-019-00327-y

Duangthongsuk, W., Wongwises, S. (2009). Heat transfer enhancement and pressure drop characteristics of TiO2–water nanofluid in a double-tube counter flow heat exchanger. *International Journal of Heat and Mass Transfer*, 52(7–8), 2059–2067.

El-Samie, M.M.A., Ju, X., Zhang, Z., Adam, A.S., Pan, X., Xu, C. (2020). Three-dimensional numerical investigation of a hybrid low concentrated photovoltaic/thermal system. *Energy*, 190, 116436. https://doi.org/10.1016/j.energy.2019.116436

Faizal, M., Saidur, R., Mekhilef, S. (2014a). Potential of size reduction of flat-plate solar collectors when applying Al2O3 nanofluid. *Advances in Materials Research*, 832, 149–153.

Faizal, M., Saidur, R., Mekhilef, S. (2013b). Potential of size reduction of flat-plate solar collectors when applying MWCNT nanofluid. *Earth and Environmental Science*, 16, 1–4.

Faizal, M., Saidur, R., Mekhilef, S., Alim, M.A. (2013a). Energy, economic and environmental analysis of metal oxides nanofluid for flat plate solar collector. *Energy Conversion and Management*, 76, 162–168.

Faizal, M., Saidur, R., Mekhilef, S., Hepbasli, A., Mahbubul, I.M. (2015). Energy, economic, and environmental analysis of a flat-plate solar collector operated with SiO2 nanofluid. *Clean Technologies and Environmental Policy*, 17(6), 1457–1473.

Gaur, A., Ménézo, C., Giroux—Julien, S. (2017). Numerical studies on thermal and electrical performance of a fully wetted absorber PVT collector with PCM as a storage medium. *Renewable Energy*, 109, 168–187. https://doi.org/10.1016/j.renene.2017.01.062

Gholampour, M., Ameri, M. (2016). Energy and exergy analyses of photovoltaic/thermal flattran spired collectors: Experimental and theoretical study. *Applied Energy*, 164, 837–856. https://doi.org/10.1016/j.apenergy.2015.12.042

Hassani, S., Taylor, R.A., Mekhilef, S., Saidur, R. (2016). A cascade nanofluid-based PV/T system with optimized optical and thermal properties. *Energy*, 112, 963–975.

Hong, T.-K., Yang, H.-S., Choi, C. (2005). Study of the enhanced thermal conductivity of Fe nanofluids. *Journal of Applied Physics*, 97(6), 064311.

Jha, P., Das, B., Gupta, R. (2019). An experimental study of a photovoltaic thermal air collector (PVTAC): A comparison of a flat and the wavy collector. *Applied Thermal Engineering*, 163, 114344. https://doi.org/10.1016/j.applthermaleng.2019.114344

Khanjari, Y., Pourfayaz, F., Kasaeian, A.B. (2016). Numerical investigation on using of nanofluid in a water-cooled photovoltaic thermal system. *Energy Conversion and Management*, 122, 263–278.

Kong, D., Wang, Y., Li, M., Keovisar, V., Huang, M., Yu, Q. (2020). Experimental study of solar photovoltaic/thermal (PV/T) air collector drying performance. *Solar Energy*, 208, 978–989. https://doi.org/10.1016/j.solener.2020.08.067

Kosan, M., Demirtaş, M., Aktaş, M., Dişli, E. (2020). Performance analyses of sustain- able PV/T assisted heat pump drying system. *Solar Energy*, 199, 657–672. https://doi.org/10.1016/j.solener.2020.02.040

Lari, M.O., Sahin, A.Z. (2017). Design, performance and economic analysis of a nanofluid-based photovoltaic/thermal system for residential applications. *Energy Conversion and Management*, 149, 467–484.

Lebbi, M., Touafek, K., Benchatti, A., Boutina, L., Khelifa, K.A., Baissi, M.T., Hassani, S. (2020). Energy performance improvement of a new hybrid PV/T Bi-fluid system using active cooling and self-cleaning: Experimental study. *Applied Thermal Engineering*. 182, 116033. https://doi.org/10.1016/j.applthermaleng.2020.116033

Ma, X., Qiu, T., Lv, J., Shi, Q., Meng, D. (2020). Performance investigation of an iron scrap filled tube-plate PV/T system. *Energy for Sustainable Development*, 58, 196–208. https://doi.org/10.1016/j.esd.2020.08.002

Manikandan, S., Rajan, K.S. (2016). Sandpropyleneglycolwaternanofluids for improved solar energy collection. *Energy*, 113, 917–929.

Othman, M.Y., Hamid, S.A., Tabook, M.A.S., Sopian, K., Roslan, M.H., Ibrahim, Z. (2016). Performance analysis of PV/T combi with water and air heating system: An experimental study. *Renewable Energy*, 86(C), 716–722.

Rahbar, K., Riasi, A., Khatam Bolouri Sangjoeei, H., Razmjoo, N. (2019). Heat recovery of nanofluid based concentrating photovoltaic thermal (CPV/T) collector with organic Rankine cycle. *Energy Conversion and Management*, 179, 373–396.

Rejeb, O., Sardarabadi, M., Ménézoc, C., Passandideh-Fard, M., Houcine, M., Jemni, A. (2016). Numerical and model validation of uncovered nanofluid sheet and tube type photovoltaic thermal solar system. *Energy Conversion and Management*, 110, 367–377. https://doi.org/10.1016/j.enconman.2015.11.063

Riaz, A., Liang, R., Zhou, C., Zhang, J. (2020). A review on the application of photo-voltaic thermal systems for building façades. *Building Services Engineering Research and Technology*, 41(1), 86–107. https://doi.org/10.1177/0143624419845117

Rubbi, F., Habib, K., Aslfattahi, N., Yahaya, S.M., Das, L. (2020). Performance optimization of a hybrid PV/T solar system using Soybean oil/MXenenanofluids as a new class of heat transfer fluids. *Solar Energy*, 208, 124–138. https://doi.org/10.1016/j.solener.2020.07.060

Sachit, F.A., Rosli, M.A.M., Tamaldin, N., Misha, S., Abdullah, A.L. (2018). Nanofluids used in photovoltaic thermal (PV/T) systems: A review. *International Journal of Engineering and Technology*, 73(20), 599–611.

Samylingam, L., Aslfattahi, N., Saidur, R., Yahya, S.M., Afzal, A., Arifutzzaman, A., Tan, K.H., Kadirgama, K. (2020). Thermal and energy performance improvement of hybrid PV/T system by using oleinpalm oil with MXeneas a new class of heat transfer fluid. *Solar Energy Materials and Solar Cells*, 218, 110754. https://doi.org/10.1016/j.solmat .2020.110754

Sardarabadi, M., Hosseinzadeh, M., Kazemian, A., Passandideh-Fard, M. (2017). Experimental investigation of the effects of using metal-oxides/water nanofluids on a photovoltaic thermal system (PVT) from energy and exergy viewpoints. *Energy*, 138, 682–695.

Sardarabadi, M., Passandideh-Fard, M., Heris, S.Z. (2014). Experimental investigation of the effects of silica/water nanofluid on PV/T (photovoltaic thermal units). *Energy*, 66, 264–272.

Saroha, S., Mittal, T., Modi, P.J., Bhalla, V., Khullar, V., Tyagi, H., Taylor, R.A., Otanicar, T.P. (2015). Theoretical analysis and testing of nanofluids-based solar photovoltaic/thermal hybrid collector. *Journal of Heat Transfer*, 137(9), 091015.

Sathe, T.M., Dhoble, A.S. (2017). A review on recent advancements in photovoltaic thermal techniques. *Renewable and Sustainable Energy Reviews*, 76, 645–672. https://doi.org /10.1016/j.rser.2017.03.075

Sokhansefat, T., Kasaeian, A.B., Kowsary, F. (2014). Heat transfer enhancement in parabolic trough collector tube using Al2O3/synthetic oil nanofluid. *Renewable and Sustainable Energy Reviews*, 33, 636–644.

Tiwari, S., Bhatti, J., Tiwari, G.N., Al-Helal, I.M. (2016). Thermal modeling of photovoltaic thermal (PVT) integrated greenhouse system for biogas heating. *Solar Energy*, 136, 639–649. https://doi.org/10.1016/j.solener.2016.07.048

Tsai, H.L. (2015). Modeling and validation of refrigerant based PVT-assisted heat pump water heating (PVTA-HPWH) system. *Solar Energy*, 122, 36–47. https://doi.org/10 .1016/j.solener.2015.08.024

Yun, C., Qunzhi, Z. (2012). Study of photovoltaic/thermal systems with Mg/O water nano-fluid flowing over silicon solar cells. In: *Power and Energy Engineering Conference (APEEC)*, Asia-Pacific Shanghai.

Zafar, S., Sahil, A., Evangelos, B. (2018). A review on performance and environmental effects of conventional and nanofluid-based thermal photovoltaics. *Renewable and Sustainable Energy Reviews*, 94, 302–316.

Zhou, J., Ma, X., Zhao, X., Yuan, Y., Yu, M., Li, J. (2020a). Numerical simulation and experi-mental validation of a micro-channel PV/T modules based direct-expansion solar heat

pump system. *Renewable Energy*, 145, 1992–2004. https://doi.org/10.1016/j.renene
.2019.07.049

Zhou, J., Zhu, Z., Zhao, X., Yuan, Y., Fan, Y., Myers, S. (2020b). Theoretical and experimental study of a novel solar in direct-expansion heat pump system employing mini channel
PV/T and thermal panels. *Renewable Energy*, 151, 674–686. https://doi.org/10.1016/j
.renene.2019.11.054

Zondag, H.A., Vries, D.W., van Helden, W.G.J., van Zolingen, R.J.C., van Steenhoven, A.A.
(2003). The thermal and electrical yield of a PV-thermal collector. *Solar Energy*, 72(2),
113–128.

4 Applications of Bio-Nanoenabled Materials and Coatings for Affordable, Reliable and Sustainable Green Energy Production

Musa S.I., Harrison Ogala, Comfort Okoji, Nathan Moses, Nathaniel Iboyi and Sheila Ojei

CONTENTS

DOI: 10.1201/9781003316374-4

4.1 INTRODUCTION

Generally, the recent advancement in technology has proven that material at a nanoscale level exhibits unique properties and functions. These unique properties and functions have made it possible for researchers in various fields to consider employing nano-substances in improving their products. Nanotechnology is a multidisciplinary subject dealing with nanosized materials (1–100 nm). This field has continued to witness increased development because of its countless uses in various science domains like biology, robotics, physics, chemistry, materials science, biotechnology, microbiology and agricultural sciences. The word nanotechnology originates from the Greek word 'nano,' which means one-billionth of a meter, and was first used by Norio Taniguchi in 1974. Nanotechnology has efficiently revolutionized the medical field by introducing nanosized particles and materials, which show excellent biocompatibility and low toxicity. A special field of nanotechnology depends on using biological substances in the production of nanoparticles, and hence, bionanotechnology became one of the most used alternatives. Bionanotechnology refers to the science of producing a one-billionth of a meter particle using a biological substance (Iqbal *et al.*, 2012). Because of the biological source of this type of nanoparticles, researchers in different energy-related disciplines are also considering the employment of nanoparticles in improving sustainable and green energy production (Iqbal *et al.*, 2012; Ridolfo *et al.*, 2021). The nanosized materials which are synthesized by biological means are known as bionanomaterials. Their extraordinarily small size helps them exhibit unique structural, chemical, physical, optical, biological, mechanical and electrical properties that differentiate them from the bulk matter. These unique properties help them find diverse roles in the biomedical domain, such as tissue engineering, drug and gene delivery, cancer treatment, neurodegenerative diseases, inflammation, etc. Apart from therapeutics, bionanomaterials are also used in the diagnosis and imaging of various substances. Furthermore, bio-nanoenabled materials have been used as coatings for efficient, affordable and sustainable green energy production. Although bionanomaterials are biocompatible, their safety, toxicity and regulation are still a major concern in both research and market aspects. However, the ongoing research has now become more organized and broader (Abah *et al.*, 2010). Bionanomaterial is defined as molecular material composed of partial or nonpartial biological molecules like proteins, antibodies, enzymes, nucleic acids, lipids, poly- and oligosaccharides, viruses, secondary metabolites, etc. These can also be utilized to fabricate complex devices through self-assembling in ecofriendly and mild conditions. The reason why bionanomaterials are usually considered is because they are a unique material combining human-made or nonbiological material with natural compounds for nonbiological use (John, 2013). Therefore, many biodegradable polymers and naturally derived nanomaterials have been extensively used in various biomedical, pharmaceutical, industrial, packaging and agriculture fields. Furthermore, the fabrication of various biosensors for use in bio-coatings and biomedical utility-based bionanomaterials is well documented.

4.2 BIO-NANOENABLED MATERIALS

A general definition of bionanomaterials would be molecular materials composed partially or completely of biological molecules (such as antibodies, proteins/enzymes, DNA, RNA, lipids, oligosaccharides, viruses and cells) and resulting in molecular structures having a nanoscale-dimension(s). The resulting bionanomaterials may have potential applications as novel fibers, sensors, adhesives, energy generating and/or harnessing materials, to mention just a few aspects. These types of systems can allow for the fabrication of complex devices by self-assembly under mild experimental conditions and in an ecofriendly manner such as at room temperature and in aqueous conditions. The nanoscale is generally meant to encompass the dimensions between 1 and 100 nanometers. A sheet of regular paper is approximately 100,000 nanometers thick, and the diameter of a gold atom is roughly one-third of a nanometer.

Generally, all living organisms have a common basic feature of being comprised of cell(s). The more developed an organism, the more the varieties and the number of the existing cells. These cells have some functions that are mandatory for their existence and subsequently for the living organisms. Their sizes are in the range of 1–100 μm, which means that their functions, such as ion transport and signal transduction, actually occur at one-tenth of their size. These facts, indeed, directed researchers to minimize the material working scale to be able to interact with these cells at their desired scale, i.e., accomplishing nanoscale products capable of speaking the same language spoken at the cellular level. Biomaterials can be defined as any material or surfaces which have the ability to interact positively or preferably synergistically, in the biological environment. There are different grades of biomaterials ranging from inert up to bioactive materials, which, in turn, not only fill gaps in the biological system, but also perform bioactive functions. The smaller the particulates from which they are synthesized, the higher the 'bio-synergy' and the benefits that could be achieved. Therefore, bionanomaterials represent an important field of research with increasing technological advancement.

4.3 CHARACTERISTICS AND PROPERTIES OF BIO-NANOENABLED MATERIALS

Bionanomaterials properties are classified into physical, chemical, biological, mechanical, optical, electric, catalytic properties, etc. These properties are detected using different sophisticated characterization techniques like spectroscopy, microscopy, etc. (Li *et al.*, 2010). The physicochemical properties of bio-nanoenabled materials include:

4.3.1 THE SIZE AND SURFACE AREA OF THE PARTICLES

Bionanomaterials' extraordinarily small size helps them exhibit unique structural, chemical, physical, optical, biological, mechanical and electrical properties that differentiate them from the bulk matter. Bionanomaterials range in size from 1 to 100 nm and surface area increases as size decreases.

4.3.2   PARTICLE SHAPE AND ASPECT RATIO

Bionanomaterials come in varied shapes including fibers, rings, tubes, spheres, planes, oval, cubic, prism and rod (Manzoor *et al.*, 2014).

4.3.3   SURFACE CHARGE

Various aspects of nanomaterials such as the selective adsorption of nanoparticles, colloidal behavior, plasma protein binding, blood-brain barrier integrity and trans-membrane permeability are primarily regulated by the surface charge of nanopar-ticles. Of note, positively charged nanoparticles show significant cellular uptake compared to negatively charged and neutral nanoparticles, owing to their enhanced opsonization by the plasma proteins (Hoshino *et al.*, 2004).

4.3.4   THE OPTICAL AND ELECTRONIC PROPERTIES

The optical and electronic properties of bio-nanoenabled materials are inter-depen-dent to a greater extent. For instance, noble metal bionanomaterials have size-depen-dent optical properties and exhibit a strong UV–visible extinction band that is not present in the spectrum of the bulk metal. This excitation band results when the incident photon frequency is constant with the collective excitation of the conduction electrons and is known as the localized surface plasma resonance (LSPR). LSPR excitation results in wavelength selection absorption with extremely large molar excitation coefficient resonance ray light scattering with efficiency equivalent to that of ten fluorophores and enhanced local electromagnetic fields near the surface of NPs with enhanced spectroscopies (Ibrahim *et al.*, 2019).

4.3.5   MECHANICAL PROPERTIES OF BIO-NANOENABLED MATERIALS

Nanomaterials have excellent mechanical properties due to the volume, surface and quantum effects of nanoparticles. As nanoparticles are added to a common mate-rial, these particles will refine the grain to a certain extent, forming an intragranular structure or an intergranular structure, thereby improving the grain boundary and promoting the mechanical properties of materials (Abdul *et al.*, 2013).

4.4   PRODUCTION AND FORMATION OF BIO-NANOENABLED MATERIALS

Bio-based materials fall under the broader category of bio products or bio-based prod-ucts which includes materials, chemicals and energy derived from renewable bio-logical resources. Phytogenic nanoparticles have privileged status in nanoscience and nanotechnology research due to their innovative nature such as unique size, shape and reduced dimensions. Over the last few decades, nanoparticles have been much explored in various research fields including medicinal, optical and catalytic activities (Kaur and Chopra, 2018). Most nanoproducts produced on an industrial scale are nanoparticles, although they also arise as byproducts in the manufacture of other materials. Helmlinger *et al.* (2016) showed that the notable properties and activities of nanoparticles are completely dependent on their size, shape, morphology and distribution. Particularly, the size and shape of the particles are factors playing determinative roles in achieving adequate results in biomedical applications (Boselli *et al.*, 2020). Such unique particles can be prepared by employing plant extract (Raza *et al.*, 2016).

Due to their noteworthy physical, chemical and biological properties, nanoparticles have superlative applications in medicinal chemistry including cancer therapy, biotechnology and water treatment depending on their particle size (Iqbal *et al.*, 2012). Nanomaterials and/or nanoparticles are used in a broad spectrum of applications. Today they are contained in many products and used in various technologies. Recently, another research group has reported that the biomaterials are used as food-packing materials due to their high efficacy antimicrobial activity (Wang *et al.*, 2021). Among the various availed synthetic routes, environmentally benign green synthetic methodology is an ecofriendly alternative to save the globe and human health from toxic chemical synthetic methods (Ovais *et al.*, 2016, Singh *et al.*, 2021). Also, using natural resources, such as plants, including fruits, barks, roots, seeds, rhizomes, leaves, agricultural waste, and by-products, to prepare metal nanoparticles is an emergent solution for cheaper and cost-effective materials (Rodriguez-Felix *et al.*, 2021). Unlike plants, the physical and chemical methods used in preparing nanoparticles may require relatively long incubation times. This triggers biosafety issues. Furthermore, the medicinal plants extract-mediated act as a reduction of ions, and is more stable and cost effective (Restrepo and Villa, 2021).

Also, using such plant extract green synthetic methods is beneficial in achieving large-scale biogenic substances without any toxic by-products. So, it is the responsibility of researchers worldwide to reduce the production of toxic residues and develop harmless and highly sustainable methods like green synthetic methods of nanoparticles. The plant extracts employed contain biodegradable antioxidant biocomponents including phenols, phenolic polysaccharides, proteins, and vitamins that simultaneously play a role both in the reduction of metal ions and in the stabilization of nanoparticles (Bagalkotar *et al.*, 2010). Also, the interesting feature of this green methodology is, despite the fluctuating concentration of biocomponents depending on the age of the natural product and differing with the growing environment, the most relevant reducing compounds remains in high concentration to reduce the metal ions (Ghimire *et al.*, 2021). Today they are contained in many products and used in various technologies.

A recent work by Ghimire et al. (2021) focused on the development of nanoparticles using *Euphorbia granulata Forst* as precursor in *in vitro* biological applications. The synthesis and stabilization of biogenic can be achieved by employing *E. granulata* plant extract, which is a widely existing weed plant in dry areas. *E. granulata* (EG) belongs to the Euphorbiaceae family and contains a considerable amount of phenolic and flavonoid compounds and possesses antioxidant capacity.

Bionanomaterials can also be extracted from algae and sea animals and synthesized by bacteria. It has been shown that bio-residues are suitable for the production of cellulose nanomaterials and that the extracted nanofibers and crystals have similar structure and properties compared with the nanomaterials extracted from primary resources, but can be produced at lower cost, and most applications require a precisely defined, narrow range of particle sizes (monodispersity).

Biomaterials are those materials (synthetic and natural; solid and sometimes liquid) that are used in contact with biological systems or in medical devices. As a field, biomaterials has seen continuous growth and utilizes various methods from materials science and engineering, chemistry, medicine and biology. Biomaterials researchers must also consider ethics, law and the health care delivery system. Mainly biomaterials are used for medical purposes, but they can also be useful in the sector of growing cells in culture, to assay for blood proteins in the clinical laboratory, in processing biomolecules in biotechnology, for fertility regulation implants in cattle, in diagnostic gene arrays, in the aquaculture of oysters and for investigational cell-silicon 'biochips.' The commonality of these applications is the interaction between biological systems and synthetic or modified natural materials.

4.5   CLASSIFICATION OF BIONANOMATERIALS

The entry of organic bionanomaterials into the biomedical domain has attracted much focus, and therefore a sudden increase in their applications in bone or tissue regeneration, the formation of scaffolds, etc. Bionanomaterials should meet certain criteria like biodegradability, biocompatibility and non-toxicity for the use of these organic bionanomaterials in the biomedical field (Gnach *et al.*, 2015; Ajalloueian et al., 2014). Some organic bionanomaterials include silk fibroin, chitosan and other biodegradable polymers such as poly (lacticco-glycolic) acid (PLGA). PLGA is considered a synthetic polymer but, as it is biocompatible, it is now counted as an organic bionanomaterial. Silk fibroin is a natural polymer with a large molecular weight obtained from various spider species and silkworms. It has various biomedical applications like in regenerative medicines to control the drug's delivery to the cell efficiently. Moreover, to enhance the applications of bionanomaterials, it can be functionalized by adding different functional groups, as this functionalized silk fibroin helps in drug and gene delivery to the cancer cells, but it also regulates the drug-releasing kinetics (Mittal *et al.*, 2013). It was observed that silk fibroin acted as a nanocarrier loaded with cisplatin, and prevented cytotoxicity and side effects of the drug on normal tissues (Gnach *et al.*, 2015). Another organic bionanomaterial is chitosan, a linear polysaccharide made up of a deacetylated unit D-glucosamine and an acetylated unit β-(1–4)-linked N-acetyl-D-glucosamine. Recently, chitosan has found uses in bone formation therapies like chitosan nanomaterials-based scaffolds for tissue regeneration, and with the help of collagen, it increased the growth of bone regeneration (Eap *et al.*, 2014). PLGA is a copolymer made up of different monomers of lactic and glycolic acid. PLGA nanomaterials could be easily fabricated with various natural and synthetic materials to produce nanofibers or nanoscaffolds, which can be used in tissue engineering and drug delivery applications. It was observed that when PLGA was combined with chito-san, it produced nanostructures that could be used in dentistry and the healing of wounds (Chronopoulou *et al.*, 2016). Various cancer-treating drugs, such as cispla-tin, paclitaxel, doxorubicin, triptorelin, xanthone, 9-nitrocamptothecin, dexameth-asone and 5-fluoracil, have been successfully incorporated into PLGA, PLA and PCL nanoparticles (NPs) (Gnach *et al.*, 2015). Polymer-based bionanomaterials have also become a new vehicle for the regulated release of drugs. Similarly, silica is considered the most multifaceted compound in the materials family, easily exist-ing with other minerals, and is found abundantly in nature. Organic silica-based bionanomaterials possess special properties like small size, chemical nature, high surface area, high absorption capacity, tunable pore volume and size, hydrophilic nature and high biocompatibility, which have made them capable of being uti-lized in the biomedical domain. Organically modified silica-based NPs have been used for the *in vivo* delivery of genes. Unlike other silica NPs, mesoporous silica NPs (MSNs) have much utility in the biomedical domain such as for loading large amounts of drugs and producing biosensing molecules. Their unique honeycomb-like porous structure, low cytotoxicity and structural stability have given MSNs an efficient role in drug delivery applications (Gnach *et al.*, 2015). Figure 4.1 provides a brief classification of bionanomaterials. Ceramic bionanomaterials predomi-nantly include zirconia, alumina, nitride, hydroxyapatite tricalcium phosphate,

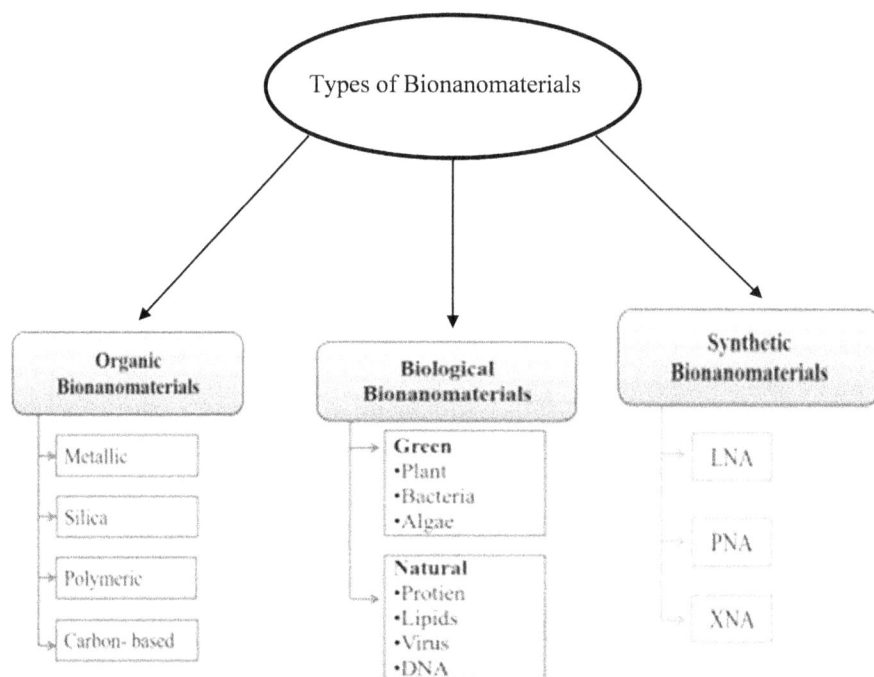

FIGURE 4.1   Schematic representations of types of bionanomaterials.

etc., as they exhibit many important properties like high resistance, good chemical stability, high biocompatibility and low density. It has been observed that nano-crystalline hydroxyapatite shows numerous applications in tissue engineering by enhancing the biocompatibility of titanium alloy, and acts as an antibody delivery agent in different bone infections and also works as an injectable paste that can be used as a substitute in producing bone structure as it exhibits great osteoconductive properties (Bigi *et al.*, 2007). Currently, bionanocomposites have gained much interest from researchers as are utilized in nanotechnology, material sciences, biology and nanobiotechnology. However, nowadays, bionanocomposites are gaining utility in the biomedical domain as well. Bionanocomposites are considered precursors of various biopolymers like polysaccharides, polypeptides, poly-nucleic acids and aliphatic polyesters (Gnach *et al.*, 2015). The size of a biopolymer lies in the range of 1–100 nm, and the method of synthesis and the unique structure of a bionanocomposite mainly decide its properties.

Generally, the biopolymers present in bionanocomposites are soluble in water, although petroleum-derived polymers show non-solubility in the water but are soluble in organic matter. The various properties of bionanocomposites like solubility in water, biodegradability, biocompatibility and thermal stability help in determining the preparation methods and applications. Bionanocomposites have potential in biomedical applications such as the antimicrobial activity of tourmaline/cellulose nano-crystal composite films against *Staphylococcus aureus*. Moreover, due to the starch's commendable properties like reduced toxicity, high biodegradability, biocompatibility and tunable mechanical properties, researchers have started focusing on starch-based bionanocomposites for the biomedical domain (Eid, 2011, Gao *et al.*, 2011). Some starch-based bionanocomposites in biomedical applications include bone regenerating treatments, drug-delivery systems, bone and tissue engineering and hydrogel production. Similarly, carbon nanotubes (CNTs) have also found their uses in medical applications, like reinforcement, to help initiate formation (Barber and Freestone, 1990). An experiment was performed by combining multi-walled carbon nanotubes (MWCNTs) with starch-based nanocomposites to study their biomedical applications; the results showed that they successfully regenerated bone and can be used for regenerating or treating bones or tissue scaffolds (Fama *et al.*, 2011). Another nanocomposite material was produced through the self-assembly of oleyl phosphate (OPh) which was stabilized with the help of iron oxide (Fe_3O_4) NPs (Domènech *et al.*, 2019). Currently, biologically derived bionanomaterials have also gained much attention from researchers in the biomedical domain field. Biologically derived bionanomaterials can be further divided into two types: natural and green-based. The natural bionanomaterials involve certain lipoproteins, DNA or RNA, peptides, etc., whereas green-based bionanomaterials include plant-based, bacteria-based, fungi-based and viral-based bionanomaterials. Some naturally derived bionanomaterials are further discussed. Lipoproteins are hydrophobic biomolecules consisting of lipids and proteins, which have a spherical shape in nanodimensioned form. Lipoprotein-based nanomaterials are biocompatible, biodegradable and stable in blood circulation, making them more suitable for biomedical applications. Lipoprotein-based nanomaterials have found a potential role in both therapeutics and imaging. Lipoproteins are widely used in drug delivery and cancer diagnosis. Similarly, peptide bionanomaterials can be structurally modified either through the self-assembly of their molecules or by binding with other nanomaterials, leading to variation in their structure (Fraysse-Ailhas *et al.*, 2007). The small size (6–10 nm) of peptide NPs' central core has found profound use in fluorescence imaging and gene delivery by attaching certain oligonucleotide sequences. It has been found that peptide-based NPs exhibit efficient antimicrobial properties, drug delivery and gene therapy (Iqbal *et al.*, 2012). Also, deoxyribonucleic acid (DNA) is a biomolecule made up of base pairs that store up the genetic information of a living body, and it is considered the most versatile biomolecule in nanotechnology. Its unique physical and chemical properties have been utilized in various biomedical applications; various forms of DNA have been utilized in the biomedical domain, such as in the production of sensors, molecules for imaging and therapeutics. DNA-based biosensors are under the spotlight as they have been applied in all domains like biological, environmental, industrial and pharmaceuticals, because DNA plays various roles in biosensors such as target, linking strand recognition motif, probe, etc. (Gnach *et al.*, 2015). Cancer is considered one of the leading causes of death worldwide, and various cures like radiotherapy or chemotherapy have a lot of side effects on the human body. Therefore, it has become a necessity to develop alternative agents for the curing of cancer. DNA nanostructures exhibit unique features and are now widely utilized to create various heterogeneous anticancer agents for targeted cancer treatments (Iqbal *et al.*, 2012). Also, a variety of DNA nanostructures have been developed for chemotherapeutic drug-loading platforms along with the aptamer-based delivery system, which will specifically decide the uptake of drugs in *in vivo* conditions (Gnach *et al.*, 2015). Doxorubicin (DOX) is a well-known anticancer drug that is widely used in chemotherapy to treat cancer, as it works by stopping macromolecular biosynthesis by communicating with DNA (Chang *et al.*, 2011). Furthermore, it is important to control the release of DOX by modifying DNA's coiling degree. Another approach was developed to overcome multi-drug resistance using targeted drug delivery. In this a combination of an aptamer along with GC-rich dsDNA was prepared which showed great stability, reduced side effects and had good payload capacity (Gnach *et al.*, 2015). Similarly, a novel self-assembled DNA nanostructure showed the capacity to particularly recognize cancer cells and to also control the release of the loaded anticancer drug maintained at pH 5.0 (Li *et al.*, 2010, Baker and Baker, 2010). With the development of nanotechnology, the attention paid to green-based bionanomaterials has increased. Certain plants, bacteria, fungi, yeasts, and viruses are used to derive bionanomaterials as they provide great cytocompatibility, less toxicity and ecofriendly techniques and are cost-friendly; therefore, these green-based bionanomaterials have found potential uses in the biomedical domain. Nowadays, it has been proved that viruses can also be utilized in the field of nano-biotechnology as they have found potential roles in various fields like bionanomaterials, bioimaging and science. Some of the viruses that have been used for these applications are cowpea chlorotic mottle virus (CCMV), hepatitis B cores, cowpea mosaic virus (CPMV), vault nanocapsules, MS2 bacteriophages and M13 bacteriophages. It was observed that the paramagnetic Gd3+-bound NPs exhibited higher relaxivity as compared to the protein-bound Gd3+ chelates (Allen *et al.*, 2005), and therefore they are now used for *in vivo* small animal MRI. Canine parvovirus (CPV) is a natural pathogen of dogs that carries a gene delivery vehicle known as adeno-associated virus (AAV), which has been utilized for targeting tumors (Gnach *et al.*, 2015). Various bionanomaterials, which are analogs of natural material, can be defined as inorganic bionanomaterials like peptide nucleic acid (PNA), aptamers, xeno nucleic acids (XNAs), etc. PNA is a synthetic analog of DNA/RNA consisting of a 2-([2-aminoethyl] amino) acetic acid backbone that results in the formation of achiral and uncharged mimic. It is generally considered DNA, but it consists of a neutral peptide backbone instead of a negatively charged sugar-phosphate backbone. This change in the PNA's backbone means it possesses unique physical and chemical properties like resistance to enzymatic degradation inside living cells, and it is under consideration for application in various fields (Gnach *et al.*, 2015). It has also shown potential applications as a biomolecular tool, biosensors, antisense and antigen agent and molecular probes. The lack of sugar-phosphate backbone in PNA makes it neutral, and therefore it lacks electrostatic repulsion, which helps it to exhibit great affinity towards the target. PNA is a polyamide-based synthetic molecule which gives it more chemical and pH stability as compared to the DNA or RNA. This property of PNA helps it to bind with more specific sequences. PNA generally consists of uracil, adenine, guanine and cytosine-N-acetic acids, making it a precursor of RNA. Moreover, PNA's ability to easily recognize specific RNA or DNA sequences using the Watson–Crick hydrogen bonding arrangement gives it distinctive ionic strength and thermal stability (Ridolfo et al., 2021). Therefore, it exhibits a broad range of applications in the biomedical domain as nucleic acid biosensors are one of the applications being tested in genetics and biomedicine using PNA specifically. It can be utilized in the development of high-performance affinity biosensors for application in DNA genotyping, although studies about specificity and sensitivity are yet to be explored. XNAs are considered artificial genetic systems which constitute only synthetic nucleotide monomers, with the change in the chemical composition of the sugar moiety. Examples include 3 acyclic L-threoninol (L-aTNA), cyclohexene (CeNA), 2′-O,4′-C-methylene linked ribose (LNA), threose (TNA), acyclic L-threoninol (L-aTNA) or a glycol (GNA) instead of the natural (2′-deoxy) ribose. The formation of homoduplexes by XNAs gives them more stability than heteroduplexes formed with complementary DNA and RNA counterparts (Iqbal *et al.*, 2012). Various bionanomaterial-based metals and their oxides used in the biomedical domain are the most widely studied elements in nanotechnology, and plenty of nanomaterials have found significant use in different domains. The metal and metal oxide nanomaterials are easy to synthesize and characterize, as they show excellent properties, making them unique and different from the bulk material. They can be easily synthesized using chemical, physical or biological routes, but generally, biologically synthesized bionanomaterials grab the attention as biological synthesis is cost-effective, they are free of chemical toxins and they do not produce any side effects in either the environment or in living cells. Various reviews and book chapters have focused on the metal NP and its applications in various domains (Ridolfo *et al.*, 2021, Fernandes, 2020). The commonly used metallic bionanomaterials are gold, silver, iron, copper, zinc and nickel, but researchers have now started to focus on different metals like cerium, selenium, magnesium, etc. The unique properties exhibited by metallic bionanomaterials like size, shape, conductivity, etc., have made them the most common bionanomaterials used in biomedical applications as they are utilized in both diagnosis and therapeutics. When these metal NPs are fabricated with other compounds their properties are enhanced, which in turn increases their utility (Iqbal *et al.*, 2012).

4.6 PRODUCTION AND FORMATION OF BIO-NANOENABLED MATERIALS (NM)

The science of nanotechnology deals with the production, design, characterization and potentiality of substances at a nano-level; therefore, it is also known as 'engineering at a molecular level.' These nano-scaled substances, when obtained from a biological origin, are regarded as bionanomaterials. Since nanotechnology has provided great benefits in various fields of biotechnology, obtaining nano-scaled materials from biological substances may give effective results. There are many approaches used in the production of nanoscale substances. Recently, biological approaches are preferred to other approaches as they are easy, convenient, eco-friendly, reduce the use of excess toxic chemicals and are cost-effective. Due to these reasons, green synthesized bionanomaterials are more preferred in the bio-medical domain, as these bionanomaterials are biocompatible and non-toxic. Biological approaches include the utilization of various living organisms like bacteria, fungi, yeast, plants and animals, as well as the use of various natural components like honey, pectin, glucose, starch, etc. The utilization of plant extracts is known as the green synthesis of bionanomaterials and is considered more than other green-based synthesis because plants show immunity to the accumulation of heavy metal ions, so there is no need to maintain the culture media of microorganisms, and rate production of the plant is fast compared to microorganisms. Furthermore, the plant-based production of nanoscale material is preferable because it exhibits slow reaction kinetics so that both the stabilization and growth of bionanomaterials can be controlled. Previous studies have discussed the influence of this product in terms of green production (Mittal *et al.*, 2013). At first, it is important to ensure the appropriate plant or plant part is selected for the production of bio-nanoenabled materials. Different plants have been observed to show different enzymes and different activity rates, phytochemical constituents, biochemical processing, etc. In this case, using a plant as a source of bionanomaterial can either be *in vivo* or *in vitro*, where *in vivo* means the synthesis of bionanoparticles inside the plant, while *in vitro* synthesis is the synthesis of bionanoparticles using the plant. These plant extracts naturally act as stabilizers and reductants; therefore, they play a vital role in synthesizing nanomaterials (Annu and Ahmed, 2018). The plant-based synthesis of bionanomaterial follows a common protocol in which plant extract is first washed, squeezed and filtered, followed by salt (metal or non-metal), which causes a visible color change. This color change is considered as the formation of NPs, and they are further characterized. Nanomaterials' unique properties have increased their role in various research fields and have also developed markets of products containing nano-objects. To understand the physicochemical properties of bionanomaterials produced from plant materials, different characterization techniques are used. Some of the factors like solvent medium and the stabilizing and reducing agent need to be considered while green synthesizing bionanomaterials. The characterization techniques are mainly used to study and improve various properties like distinct size, shape, molecular weight, purity, solubility, chemical composition and stability of the green synthesized bionanomaterials, as these properties help in determining their applications in various domains like biomedicine, sensing, agriculture, the environment, etc. (Biao *et al.*, 2018; Iqbal *et al.*, 2012). The characterization techniques are divided into two types: structural characterization and chemical characterization. Structural characterization is used to detect the structural morphology of a bionanomaterial, that is, the size, shape and crystallinity, whereas chemical characterization is used to detect various surface and chemical atoms and the spatial conjugation of the bionanomaterial.

The characterization techniques are mainly used to study and improve various properties like distinct size, shape, molecular weight, purity, solubility, chemical composition and stability of the green synthesized bionanomaterials, as these properties help in determining their applications in various fields (Biao *et al.*, 2018, Jain *et al.*, 2011). The characterization techniques are divided into two types: structural characterization and chemical characterization. Structural characterization is used to detect the structural morphology of a bionanomaterial, that is, the size, shape and crystallinity, whereas chemical characterization is used to detect various surface and chemical atoms and the spatial conjugation of the bionanomaterial. The characterization techniques use highly sophisticated instruments. Some structural characterization instruments are X-ray photoelectron spectroscopy (XPS) which is also known as electron spectroscopy used generally for chemical analysis (ESCA) and is required to study the physical and chemical properties of the nanomaterials. XPS works on the principle of photoionization and is widely utilized to study the surface properties of bionanomaterials which helps in studying their chemical state and energy distribution. XPS was utilized to study the interaction effects of proteins extracted from *Capsicum annuum* in the formation of silver NPs (Li et al., 2010). Scanning electron microscopy (SEM) and transmission electron microscopy (TEM) are among the most highly utilized tools in analyzing green synthesized bionanomaterials because they are simple and convenient to use. They are used to study the morphological characteristics like the size and shape and size distribution of green synthesized bionanomaterials. SEM analysis was used to study the ultra-thin coating of electrically conductive material and various other green synthesized bionanomaterials, but in contrast to SEM, TEM is considered a more sophisticated technique as it gives details about the chemical composition of the material and it also produces a direct image of high spatial resolution within atomic range. There is not much difference between SEM and TEM except for the sample thickness and the means of data collection from the material. To study detailed information about the chemical composition of bionanomaterials, X-ray spectroscopy can be combined with TEM and SEM (Gnach *et al.*, 2015). TEM has been utilized for the morphological study of various green synthesized bionanomaterials like silver, copper, nickel, etc. (Li *et al.*, 2010). Another technique used for structural detection is atomic force microscopy (AFM), which is considered the most modern, promising and advanced technology as it produces nano and atomic scale images of the surface, and it can be performed in different surroundings like air, vacuum, liquid, etc. AFM is used to detect and analyze the characteristic properties of many green synthesized bionanomaterials (Bhattacherjee *et al.*, 2018). Energy dispersive X-ray (EDX) is one of the most sophisticated techniques used to study bionano-materials' composition. SEM and TEM combined with EDX can efficiently detect elemental composition, microstructure and spectra (Anake *et al.*, 2018). EDX can give qualitative, quantitative or semiquantitative analysis by mapping the spatial distribution of components. EDX is considered one of the most powerful tools for studying various physical, compositional and chemical properties of green syn-thesized bionanomaterials like gold and silver nanomaterials (Gangadoo *et al.*, 2015), and similarly, the X-ray diffraction (XRD) method is one of the important techniques used in the nanotechnology domain, as it determines the crystal structure, crystallinity, crystal defects and atomic spacing of the crystalline material. XRD is used to characterize many green synthesized bionanoparticles like silver NPs (Li *et al.*, 2010). To study the chemical composition of the green synthesized bionanomaterials, various techniques are utilized like UV–visible spectroscopy, a basic technique utilized for both the quantitative and qualitative characteriza-tion of all biochemical compounds. It is generally used to analyze physical and chemical properties such as size, aggregation and molecules' concentration. It is used to study the optoelectronic properties and the structural conformation of the bionanomaterials. UV–visible spectroscopy works on the principle of the Beer–Lambert law by applying linear relations between absorbance and concentrations. Many green synthesized bionanomaterials like silver, gold, copper, cerium, sele-nium, etc., have been characterized and used to study the structural properties of green bionanomaterials. Similarly, Fourier transform infrared spectroscopy (FT-IR) is considered the most sensitive and rapid analysis technique to characterize bionanomaterials. It works by analyzing specific vibration patterns of individual molecules that help determine the functional groups like aldehydes, ketones, alco-hols, carboxylic acid and terpenoids present in the bionanomaterial. Therefore, it is used for both quantitative and qualitative analysis of solid and aqueous samples. Additionally, it is also used to detect the structural composition, dynamics, stabil-ity, aggregation, etc. Several green synthesized bionanomaterials like silver NPs, gold NPs, nickel NPs, iron oxide NPs, etc., have been analyzed with the help of FT-IR (Arunachalam and Annamalai, 2013, Davar *et al.*, 2009). Moreover, nuclear magnetic resonance (NMR) spectroscopy is also a non-invasive method used for analyzing various constituents present in samples. NMR is considered an efficient analyzing method as it provides high output and high reproducibility, and it is easy to modify data. NMR is used to analyze the properties of various polymers, amor-phous materials, biomolecules and bionanomaterials.

4.7 APPLICATIONS OF BIO-NANOENABLED MATERIALS

Nanobiomaterials have become a useful tool for medical applications for several reasons including compatibility and novel effects due to nanoscale. The possibilities of adding biomaterials to the development of nanostructures have opened the door for innovative applications in several fields (Karagkiozaki *et al.*, 2012). One of the most important applications of bio-nanoenabled materials is the use in modern medicine. . This has been used in solving complex problems. Modern medicine has incorporated nanotechnology into two major areas of general concern, tissue engineering and novel viruses, which were chosen in this review due to significant cases and impact (Torres-Sangiao *et al.*, 2016; Kapat *et al.*, 2020).

Biomaterials are a group of substances that are either produced by living organisms or highly compatible. In this matter, a great deal of research work has been done to test and use the materials in modern medicine. An extensive list of medical applications can be found elsewhere (Nune and Misra, 2016; Gim *et al.*, 2019). The property of biocompatibility invited the scientific community to explore the characteristics of nanotechnology (Bayda *et al.*, 2020). Nanotechnology is an emerging research field based that utilizes the smallest material scale which can be applied in solving problems from all scientific fields. The use of this small material scale in areas such as biotechnology, genetic engineering and other disciplines allowed nanometer-scale manipulation (Wong *et al.*, 2013). The characteristics of novel biomaterials at the nanoscale make them a powerful tool for achieving precise and smart functions, e.g., drug delivery, localized effects dependent on size, features triggered by stimuli (Lombardo *et al.*, 2019).

The presented work explores the most prominent alternatives focused on modern medicine applied in tissue engineering, as well as therapeutics and vaccines for COVID-19 (Tang *et al.*, 2016). First, a detailed description of nanostructures is given to in order to understand the principles of nanomaterials. Then, a description of the administrative mechanisms is given to provide an understanding of the conditions for the materials and characteristics of the structures used depending on the application. Tissue engineering can help a vast number of diseases, including the current COVID-19 pandemic, from the angle of regenerative medicine. In general, multidimensional applications refer to the inclusion of several materials, drugs, geometries and other characteristics in the nanobiomaterial for desired multipurpose (Palestino *et al.*, 2020). Applications of nanomaterials in biology or medicine are as follows:

- Used in fluorescent biological labels (Mittal *et al.*, 2013).
- Used in drug and gene delivery (Parak *et al.*, 2002).
- Used in the bio detection of pathogens (Zhang *et al.*, 2002).
- Used in the detection of proteins (Iqbal *et al.*, 2012).
- Used in the probing of the DNA structure (Mittal *et al.*, 2013).
- Used in tissue engineering (Sinani *et al.*, 2003).
- Used in tumor destruction via heating (hyperthermia) (Nam *et al.*, 2003).
- Used in the separation and purification of biological molecules and cells.
- Bio-based materials are often biodegradable, but this is not always the case.

Examples include:

1. Cellulose fibers – fibers made from reconstituted cellulose.
2. Casein – a phosphoprotein extracted from milk during the process of creating low-fat milk, it is processed in various ways to make plastic, dietary supplements for body builders, glue, cotton candy, protective coatings and paints, and occurs naturally in cheese, giving it a creamy texture.
3. Polylactic acid – a polymer produced by industrial fermentation.
4. Bioplastics – including a soy oil-based plastic now being used to make body panels for John Deere tractors.
5. Engineered wood – products such as oriented strand board and particle board.
6. Zein – a natural biopolymer which is the most abundant corn protein.
7. Cornstarch – the starch of the maize grain, used to make packing pellets.
8. Grease – lubricants made from vegetable oils, including soybean oil, that can replace petroleum-based lubricants.

As mentioned above, the fact that nanoparticles exist in the same size domain as proteins makes nanomaterials suitable for bio tagging or labeling. However, size is just one of many characteristics of nanoparticles and is rarely sufficient if one is to use nanoparticles as biological tags. In order to interact with a biological target, a biological or molecular coating or layer acting as a bioinorganic interface should be attached to the nanoparticle. Examples of biological coatings may include antibodies, biopolymers like collagen (Bruchez *et al.*, 1998) or monolayers of small molecules that make the nanoparticles biocompatible (Mittal *et al.*, 2013). In addition, as optical detection techniques are widespread in biological research, nanoparticles should either fluoresce or change their optical properties.

Nanoparticles usually form the core of nanobiomaterial. It can be used as a convenient surface for molecular assembly, and may be composed of inorganic or polymeric materials. It can also be in the form of a nano-vesicle surrounded by a membrane or a layer. The shape is most often spherical, but cylindrical, plate-like and other shapes are possible. The size and size distribution might be important in some cases, for example if penetration through a pore structure of a cellular membrane is required (Elmowafy *et al.*, 2019). The size and size distribution are becoming extremely critical when quantum-sized effects are used to control material properties. A tight control of the average particle size and a narrow distribution of sizes allow the creation of very efficient fluorescent probes that emit small amount of light in a very wide range of wavelengths. This helps with creating biomarkers with many and well-distinguished colors. The core itself might have several layers and can be multifunctional. For example, combining magnetic and luminescent layers one can both detect and manipulate the particles.

The core particle is often protected by several monolayers of inert material, for example silica. Organic molecules that are adsorbed or chemisorbed on the surface of the particle are also used for this purpose (Song *et al.*, 2020). The same layer might act as a biocompatible material. However, more often an additional layer of linker molecules is required to proceed with further functionalization. This linear linker molecule has reactive groups at both ends. One group is aimed at attaching the linker to the nanoparticle surface, and the other is used to bind various moieties. Bionanomaterials are utilized in biomedical applications, including tissue engineering, drug delivery, nanomedicines and the diagnosis of various diseases, as shown in Figure 4.2. Various other applications have been described below.

4.7.1 Tissue Engineering

Tissue engineering has allowed a great advance in surgical and therapeutic techniques as it helps develop and identify the desired cell or tissue behavior, functions and structure. It is considered a promising technology that helps to build on defective or damaged tissues and aims to restore partial or full functionality (Cao et al., 2007). Tissue engineering utilizes various combinations of scaffolds, biologically active molecules and cells to construct, maintain or improve on the damaged tissue or organ functions. In order to produce an ideal scaffold, it should be able to carry active biomolecules and initiate physiological signals and should have mechanical properties similar to the original tissue (Iqbal et al., 2012). With the introduction of bionanomaterial, tissue engineering's potential has enhanced and developed, making it more

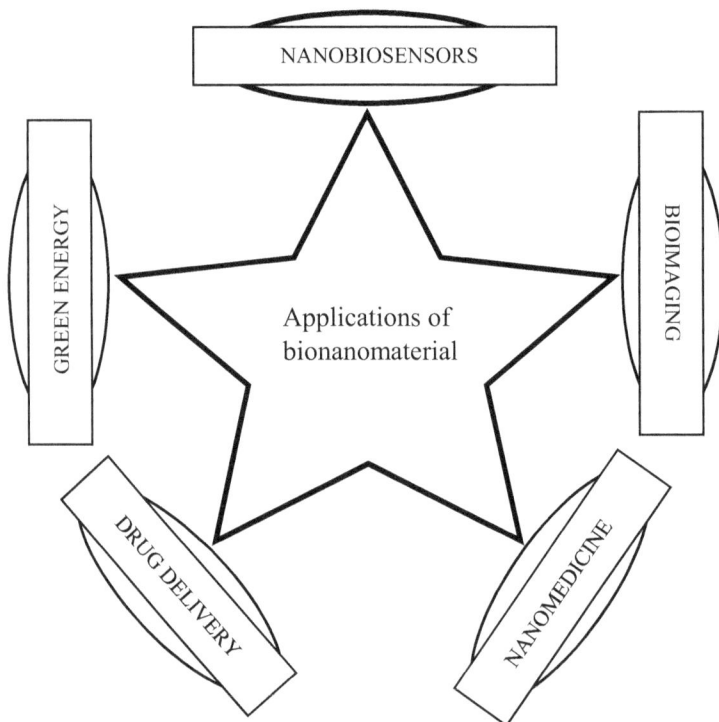

FIGURE 4.2   Fields of application of bionanomaterials.

efficient and more often considered in therapeutic applications. Bionanomaterials possess a superfine structure, which helps them directly interact with the cell or tissue surface receptors, creating a microenvironment to form tissue regeneration (An *et al.*, 2013). Further, bionanomaterials produce a wide range of basic structural units like NPs, nanofibers, nanofilms and nanotubes essential in replacing damaged or defective tissue or organs in the body. Vascular tissue engineering is one of the leading applications in tissue engineering, which replaces or repairs the damaged vessels. Several *in vivo* studies have shown that vascular tissue can efficiently interact with NPs when there is a dire need to create an appropriate method to repair or replace the damaged vessel. In recent times, the formation of allografts or autografts for repairing damaged vascular tissues has not gained much attention because of the formation of compliance incongruity and thrombus. Therefore, several bionanomaterials have been targeted, assembled and toned to overcome these problems and improve the function of vascular endothelial cells. Comparative studies have concluded that the nano-structured titanium surface provides great adhesion and expansion of vascular endothelial cells as compared to bulk titanium; therefore, this can be used to enhance the vascular endothelial cells (Choudhary *et al.*, 2007). It was observed that nanometer PLGA vascular grafts helped in triggering the proliferation of both smooth muscle cells (SMCs) and vascular endothelial cells, and it also enhanced the absorption of both fibronectin and vitronectin from the serum (Iqbal *et al.*, 2012). Similarly, the repair and restoration of damage in the nervous system are dangerous to life and more complicated and challenging than vascular tissue engineering. However, many biomaterials have been considered promising to date and have been widely explored; complete recovery is not possible, but full function of the damaged nerve is restored. Unlike vascular tissue damage, nervous tissue damage is considered fatal and dangerous for life because the nervous system is a sophisticated network that works by receiving, analyzing and responding to information. Slight damage in the nervous system could lead to serious illness and in some cases death, and for this reason, reconstruction of nervous system injuries is still a challenge. Autograft and allograft are considered important approaches in fixing nervous system repairs, but the complete recovery of tissue is still not possible, and the allograft causes inflammation reactions, tumor formation and the generation of infections in the body. Another common problem that allograft and autograft possess is the regular intake of immunosuppressors to prevent graft rejection (Mittal *et al.*, 2013). Several bionanomaterials have been studied and explored to overcome these problems, as they exhibit excellent electrical conductivity, phase separation, electrospinning, self-assembly and cytocompatibility, which help them promote neural regeneration and enhance the regeneration of axons (Mittal *et al.*, 2013). It has been found that several piezoelectric bionanomaterials like Poled PVDF, GO nanosheets, Electrospun PVDF and P(VDF-TrFE), etc., can be used in the regeneration of nervous tissue grafts (Kapat *et al.*, 2020, Valentini *et al.*, 1992). Bionanomaterials possess a great future and potential applications in tissue implantation, and more research is needed in this field. A detailed analysis is presented in further chapters, giving plenty of room to the reader to explore the potential of bionanomaterials in tissue engineering.

4.7.2 NANOMEDICINE

Nanomedicine promises a revolution in traditional clinical practices, as the introduction of novel medicines in both diagnosis and treatment has made it possible to meet some previously impossible medical needs, like the integration of highly toxic but effective molecules like Mepact, increasing the efficacy and bioavailability and decreasing dose toxicity and side effects, while increasing targeting of the drug and the controlled and site-specific release of the drug. It also enhances transport across biological barriers (Chan, 2006, Nano on reflection, 2016). This is possible due to bionanomaterials' unique properties, which have brought many pharmaceutical and biomedical development benefits. The extremely small size and high surface area-to-volume helps them in adsorbing biomolecules, like proteins, lipids, etc., when they are in contact with biological fluids. This has gained them a potential position in the biomedical field. One of the most important interactions within living matter is the plasma/serum biomolecule adsorption layer, also known as 'corona,' which is formed on the surface of colloidal NPs (Gnach et al., 2015). Its composition depends on the point of entry inside the body and the particular fluid that the NPs come across like blood, lung fluid and saliva. Additional dynamic changes can influence the 'corona' constitution as the NP crosses from one biological compartment to another (Iqbal et al., 2012). Furthermore, other properties like electrical, optical and magnetic properties can be modified through electron confinement in nanomaterials. Various polymeric bionanomaterials are considered excellent carriers of the required drugs in the treatment of cancer treatment. Synthetic polymeric bionanomaterials include poly (lactic acid), PLGA and polyethyleneimine, while the natural ones are albumin and collagen chitosan and gelatin. Various anticancer drugs like cisplatin, doxorubicin, paclitaxel, 5-fluorouracil, triptorelin, 9-nitrocamptothecin, xanthone, dexamethasone, etc., have been effectively conjugated with PLGA and have shown effective drug delivery to cancer cells (Ali et al., 2020).

4.7.3 BIONANOMATERIALS FOR DRUG DELIVERY

Bionanomaterials have been found to enhance pharmacological effects along with minimizing the negative side effects, and giving plenty of room to researchers to explore their potential in drug delivery (Bayda et al., 2017). Various criteria should be kept in mind while preparing an efficient drug delivery material, for example, it should be able to enhance the water solubility of hydrophobic drugs, control drug release and also enhance the rate of drug delivery. Carbon-based, silicon-based and polymeric-based bionanomaterials have gained a lot of attention in the past few decades; of these, CNTs have attracted a lot of exploration, because they exhibit unique enclosed nanochannels making them a potential material for drug-delivery applications. It was observed that single-walled carbon nanotubes (SWCNTs), when functionalized with different phospholipids along with a folic acid (FA) terminal group and polyethylene glycol (PEG) chain segment (SWCNTs-(PL-PEG-FA)), selectively killed the cancer cells without causing any harm to the normal cells. This experiment proved that CNTs can be used as an efficient transport material for carrying drug (Chen *et al.*, 2014). Similarly, CNTs covalently bonded with amphotericin B (AmB) were easily taken up by mammalian cells (Xu *et al.*, 2016). Apart from various types of CNTs, graphene-based bionanomaterials have also achieved a lot of success in drug delivery applications (Iqbal *et al.*, 2012). It was observed that when graphene nanosheets (GNS) were attached to the anticancer drug, it showed high loading capacity which was also found to be useful in cellular imaging (Mittal *et al.*, 2013). Mesoporous silica NPs (MSNs) exhibiting unique properties like adjustable and uniform pore size are suitable as a drug delivery material and an excellent pharmacological agent (Li *et al.*, 2010). It was observed that MSNs when coated with a lipid bi-layer (LMSNs) showed excellent biocompatibility and also exhibited great cellular uptake in cancer cells as compared to other cancer-treating drugs, therefore making LMSNs effective in cancer treatment (Gnach *et al.*, 2015). Besides carbon and silica nanomaterials, polymer-based nanomaterials have also proved their potential in drug delivery applications by protecting drugs from premature degradation, controlling the release of drug and reducing the drug's side effects (Allhoff, 2007). Natural polymer-based bionanomaterials can easily be loaded with the drug and show a great adsorption pattern which can be used in the development of various devices (de Oliveira *et al.*, 2013). On the other hand, synthetic polymer-based bionanomaterials show continuous drug release *in vitro* along with high loading efficiency (Cao *et al.*, 2007). Recently, the drug delivery system has been enhanced and is now shifting towards three-dimensional (3D) technology as it efficiently controlled the drug release for more than four months (Kudgus *et al.*, 2011). The introduction of bionanomaterials in the biomedicine domain has efficiently revolutionized drug delivery applications and has also decreased the toxic effects of many harmful drugs to living cells.

4.7.4 BIONANOMATERIALS FOR NANOBIOSENSING

In the past decades, biosensors have experienced remarkable growth and have revolutionized the biomedical domain by enhancing diagnostic and therapeutic applications. A biosensor is an analytical device which is used to detect an analyte or to measure physiological signals by combining various biological elements and transducers for converting electrical, optical and thermal signals (Gnach *et al.*, 2015). A biosensor is primarily made up of two components: firstly, a probe which is used to detect various biomolecules like DNA, RNA, antibodies, aptamers, enzymes, proteins, etc., and secondly a transducer which is used to transform physiochemical information into detectable magnetic, electrical, optical, thermal and electrochemical signals. Various types of bionanomaterials have been utilized in the development of different nanobiosensors as they possess unique properties like greater surface area, tunable size, etc. Based on different types of bionanomaterials, like zero-dimensioned, one dimensioned, two-dimensioned and three-dimensioned biosensors, optical and electrical biosensors have been produced, as different types of material exhibit different configurations, helping them to enable the enhanced integration of biosensors in the human body. Bionanomaterials have made the production of ultrasensitive biosensors possible, as they provide high surface area, excellent biocompatibility, electrocatalytic activity and electronic properties. They have made it possible to directly wire the enzymes with the electrode surface, to promote an electrochemical reaction and to amplify a signal produced by biorecognition reactions. It was observed that CNT-modified electrodes can easily detect NADH at low potential. Similarly, gold NPs were used in electrochemical immunosensors, carbon nanotube-based sensors (Agui *et al.*, 2008, Umasankar and Chen, 2008) and NPs-based biosensing (Iqbal *et al.*, 2012). Bionanomaterials have not just shown their potential applications in therapeutics, but they have also exhibited great potential in diagnosis. Various bionanomaterials are utilized in sensor-based applications, computed tomography (CT) imaging, magnetic resonance imaging (MRI) and photo-thermal therapy (PTT). Nanotechnology is considered an emerging technology in diagnostic applications as it enhances and simplifies the detection of various biomolecules like DNA, RNA, proteins, enzymes, antibodies, etc., by functionalizing NPs to monitor these molecules at the molecular level. Among many bionanomaterials, quantum dots have gained much popularity due to their semiconductive nature and high resistance to chemical degradation. Quantum dots are also stable, specific and sensitive, which means they are considered to be more diagnostic than other bionanomaterials. Quantum dots have provided a strong base for utilizing fluorescence resonance energy transfer (FRET)-based nanobiosensors to monitor and detect various biological responses; for example, it was observed in FRET-based biosensors that when quantum dots were linked with DNA probes to catch DNA targets, they efficiently monitored the low concentration of DNA in a free-separation format. It was also observed that quantum dots in the nano-range efficiently amplified the target signals (Gnach *et al.*, 2015).

4.7.5 BIOIMAGING

The diagnostic branch also includes molecular imaging that uses both nuclear medicines and radiology to check the changes in molecules during the living body's disease process (Mittal *et al.*, 2013).

Other applications include:

1. Electrospun nanofibers are employed to induce fast hemostasis and cell proliferation in wound healing because of their high porosity (Li *et al.*, 2010). Also, in 3D cell culture and tissue repair, they promote cell proliferation and differentiation (Mittal *et al.*, 2013).
2. Calcium phosphate nanorods, particularly hydroxyapatite, have been employed in composite scaffolds as nanocarriers as well as for bone tissue regeneration (Allhoff, 2007; Li *et al.*, 2010).
3. In tissue engineering, carbon nanotubes can support and promote the proliferation of various tissues, particularly neural and cardiac cells. Recently, they have been recognized for their use in stem cell culture, where they have been shown to modulate the proliferation and differentiation of a variety of stem cells (Xu *et al.*, 2004), such as mouse neural stem cells to neurons and oligodendrocytes (Jan and Kotov, 2007).

4. Nanotubes are used in the manufacture of biosensors, as a contrast agent in computed tomography (Iqbal *et al.*, 2012), as nanocarriers (the right candidate for DNA or RNA attachment) for use in gene therapy (Freestone *et al.*, 2007) and as neurodegenerative disease therapy, because they can cross the blood-brain barrier.

5. Halloysite nanotubes (HNTs) are made up of aluminosilicate layers and have great mechanical strength, outstanding biocompatibility and homeostatic qualities, making them ideal for drug delivery, tissue engineering, wound healing and imaging (Iqbal *et al.*, 2012).

6. Cancer cell protein biomarkers have been identified using two-dimensional nanomaterials such as a thick graphene oxide nanosheet covered with gold nanoparticles (Gnach *et al.*, 2015). Biological characteristics of polymer-coated graphene oxide nanosheets against Gram-positive and Gram-negative bacteria have been established (Cao *et al.*, 2007).

7. The organic and inorganic components determine the physicochemical, thermal and mechanical properties of organic/inorganic hybrid nanocomposites, where porosity allows drug delivery with applications as scaffolds for tissue regeneration containing chemicals to stimulate cell differentiation (Iqbal *et al.*, 2012).

8. GNPs and titanium dioxide (TiO_2) nanoparticles, respectively, have been employed to boost cell proliferation rates for bone and heart tissue regeneration. GNPs were first discovered to stimulate osteogenic development of an osteoblast precursor cell line, MC3T3-E1, in bone TE (Gnach et al., 2015). Furthermore, these nanoparticles influenced the formation of osteoclasts (or bone resorbing cells) from hematopoietic cells while also protecting osteoblastic cells from mitochondrial dysfunction (Gnach et al., 2015).

9. Composite scaffolds with silver nanoparticles, including collagen–silver nanoparticle scaffolds for skin restoration, particularly for burn sufferers, and composite HA–silver nanoparticle scaffolds and HA/collagen–silver nanoparticle composite scaffolds for bone transplant materials have been developed (Iqbal *et al.*, 2012). Iron oxide nanoparticles have shown great potential in destroying bacteria after they have formed a biofilm (Gnach *et al.*, 2015).

4.8 CONCLUSION

The emergence of biotechnology has revolutionized scientific studies over the last decade. This advancement has developed to form a special branch which is nanotechnology. The application of nanotechnology in various fields has proved effective; however, its synthesis has become an environmental challenge. In order to improve its synthesis and function, biological organisms or substances were considered as sources of nanoparticles and that led to the study of bio-nanoenabled substances. Various bio-nanoenabled substances have been used as efficient and more ecofriendly alternatives to other methods of synthesis. These bionanomaterials exhibit great biocompatibility, which gives them biomedical domain potentialities in therapeutics, diagnosis and imaging. Hence, this chapter briefly describes the fundamentals of bionanomaterials along with their types and how they have been used for affordable, reliable and sustainable green energy production.

REFERENCES

Abah, J., Ishaq, M. N. and Wada, A. C. (2010). The role of biotechnology in ensuring food security and sustainable agriculture. *Afr. J. Biotechnol.* 9(52): 8896–8900.

Abdul, K. S., Fizree, H. M., Bhat, A. H., Jawaid, M. and Abdullah, C. K. (2013). Development and characterization of epoxy nanocomposites based on nano-structured oil palm ash Composite. *Eng. J.* 53: 324–333.

Agui, L., Yanez-Sedeno, P. and Pingarrón, J. M. (2008). Role of carbon nanotubes in electro-analytical chemistry: A review. *Anal. Chim. Acta* 622(1–2): 11–47.

Ajalloueian, F., Tavanai, H., Hilborn, J., Donzel-Gargand, O., Leifer, K. and Wickham, A. (2014). Emulsion electrospinning as an approach to fabricate PLGA/chitosan nanofibers for biomedical applications. *Biomed. Res. Int.* 2014:475280.

Ali, I., Alsehli, M., Scotti, L., Tullius, M., Tsai, S. and Yu, R. (2020). Progress in polymeric nano-medicines for theranostic cancer treatment. *Polymers* 12(3): 598–590.

Allen, M., Bulte, J., Liepold, L., Basu, G., Zywicke, H. and Frank, J. (2005). Paramagnetic viral nanoparticles as potential high-relaxivity magnetic resonance contrast agents. *Magn. Reson. Med.* 54(4): 807–812.

Allhoff, F. (2007). On the autonomy and justification of nanoethics. *Nanoethics.* 1: 185–210.

An, J., Chua, C., Yu, T., Li, H. and Tan, L. (2013). Advanced nanobiomaterial strategies for the development of organized tissue engineering constructs. *Nanomedicine (London)* 8(4): 591–602.

Anake, W., Ana, G. and Benson, N. (2018). Study of surface morphology, elemental composition and sources of airborne fine particulate matter in Agbara industrial estate, Nigeria. *Int. J. Appl. Env. Sci.* 11: 881–890.

Annu, A. and Ahmed, S. (2018). Green synthesis of metal, metal oxide nanoparticles, and their various applications. In *Handbook of Ecomaterials*, Cham: Springer. https://doi .org/10.1007/978-3-319-48281-1_115-1.

Arunachalam, K., Annamalai, S. and Hari, S. (2013). One-step green synthesis and characterization of leaf extract-mediated biocompatible silver and gold nanoparticles from *Memecylon umbellatum. Int. J. Nanomed.* 8: 137–147.

Bagalkotkar, G., Sagineedu, S. R., Saad, M. S. and Stanslas, J. (2010). Phytochemicals from *Phyllanthus niruri* Linn. And their pharmacological properties: A review. *J. Pharm. Pharmacol.* 58(12): 1559–1570.

Baker, S. N. and Baker, G. A. (2010). Luminescent carbon nanodots: Emergent nanolights. *Angew. Chem. Int. Ed. Engl.* 49: 6726–6744.

Barber, D. J. and Freestone, I. C. (1990). An investigation of the origin of the colour of the Lycurgus Cup by analytical transmission electron microscopy. *Archaeometry.* 32: 33–45.

Bayda, S., Hadla, M., Palazzolo, S., Kumar, V., Caligiuri, I., Ambrosi, E., Pontoglio, E., Agostini, M., Tuccinardi, T. and Benedetti, A. (2017). Bottom-up synthesis of carbon nanoparticles with higher doxorubicin efficacy. *J. Control. Release.* 248: 144–152.

Bayda, S., Muhammad, A., Tiziano, T., Marco, C. and Flavio, R. (2020).The history of nano-science and nanotechnology: From chemical–physical applications to nanomedicine. *Molecules.* 25: 112–122.

Bhattacherjee, A., Ghosh, T. and Datta, A. (2018). Green synthesis and characterization of antioxidant-tagged gold nanoparticle (X-GNP) and studies on its potent antimicrobial activity. *J. Exp. Nanosci.* 13(1): 50–61.

Biao, L., Tan, S., Meng, Q., Gao, J., Zhang, X., Liu, Z. and Fu, Y. (2018). Green synthesis, characterization and application of proanthocyanidins-functionalized gold nanoparticles. *Nanomaterials (Basel)* 8(1): 53–59.

Bigi, A., Nicoli-Aldini, N., Bracci, B., Zavan, B., Boanini, E. and Sbaiz, F. (2007). In vitro culture of mesenchymal cells onto nanocrystalline hydroxyapatite-coated Ti13Nb13Zr alloy. *J. Biomed. Mater. Res.* 82: 213–222.

Boselli, L., Lopez, W. and Zhang, H. (2020). Classification and biological identity of complex Nano shapes. *Commun. Mater.* 1(1): 35.

Bruchez, M., Moronne, M., Gin, P., Weiss, S. and Alivisatos, A. P. (1998). Semiconductor nanocrystals as fluorescent biological labels. *Science* 281(5385): 2013–2016. https:doi .org/10.1126/science.281.5385.2013.

Chan, V. (2006). Nanomedicine: An unresolved regulatory issue. *Regul. Toxicol. Pharmacol.* 46(3): 218–240.

Chang, M., Yang, C. and Huang, D. (2011). Aptamer-conjugated DNA icosahedral nanoparticles as a carrier of doxorubicin for cancer therapy. *ACS Nano* 5: 66–69.

Cao, L., Wang, X., Meziani, M. J., Lu, F., Wang, H., Luo, P. G., Lin, Y., Harruff, B. A., Veca, L. M. and Murray, D. (2007). Carbon dots for multiphoton bioimaging. *J. Am. Chem. Soc.* 129: 11318–11319.

Chen, J., Shi, M., Liu, P., Ko, A., Zhong, W., Liao, W. and Xing, M. M. (2014). Reducible polyamidoaminemagnetic iron oxide self-assembled nanoparticles for doxorubicin delivery. *Biomaterials* 35(4): 240–245.

Choudhary, S., Haberstroh, K. and Webster, T. (2007). Enhanced functions of vascular cells on nanostructured Ti for improved stent applications. *Tissue Eng.* 7: 21–30.

Chronopoulou, L., Nocca, G., Castagnola, M., Paludetti, G., Ortaggi, G. and Sciubba, F. (2016). Chitosan based nanoparticles functionalized with peptidomimetic derivatives for oral drug delivery. *Nat. Biotechnol.* 33(1): 23–31.

Davar, F., Fereshteh, Z. and Salavati-Niasari, M. (2009). Nanoparticles Ni and NiO: Synthesis, characterization and magnetic properties. *J. Alloys Compd.* 476(1–2): 797–801.

de Oliveira, R., Zhao, P., Li, N., de Santa Maria, L., Vergnaud, J. and Ruiz, J. (2013). Synthesis and in vitro studies of gold nanoparticles loaded with docetaxel. *Int. J. Pharm.* 454(2): 703–711.

Domènech, B., Kampferbeck, M., Larsson, E., Krekeler, T., Bor, B. and Giuntini, D. (2019). Hierarchical supercrystalline nanocomposites through the self-assembly of organicallymodified ceramic nanoparticles. *Sci. Rep.* 9(1): 3435.

Eap, S., Ferrand, A., Schiavi, J., Keller, L., Kokten, T. and Fioretti, F. (2014). Collagen implants equipped with 'fish scale'-like nanoreservoirs of growth factors for bone regeneration. *Nanomedicine (London)* 9(8): 1253–1261.

Eid, M. (2011). Gamma radiation synthesis and characterization of starch based polyelectrolyte hydrogels loaded silver nanoparticles. *J. Inorg. Organomet. Polym. Mater.* 21(2): 297–305.

Elmowafy, E., Abdal-Hay, A., Skouras, A., Tiboni, M., Casettari, L. and Guarino, V. (2019). Polyhydroxyalkanoate (PHA): Applications in drug delivery and tissue engineering. *Expert Rev. Med. Devices* 16(6): 467–482. https://doi.org/10.1080/17434440.2019.1615439.

Famá, L. M., Pettarin, V., Goyanes, S. N. and Bernal, C. R. (2011). Starch/multi-walled carbon nanotubes composites with improved mechanical properties. *Carbohydr. Polym.* 83(3): 1226–1231.

Fernandes, M. (2020). Recent applications of magnesium oxide (MgO) nanoparticles in various domains. *Adv. Mater. Lett.* 11(2): 54–63.

Fraysse-Ailhas, C., Graff-Meyer, A., Per Rigler, C., Mittelhozer, S., Raman, U. and Aebi, P. (2007). Peptide nanoparticles for drug delivery applications. *Eur. Cells Mater.* 14: 115–119.

Gangadoo, S., Taylor-Robinson, A. and Chapman, J. (2015). Nanoparticle and biomaterial characterization techniques. *Mater. Technol.* 30(sup5): 44–56.

Gao, X., Wei, L., Yan, H. and Xu, B. (2011). Green synthesis and characteristic of core–shell structure silver/starch nanoparticles. *Mater. Lett.* 65(2): 3–5.

Ghimire, B. K., Seo, J.-W., Kim, S.-H., Ghimire, B., Lee, J., Yu, C. and Chung, I. (2021). Influence of harvesting time on phenolic and mineral profiles and their association with the antioxidant and cytotoxic effects of *Atractylodes japonica* Koidz. *Agronomy* 11(7): 1327.

Gim, S., Zhu, Y., Seeberger, P. H. and Delbianco, M. (2019). Carbohydrate-based nanomaterials for biomedical applications. *Wiley Interdiscipl Rev. Nanomed. Nanobiotechnol.* 11(5): e1558.

Gnach, A., Lipinski, T., Bednarkiewicz, A., Rybka, J., Capobianco, J. A. (2015). Upconverting nanoparticles: Assessing the toxicity. *Chem. Soc. Rev.* 44: 1561–1584.

Helmlinger, J., Sengstock, C., Groß-Heitfeld, C., Mayer, C., Schildhauer, T. A., Köller, M. and Epple, M. (2016). Silver nanoparticles with different size and shape: Equal cytotoxicity, but different antibacterial effects. *RSC Adv.* 6(22): 18490–18501.

Hoshino, A. K., Fujioka, T. and Oku, M. (2004). Physicochemical properties and cellular toxicity of nanocrystal quantum dots depend on their surface modification. *Nano Lett.* 4(11): 2163–2169.

Ibrahim, K., Khalid, S. and Idrees, K. (2019). Nanoparticles; properties, applications and toxicities. *Arab. J. Chem.* 12(7): 908–931.

Iqbal, P., Preece, J. A. and Mendes, P. M. (2012). Nanotechnology: The "top-down" and "bottom-up" approaches. In *Supramolecular Chemistry*. Chichester: John Wiley & Sons, Ltd.

Jain, N., Bhargava, A., Majumdar, S., Tarafdar, J. C. and Panwar, J. (2011). Extracellular biosynthesis and characterization of silver nanoparticles using *Aspergillus flavus* NJP08: A mechanism perspective. *Nanoscale* 3(2): 635–664.

Honek, J. F. (2013). Bionanotechnology and bionanomaterials: John Honek explains the good things that can come in very small packages. *BMC Biochem.* 14: 29.

Kapat, K., Shubhra, Q. T., Zhou, M. and Leeuwenburgh, S. (2020). Piezoelectric nano-biomaterials for biomedicine and tissue regeneration. *Adv. Funct. Mater.* 30(44): 1909045. https://doi.org/10.1002/adfm.201909045.

Karagkiozaki, V., Karagiannidis, P. G., Kalfagiannis, N., Kavatzikidou, P., Patsalas, P., Georgiou, D. and Logothetidis, S. (2012). Novelnanostructured biomaterials: Implications for coronary stent thrombosis. *Int. J. Nanomed.* 7: 6063. https://doi.org/10.2147/ijn.s34320.

Kaur, M. and Chopra, S. (2018). Green synthesis of iron nanoparticles for biomedical applications. *Glob. J. Nanomed.* 4(4): 1–10.

Li, Q., Ohulchanskyy, T. Y., Liu, R., Koynov, K., Wu, D., Best, A., Kumar, R., Bonoiu, A. and Prasad, P. N. (2010). Photoluminescent carbon dots as biocompatible nanoprobes for targeting cancer cells in vitro. *J. Phys. Chem.* 114: 12062–12068.

Lombardo, D., Kiselev, M. A. and Caccamo, M. T. (2019). Smart nanoparticles for drug delivery application: Development of versatile nanocarrier platforms in biotechnology and nanomedicine. *J. Nanomater.* 2019: 3702518.

Manzoor, A. G., Sufia, N., Mir, Y. A., Ayaz, M. D., Khusro, Q. and Swaleha, Z. (2014). *Physicochemical Properties of Nanomaterials: Implication in Associated Toxic Manifestations. BioMedical Research International.* Hindawi Publishing Corporation, pp. 1–8.

Nune, K. C. and Misra, R. D. (2016). Biological activity of nanostructured metallic materials for biomedical applications. *Mater. Technol.* 31(13): 772–781. https://doi.org/10.1080/10667857.2016.1225148.

Ovais, M., Khalil, A. T., Raza, A., et al. (2016). Green synthesis of silver nanoparticles via plant extracts: Beginning a new era in cancer theranostics. *Nanomedicine (London)* 11(23): 3157–3177.

Palestino, G., García-Silva, I., González-Ortega, O. and Rosales-Mendoza, S. (2020). Can nanotechnology help in the fight against COVID-19? *Expert Rev. Anti-Infect. Ther.* 18(9): 849–864.

Parak, W. J., Boudreau, R., Gros, M. L., Gerion, D., Zanchet, D., Micheel, C. M., Williams, S. C., Alivisatos, A. P. and Larabell, C.A. (2002). Cell motility and metastatic potential studies based on quantum dot imaging of phagokinetic tracks. *Adv. Mater.* 14(12): 882–885.

Raza, M. A., Kanwal, Z., Rauf, A., Sabri, A. N., Riaz, S. and Naseem, S. (2016). Size- and shape-dependent antibacterial studies of silver nanoparticles synthesized by wet chemical routes. *Nanomaterials (Basel)* 6(4): 74.

Restrepo, C. V. and Villa, C. C. (2021). Synthesis of silver nanoparticles, influence of capping agents, and dependence on size and shape: A review. *Environ. Nanotechnol. Monit. Manag.* 15: 100428.

Ridolfo, R., Tavakoli, S., Junnuthula, V., Williams, D. S., Urtti, A. and Van Hest, J. C. M. (2021). Exploring the impact of morphology on the properties of biodegradable nanoparticles and their diffusion in complex biological medium. *Biomacromolecules* 22(1): 126–133.

Rodríguez-Félix, F., López-Cota, A. G., Moreno-Vásquez, M. J., Graciano-Verdugo, A. Z., Quintero-Reyes, I. E., Del-Toro-Sánchez, C. L. and Tapia-Hernández, J. A. (2021). Sustainable-green synthesis of silver nanoparticles using safflower (*Carthamus tinctorius* L.) waste extract and its antibacterial activity. *Heliyon* 7(4): e06923.

Sinani, V. A., Koktysh, D. S., Yun, B. G., Matts, R.L., Pappas, T.C., Motamedi, M., Thomas, S.N. and Kotov, N.A. (2003). Collagen coating promotes biocompatibility of semiconductor nanoparticles in stratified LBL films. *Nano Lett.* 3(9): 1177–1182.

Singh, K. R. B., Nayak, V. and Singh, R. P. (2021). Introduction to bionanomaterials: An overview. In *Bionanomaterials; Fundamentals and biomedical applications.* Singh, Ravindra Pratap, Singh, Kshitij RB. IOP Publishing, pp. 1–21.

Song, S., Shen, H., Wang, Y., Chu, X., Xie, J., Zhou, N., et al. (2020). Biomedical application of graphene: from drug delivery, tumor therapy, to theranostics. *Coll. Surf. B* 185: 110596.

Tang, Z., He, C., Tian, H., Ding, J., Hsiao, B. S., Chu, B., et al. (2016). Polymeric nanostructured materials for biomedical applications. *Prog. Polym. Sci.* 60: 86–128.

Torres-Sangiao, E., Holban, A. M. and Gestal, M. C. (2016). Advanced nanobiomaterials: Vaccines, diagnosis and treatment of infectious diseases. *Molecules* 21(7): 867. https://doi.org/10.3390/molecules21070867.

Wang, L., Periyasami, G., Aldalbahi, A. and Fogliano, V. (2021). The antimicrobial activity of silver nanoparticles biocomposite films depends on the silver ions release behaviour. *Food Chem.* 359: 129859.

Wong, I. Y., Bhatia, S. N. and Toner, M. (2013). Nanotechnology: Emerging tools for biology and medicine. *Genes Dev.* 27(22): 2397–2408. https://doi.org/10.1101/gad.226837.113.

Xu, X., Ray, R., Gu, Y., Ploehn, H. J., Gearheart, L., Raker, K. and Scrivens, W. (2004). A electrophoretic analysis and purification of fluorescent single-walled carbon nanotube fragments. *J. Am. Chem. Soc.* 126:12736–12737.

Zhang, Y., Kohler, N. and Zhang, M. (2002). Surface modification of superparamagnetic magnetite nanoparticles and their intracellular uptake. *Biomaterials* 23(7): 1553–1561.

5 Bionanocatalysts as Green Catalysts
Design and Application for Production of Renewable Biofuels

Disha Tandulkar[a], Sonam Paliya,[a,b] Ashootosh Mandpe[c], Manukonda Suresh Kumar[b] and Sunil Kumar[a,b]

[a]CSIR-National Environmental Engineering Research Institute (CSIR-NEERI), Nehru Marg, Nagpur 440 020, India

[b]Department of Biological Sciences and Bioengineering, Indian Institute of Technology Indore, Indore 453 552, India

[c]Department of Civil Engineering, Indian Institute of Technology Indore, Indore 453 552, India

CONTENTS

DOI: 10.1201/9781003316374-5

5.1   INTRODUCTION

The ever-increasing energy demand increases fossil fuel usage, which will lead to their shortage in the near future and also causes detrimental effects to the environment due to the release of greenhouse gases (GHGs). These emissions are allied with global warming and climate change (Selvaraj et al., 2019). Therefore, the need to find a substitute for renewable energy sources (such as biofuels) has been a research interest of recent studies. To satisfy global energy demand, biofuel production from renewable and sustainable technology is important (Singhvi et al., 2020). To that end, nanotechnology is a highly evolved research area where nanocatalysts are exploited and optimized to produce biofuels.

5.1.1   NANOCATALYST

Catalysis is a sustainable, economically sound and efficient approach to the development of renewable energy fuels. Catalysts can reduce the net emissions of GHGs and their adverse effects on the environment. Green chemistry is one of the vital research areas dedicated to tackling such challenges (Hunt & Farmer, 2015). Over the last few decades, researchers have been focusing on nanocatalyst applications in green chemistry.

Nanocatalysts' (and nanomaterials') unique properties like large surface area, catalytic activity, crystal structure, durability, energy storage capacity, etc., make them ideal for biofuel production (Hunt & Farmer, 2015). Nanocatalyst morphology, shape and size play a vital role in their performance. They act as the interface between homogeneous as well as heterogeneous catalysts in terms of catalytic activity, selectivity, efficacy, recovery and re-usability (Roy et al., 2021).

The reason for such unique behavior is their surface-to-volume ratio. For example, the efficiency of silver nanomaterials' antibacterial properties is increased at a higher aspect ratio. Thus, the surface-to-volume aspect ratio of nanomaterials is responsible for their wide applicability in different fields, such as silver nanomaterials having diverse applications in medicine, agriculture, textiles, imaging, cosmetics, electronics, biosensing, biotechnology and so on (Raina et al., 2020).

Though the fabrication of nanocatalysts using chemical and physical methods is proven to be useful, it comes with certain drawbacks such as particle aggregation when reacted for a long time, instability of product formed and inappropriate crystal growth. Moreover, these processes are not cost-effective, consume a lot of energy and are also not eco-friendly as they generate toxic by-products in huge amounts. Therefore, the development of new methods which are economically feasible, eco-friendly, clean and energy-efficient is necessary (Roy et al., 2021).

5.1.2 BIONANOCATALYSTS (BNCS)

Earlier, biocatalysts were used as a green and sustainable approach for the synthesis of novel compounds and biofuel production in the industry. They were best suited and profitable due to the activity, specificity and selectivity of enzymes (Gkantzou et al., 2021). However, disadvantages related to enzyme-catalyzed bioprocesses are high operational costs, less durability and non-reusability. The immobilization of enzymes on solid support materials attracted research interest as they can be recovered and reused, and also can be protected from degradation by chemical and environmental attacks (Iravani & Varma, 2020). Therefore, the development of enzyme carriers with biocompatibility and a robust nature is a focus of research.

In recent times, advancements in nanotechnology have provided various nanoscale carriers which can act as enzyme carriers for their immobilization. BNCs were developed in which enzyme immobilization on nanomaterials (Figure 5.1) became a promising method to improve enzyme performance. BNCs involve the fusion of nanotechnology and biotechnology. They are greener, sustainable, biodegradable, eco-friendly, require mild reaction conditions and the energy consumption is also very low (Misson et al., 2015). Thus, BNCs are cost-effective and energy-efficient.

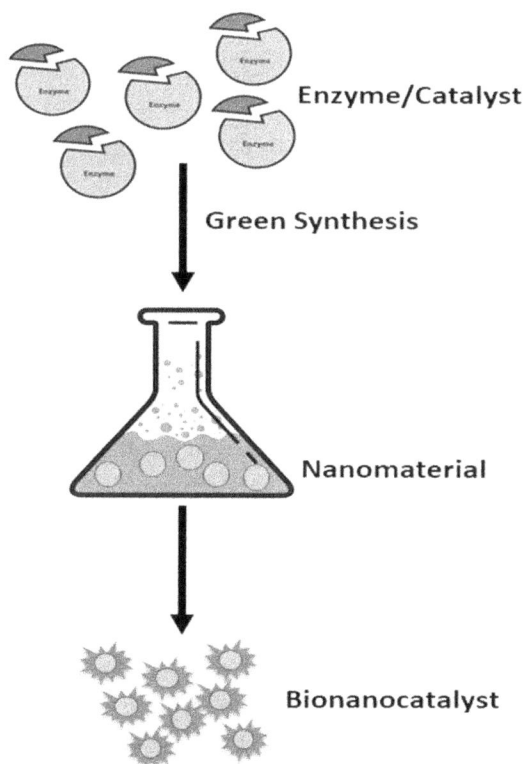

FIGURE 5.1 Mechanism of synthesis of a bionanocatalyst.

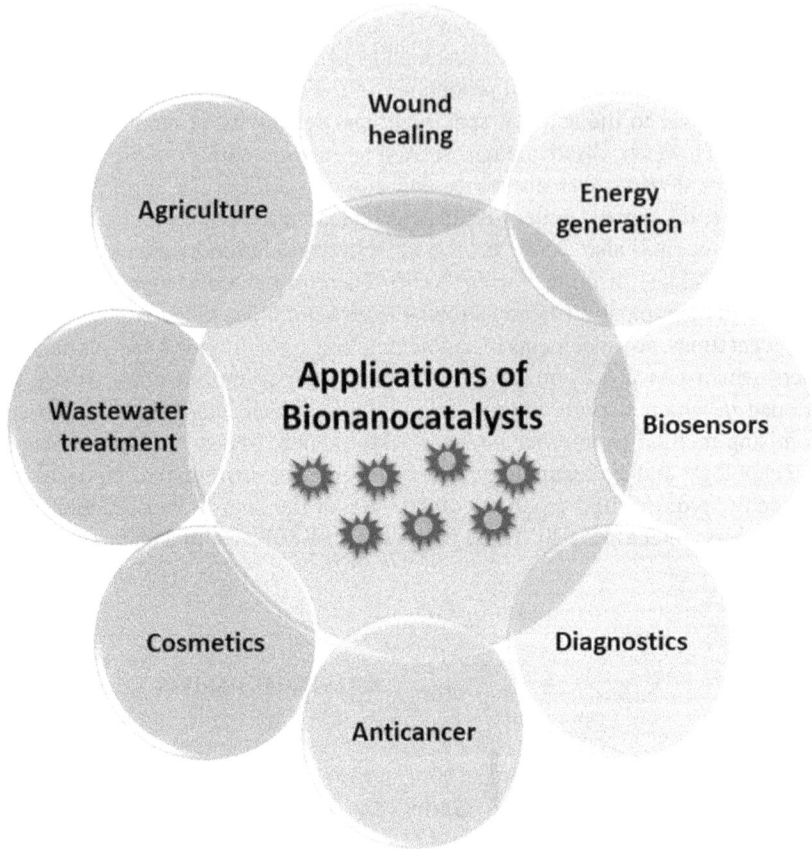

FIGURE 5.2 Application of bionanocatalysts in different sectors.

BNCs were engineered for their applications in the fields of energy, synthesis, cosmetics, therapeutics, agriculture, medicine, etc. (Figure 5.2). The fabrication of BNCs involves assembling enzyme molecules on the nanomaterial carriers in such a way that interaction with the substrate increases chemical kinetics and selectivity. The literature review shows that nanomaterials used in BNCs provide a large surface area that increases enzyme loading and reduces mass transfer resistance for substrates (Misson et al., 2015). BNCs are developed to improve enzyme stability, activity, efficiency and performance. The cost of the biocatalytic process is also decreased by the use of BNCs as they increase the life cycle of the biocatalyst. Betancor and Luckarift explained that enzyme immobilization on the nanomaterials is a versatile technology, as it is cost-effective, increases reaction rate and has mild reaction conditions, robust nature, high enzyme loading and minimum diffusional limitations and stability (Betancor & Luckarift, 2008).

The unique properties of nanomaterials can help to develop a new modified class of biocatalyst. The driving interest in BNCs forced the fabrication of nanomaterial carriers with inimitable properties and structures. The development of nanomaterials as enzyme carriers, such as nanofibers, nanotubes, nanocomposites, etc., has been used to enhance the performance and retain the functionality of the enzymes. Over the last decade, to modify and enhance the performance of BNCs, researchers have been developing active nanomaterials as enzyme carriers, working on enzyme immobilization techniques and fabricating BNCs, and their various applications in bioprocesses have been extensively studied (Verma et al., 2013).

5.2 ENZYME IMMOBILIZATION ON NANOMATERIALS FOR DESIGNING BNCs

Enzymes are natural biocatalysts possessing excellent catalytic properties, but their activity is hampered when used for industrial applications. In the last decade, the immobilization of enzymes on nanomaterial supports has gained a lot of scientific attention after the introduction of BNCs. Enzyme immobilization has various benefits in industrial applications, such as increased enzyme thermal stability, catalytic activity and enzyme loading due to large surface area, tolerance to different pH conditions and anhydrous organic solvents, reusability of the enzyme and resistance against irradiation, and products formed can be easily separated from the reaction mixture. The need for immobilization is to provide biocompatibility and an inert environment for the enzyme to avoid interference with the protein structure and its biological activity (Mitchell et al., 2002). Immobilization is needed to prevent loss of enzyme activity by altering the chemical nature or binding site of the enzyme. The protection of the binding site is important; this is done by blocking it with the help of a substrate or competitive inhibitor of the enzyme or by attachment of the protecting groups. The common methods employed for enzyme immobilization on nanomaterials are adsorption, covalent binding, entrapment and cross-linking (Ahmad et al., 2015). This section focuses on some of the nanomaterials used for enzyme immobilization.

5.2.1 CARBON-BASED NANOMATERIALS

Carbon-based nanomaterials (CBNs) have attracted the scientific community due to their unique properties, such as excellent electrical conductivity and chemical, optical and mechanical properties. Due to the presence of pure carbon in CBNs, they are less toxic, highly stable and eco-friendly. Their nanostructure gives them high surface area, linear geometry and strong anisotropic thermal conductivity. All these properties make CBNs a novel material with wide applications in biomedical fields like drug and gene delivery, bioimaging, biofuel production for energy storage, biosensors and cancer therapy (Misson et al., 2015).

CBNs are unique carriers for enzyme immobilization due to their properties like inertness, biocompatibility and thermal stability. For the fabrication of BNCs, newly developed carbon nanomaterials that act as carriers for enzyme immobilization are carbon nanotubes (CNTs), graphene (G) and its derivatives (graphene oxide (GO) and reduced graphene oxide (rGO)), nano-diamond, carbon nanoparticles and carbon nanofibers (Gkantzou et al., 2021; Misson et al., 2015).

CNTs are one-dimensional hollow tubes made up of carbon atoms. They have exceptional electrical and mechanical properties with rapid electron transfer rates. They are classified into single-walled carbon nanotubes (SWCNTs) and multi-walled carbon nanotubes (MWCNTs). Earlier SWNTs were used, and later MWNTs were used for designing BNCs. Adsorption on SWNTs and unmodified MWNTs is by simple physical adsorption. The completion of enzyme adsorption is dependent on the ionic strength of the medium (Huang et al., 2015). Enzymes are immobilized on CNTs also by forming a covalent or non-covalent bond with the hydrophobic nanotube surface. This interaction on the surface of the nanotube forms free structures of enzyme molecules and their functional group. Different charges on the surface of the enzyme and nanotube enhance binding. The free amine groups of enzymes react with the carboxylic groups on the nanotube surface which are responsible for the oxidation of CNTs and activation of the carbodiimide. This method retains enzyme bioactivity. It is also used to immobilize various enzymes such as α-amylase, urease, glucose oxidase, lipase, hydrogenase, peroxidase and other enzymes successfully on the CNTs. The enzyme loading is increased up to four times the weight of the nanotubes and also increases enzyme catalytic activity, reusability and stability. It is a co-immobilization technique where a strong covalent bond is formed. This bonding with MWCNTs surface change lipase conformation to an "open lid" structure which is surrounded to the active catalytic centers and regulates substrate towards enzyme active site. Thus, helps to increase lipase activity (Misson et al., 2015).

Graphene (G) has a two-dimensional honeycomb network of carbon atoms. Its features, such as a large surface area, high mechanical strength and exceptional electrical conductivity, gained attention in diverse fields. These remarkable properties make it a perfect carrier for enzyme immobilization. Enzyme glucose oxidase was immobilized on G-polyethyleneimine-functionalized ionic liquid (G-PFIL) by covalent attachment which enhanced stability and solubility. It also increases the enzyme's biocompatibility and the counter-anion favors electrostatic adsorption. Its derivative graphene oxide (GO) possesses functional groups such as carboxylic, epoxy and hydroxyl groups which are important for enzyme immobilization, whereas the reduced graphene oxide (rGO) does not possess surface functional groups. For enzyme adsorption on GO, electrostatic, covalent or crosslinked methods are used whereas the hydrophobic method is used for rGO. It is observed that catalytic behavior was affected by the electrostatic interaction between enzyme and GO. But the catalytic activity of the enzyme is only slightly affected in rGO due to the lack of functional groups and hydrophobic interaction. Thus, enzymes immobilized on rGO were more stable, had higher loading and had better activity (Seelajaroen et al., 2020).

Other CBNs show similar chemical and physical properties, and similar methods can be applied for immobilizing enzymes onto these materials.

5.2.2 Metal-based Nanomaterials

Metal-based nanomaterials (MBNs) are used for the immobilization of enzymes for the fabrication of BNCs. They are emerging green catalysts. Metal nanomaterials like gold nanoparticles (AuNPs) are extensively used for enzyme immobilization. The immobilization of enzymes on AuNPs adsorption by chemical, physical, cross-linking or covalent methods are used. Chemisorption increases the operational and storage stability of the prepared biosensor whereas physical adsorption improves storage, thermal stability and tolerance in extreme pH conditions. Enzymes like glucose oxidase and α-amylase were adsorbed on the AuNPs (Alshammari et al., 2016).

Metal oxide nanomaterials like Fe_3O_4, TiO_2 and ZnO are used for enzyme immobilization. Similar techniques are used to immobilize enzymes. Fe_3O_4 nanoparticles show robust silane linkages as carriers for enzymes like keratinase, lipase, laccase and cholesterol oxidase. They are magnetic nanoparticles with good dispersion ability and high recyclability for biocatalysts. They showed binding efficiency of 98% and enhanced biocatalytic performance. Therefore, they have wide biological and medical applications. TiO_2 nanoparticles improve pH and thermal stability with 80% retention in their catalytic activity. Enzymes like sucrose isomerase, adenosine deaminase and dopamine are immobilized on TiO_2 nanoparticles. Due to their high biocompatibility, low cost, huge surface area and chemical stability, ZnO nanoparticles are extensively explored for enzyme immobilization (Mauro et al., 2017). Enzymes are adsorbed on ZnO by electrostatic adsorption. A glucose biosensor was constructed using a single ZnO nanofiber, and the glucose oxidase enzyme was immobilized on the nanofiber (An et al., 2014). The biosensor showed enhanced storage stability. Enzyme α-amylase was immobilized electrostatically on ZnO nanoparticles (Pascariu & Homocianu, 2019). Metal-based nanomaterials can be merged into carbon nanotubes, polymer nanogels and mesoporous nanomaterials to synthesize BNCs.

5.2.3 SILICA-BASED NANOMATERIALS

The inorganic nanoparticles most frequently utilized as carriers for enzyme immobilization are silica-based nanomaterials. They are physically adsorbed into mesoporous silica NPs. Their mechanical robustness, excellent chemical and heat resistance make them a suitable carrier (Chen et al., 2019). The porous nature of silica improves enzyme adsorption by enhancing binding strength and lowering diffusion barriers. It also enhances enzyme stability and tolerance in organic solvents and protects from proteolysis (Yang et al., 2019). The silica nanoparticles have hydroxyl groups on their surface which help them to operate better and result in targeted covalent attachment of enzymes to the nano-support. Silica nanomaterials are mesoporous materials. Enzymes entrapped are α-chymotrypsin, trypsin, catalase, protease, peroxidase, etc. (Yang et al., 2020).

5.2.4 OTHER NANOMATERIALS

In addition to the nanomaterials discussed above, other nanomaterials like chitosan, cellulose and clay, as well as metal-organic frameworks (MOFs) have also been applied as nanocarriers for enzyme immobilization. Chitosan nanocarriers are coupled with CNTs, G and some magnetic nanoparticles as functional support for enzyme immobilization. On the other hand, cellulose-based nanomaterials provide external support for other nanomaterials. A novel nanocomposite was prepared by using AuNPs supported on cellulose nanocrystals to immobilize the enzyme cyclodextrin glycosyltransferase. In last decade, green nanotechnology has paved the way for bio-nanomaterials in the field of bionanotechnology by introducing some new nanocarriers such as nanoflowers, biogenic nanoparticles and biopolymers (Liu et al., 2019; Verma et al., 2020). These emerging bio-nanocarriers are not only eco-friendly and economically sound but also possess unique physicochemical properties. Therefore, exploring more about their characteristics can be useful to develop sustainable and efficient BNCs.

5.3   RECENT APPLICATIONS OF BNCs

In recent years, advancement in branches of biotechnology, molecular biology, nanotechnology and engineering has resulted in the development of nanocatalysts. Nanocatalyst synthesis and its optimization studies are being advanced at multiple scales such as the structural optimization of protein molecular structures, nanoscale environments of the enzymes, the properties of nanoparticles and the working of macroscale reactors (Liu et al., 2019). When comparing enzymes with and without the use of nanocatalysts as support, it can be seen that the use of nanocatalysts has significantly increased enzyme loading capacity and also enhanced mass transfer efficiency.

The use of immobilized enzymes is important for large-scale industrial operations, such as glucose isomerase for the production of fructose corn syrup, lipase for transesterification of food oils and penicillin G acylase for antibiotic modification showing good production and profit for industries. The reported studies on the development and implementation of nanocarrier-based BNCs for biological processes are still being conducted at the laboratory scale. So far, there has not been a successful use of BNCs in industrial bioprocesses that has been documented in the literature.

5.3.1   Food Industry

Enzyme application, especially in the food industry, is well reported. For many decades, food companies have relied on enzymes such as lipase, trypsin, amylase, glucose isomerase, papain and pectinase. As food safety is a major concern, only such microorganisms and products derived from them as are recognized as safe are used in the food industry.

Nanoparticles are employed for various applications like enzyme stabilization and pigment extraction from waste material. The nano-immobilization of enzymes was achieved for higher pigment extraction than free enzymes. Moreover, it is seen that Robust BNCs retained the 85% original activity even after the three recycles of real substrate hydrolysis (Verma et al., 2020). The affinity was increased and also a lower Km value was seen with the use of nanocarrier-bound enzyme compared to the natural enzyme. A nanocarrier such as graphene nanosheet was used to immobilize α-galactosidase, and gold nanoparticles were covalently immobilized on β-Galactosidase sourced from chickpea (Nguyen et al., 2019). BNCs were used to hydrolyze raffinose, which is the reason for flatulence in soybean-derived foods (Misson et al., 2015). Thus, BNCs show excellent results due to their enhanced thermal stability, operational stability, catalytic activity and resistance to extreme temperature, which will lead to their use in the food industry in the near future.

5.3.2 BIOMEDICAL AND THERAPEUTIC SYSTEMS

In the pharmacy sector, BNCs have gained greater significance than alternative enzyme immobilization technologies because of the easy recovery of products, biological compatibility and eco-friendly nature, among other factors. Recently, the enzyme penicillin G acylase (PGA) was successfully immobilized covalently using magnetic porous nano-flowers. Because of this, it was necessary to ensure inexpensive penicillin G cleavage on a wide scale (Lv et al., 2019). By utilizing such supports, diffusional barriers can be reduced, enabling the optimum accessibility of the user interface to the active catalytic sites and hence increased antibiotic production. When compared to commercial methods, PGA encapsulated in magnetic silica nanoparticles aided in the synthesis of cephalexin, demonstrating increased operational efficiency and stability. The use of click chemistry to create a magnetically switchable bio-electrocatalyst via ferrocene imbedded Fe_2O_3 on magnetic mesocellular carbon foam was completed by combining protein engineering with an enzyme immobilization technique; it is helpful to increase the efficiency of BNCs and their industrial use (Madhavan et al., 2021). Many engineered enzymes, such as purine 2′-deoxyribosyl transferase, have been modified to improve catalytic action in contrast to synthetic nucleosides by employing a pyramid strategy. The his-tagged enzyme is adsorbed on commercially synthesized Ni^{2+} chelate which is a magnetic metal oxide with porous nanospheres, which was previously reported (Arco et al., 2019).

5.3.3 BIOREMEDIATION

Bioremediation based on enzymes is highly successful and efficient compared to other approaches. BNCs can be employed in a variety of sectors to safeguard the environment, including pollutant degradation, wastewater treatment and decolorizing dyes in contaminated waters. These BNCs are effective tools for bioremediation, and they provide a variety of advantages, including high efficiency, a high degree of stability, less toxicity, the ability to work at low temperatures and a reduction in sludge production.

The direct interaction of nanomaterial supports with monolithic microreactors enables the adsorption of pollutants, which, paired with biocatalytic degradation, results in enhanced bioremediation effectiveness (Zdarta et al., 2021). In recent years, research has shown that BNCs, specifically laccase-mCLEAs, are excellent and affordable oxido-reductases for the degradation of pollutants in aqueous solutions. They have been employed to eliminate non-phenolic and phenolic chemicals from aqueous media (Sadeghzadeh et al., 2020).

BNCs can also be used in wastewater treatment bioreactors involving enzymes such as laccases, tyrosinases, lignin and phenol peroxidases, which seem to play a crucial part in preventing infection from a range of sources, including a wide array of dyes and phenolic chemicals.

Several emerging strategies for solid waste remediation employing BNCs include immobilized cutinase, PETase and lipase, as well as the remediation of solid municipal wastes using lipase, protease and other hydrolases. For example, using surface solid support carriers like FeO nanoparticles and PETase results in a BNC which increases enzyme loading and its affinity for degrading substrates of PET. These BNCs, by preserving their natural enzyme activity, can be recycled magnetically after about ten cycles (Schwaminger et al., 2021). Cost-effective renewable feedstocks can be implemented to create magnetic biocomposites derived from carbon mesopores. Laccase adsorption was accomplished by employing magnetic carbon-based nano carriers derived from luffa sponges, to eradicate hazardous contaminants like bisphenol A through enzymes (Yavaşer and Karagözler, 2021).

Congo red, an organic textile dye which is toxic to many organisms due its carcinogenic and mutagenic properties, has also been removed by employing BNCs made of laccase and an amino group graphene oxide nanosheet for CotA laccase adsorption. Lin et al. (2015) covalently attached laccase onto chitosan/CeO$_2$ microspheres for the purpose of decolorizing orange II and methyl red dyes (Lin et al., 2015). BNCs are adaptable to various bioremediation approaches, opening the door to a plethora of possible advancements in the field of enzyme technology.

5.3.4 BIOTRANSFORMATION

Enzymatic biotransformation has been investigated for the generation of useful goods such as medication intermediates and functional food additives. Enzyme immobilization on traditional supports does not function well in bioprocessing. Nanocarriers were introduced to enhance their potential in relation to the loading capacity and enzyme activity. Generally, ß-galactosidase is employed on lactose-containing dairy waste to convert it to galacto-oligosaccharide (GOS), lactulose and lactosucrose. Liu et al. (2019) employed covalent immobilization of the enzyme onto magnetic poly (GMA-EDGMA-HEMA) nanospheres to increase the rate of GOS synthesis. By following a condensation technique, a reaction between the epoxy groups on the nanosphere surfaces and nucleophilic enzyme was carried out. Up to 15 cycles of reactions, the latter preserved around 84.6% of its starting activity, but the former retained just 81.5% after 10 operations (Klaochanpong et al., 2015). Glutaraldehyde was used to link ß-galactosidase to the NT microchannel on the surface during continuous lactulose production. The technology successfully mirrored current industrial application at 78.3% of lactose conversion rate. BNCs enable one-step catalytic cascade reactions in non-aqueous media, primarily in the manufacture of chiral pharmaceuticals and their precursors (Galanakis, 2012). Malate dehydrogenase (MDHase) and citrate synthase (CSase) have been co-immobilized on 30 nm Au NPs for malate conversion. The bioconjugates were generated in one of two ways: either by directly adsorbing MDHase followed by CSase, or vice versa, or by co-adsorbing the two enzymes in the same solution. By the bioconjugation of CSase to the NPs prior to MDHase, higher specific activity and favorable kinetic parameters were achieved. The approach facilitated efficient sequential reactions through one-step biotransformation, with the MDHase products serving as substrates for the subsequent CSasereaction (Keighron and Keating, 2010). Due to the significant reduction in the use and release of organic solvents, this co-enzyme immobilization results in a low-cost and environmentally friendly approach.

5.3.5 BIOFUEL PRODUCTION

Globally, energy demand is increasing; therefore, the consumption of fossil fuels has increased. Due to limited natural resources (such as fossil fuels) and rising energy demand led to the production of Biofuel using enzymatic technology, which is proven to be more environmentally friendly. Compared to conventional alkaline catalyzed techniques, processing biodiesel especially, bioprocessing based on lipase enzyme has high efficiency as it requires very little energy and is more eco-friendly. Another potential use is cellulose hydrolysis which produces fermentable sugars to generate bioethanol utilizing BNCs. To immobilize lipase as a nanocarrier host, different nanocarriers such as GO, nano-silica, MOFs, carbon nanotubes and nano-flowers have been developed. These nanocarrier substrates have a high level of enzyme activity and loading capacity (Lv et al., 2019). Immobilization of lipase in *Pseudomonas cepacian* via physical adsorption using polyacrylonitrile biohybrid NFs boosted the production of biodiesel and immobilized lipase catalytically active increased by 23 times. Ion-exchange supports were produced for lipase immobilization from *Thermomyces lanuginosus* through adsorption, by progressively functionalized Si-based NPs activated with glycine and (3-glycidyloxypropyl) trimethoxysilane (Bolina et al., 2018). Excellent results obtained from these laboratory findings paved the way for the commercialization of this technology. It has been demonstrated that increasing the immobilization of the ß-glucosidase enzyme reduces cellulosic bioethanol production by inhibiting the conversion of cellobiose to glucose via enzyme aggregation. In comparison to covalent binding the aggregation approach enhanced bioethanol production up to four-fold (Deng et al., 2020). While BNCs are still in their infancy, their role in biofuel production has become a critical aspect of future engineering bioprocess research and innovation.

5.4 BNCS USED FOR THE PRODUCTION OF RENEWABLE BIOFUELS

5.4.1 BIOETHANOL

Major countries like the USA and Brazil use food crops (like sugarcane, maize, sugar beet, sweet sorghum), lignocellulosic biomass (agricultural waste and grasses) and algae (macroalgae and microalgae) to produce bioethanol. The steps involved in bioethanol production are pre-treatment, enzymatic hydrolysis, fermentation and separation. In the pre-treatment step, the use of nanotechnology on biomass is an inexpensive and eco-friendly method. The goal of using nanotechnology in pre-treatment is to change the biomass structure so that cellulolytic enzymes can access it, as well as to improve process efficiency. With the use of nanoscale catalyzers, nanotechnology aids in the improvement of pre-treatment methods. Wang et al. (2013) produced biofuel on corn stover using the NSHA pre-treatment method. In the study, a cellulose conversion rate of over 70% was achieved using a 12,500 s^{-1} shear rate on biomass and 1:1 NaOH at room temperature for 2 mins, showing excellent results with high shearing mechanism (Devi et al., 2021).

Microalgae bioproducts are divided into two categories: high value-low volume and low value-high volume products. Examples of high value-low volume products are biopolymers and renewable chemicals which include propanediol and lactic acid, whereas examples of low value-high volume products are biofuels like biodiesel, biohydrogen, bioethanol and biogas. Microalgae are a great source of fermentable carbohydrates, which could be used to make bioethanol. Nearly 3.83 g/L of bioethanol was obtained from 10 g/L of lipid-extracted *Chlorococcum* sp. debris (Harun et al., 2010), and a maximum of 0.3 g of ethanol was obtained per gram of total carbohydrate from wet algae *Nannochloropsis* sp. in another study (Karpagam et al., 2021).

5.4.2 BIODIESEL

Biodiesel is produced by transesterification from vegetable oils, animal fats or yellow grease. It is made up primarily of fatty acid ethyl esters (FAEE) and fatty acid methyl esters (FAME). It is regarded as among the most ecologically friendly and clean energy sources. The esterification and transesterification of fats and oils in the presence of acid-base catalysts produce biodiesel. Examples of homogenous catalysts for esterification and transesterification are H_2SO_4 and sodium methylate respectively, whereas heterogeneous catalysts such as zeolites, metal-based oxides and supported alkali metal/metal ions have been shown to be highly efficient due to low-cost production, and they are environmentally friendly in the production of biodiesel (Basumatary, 2013). However, the use of catalysts can have some drawbacks: they have low strength, are expensive and difficult to synthesize, have smaller surface area, have poor catalytic activity, are sensitive to pollution, also require high temperature and pressure and depend on alcohol to oil ratios. These catalysts work well for the transesterification reaction to produce biodiesel. The raw materials like fats or oils containing a particular percentage of free fatty acids (FFA) are regarded as important in these catalytic processes; otherwise, base catalyst poisoning occurs, which impairs the biodiesel quality (Chang et al., 2014). When molar ratio of (1:50) soybean oil to methanol was transesterified, a 6% CaO–MoO3–SBA-15 catalyst 83.2% biodiesel was produced. The acid-base interaction between MoO_3 and CaO may create a broad dispersion of catalytic active sites, hence increasing the catalyst's stability (Thangaraj et al., 2019). Karpagam et al. (2020) employed Biowaste (waste seashells) derived BNCs which are regarded as heterogeneous catalysts for whole-cell transesterification of wet microalgal biomass of Coelastrella sp. M-60 had yielded (Fatty Acids Methyl Esters) FAME. This shows that BNCs has a great potential and economical way for biodiesel production and also serve as an alernative method to be applied instead of using conventional corrosive acid catalysts. The major FAME products obtained from this direct transesterification reaction were C 16:0 (7.5–22.3%) and C 18:1 (63.2–78%).

In comparison to conventional acid catalysis, bio-nano $CaCO_3$ and bio-nano CaO produced a 1.29-fold and 1.02-fold increase in FAME proportions, respectively. As a consequence, this work demonstrates the potential of BNCs for the direct transesterification of *Coelastrella* sp. M-60 as a more environmentally friendly method of biodiesel synthesis (Karpagam et al., 2020).

5.5 CONCLUSION

The novel concept of utilizing biological agents coupled with nanomaterials for the production of biofuels is a new insight into the field of sustainable energy production. BNCs show enormous potential for widespread use and versatility. The conventional method of employing normal catalysts for the manufacturing of biofuels is not cost-effective and exhibits other drawbacks such as less selectivity, stability and complexity in removal from the developed product; therefore, the use of catalysts integrated with nanomaterial has emerged as a new possibility with implications for different sectors. However, attempts are being made to gain a better understanding of the nature of the changes that occur during the immobilization process on nano-support materials through the use of biochemical and morphological characterization approaches.

5.6 FUTURE PERSPECTIVE

The creation of BNCs is dependent on the interaction of various enzymes with nanoparticles. As a result, selective immobilization techniques are required that enable site-specific modification and orientational control of the support surface. The motivation for correctly designing and characterizing a bionanocatalyst is to produce robust and reusable systems for the selective biotransformation of industrially significant compounds. The bionanocatalysts have been effectively utilized for the production of high-density green fuels. Remarkably, bionanocatalysts have also been employed for the production of biofuels using plastic waste as feedstock in contrast to using biomass as raw material, which offers new possibilities for the sustainable utilization of waste for energy generation or the biovalorization of waste in an environmentally friendly manner for the generation of valuable products. Therefore, the adaptation of these green and effective catalysts for green energy production is the need of the hour, which has immense potential to achieve the sustainable development goals of affordable and green energy. However, BNCs should be evaluated for their long-term operation, stability, activity, leakage and mechanical strength. Second, bioprocess engineers should play a significant role in bringing these lab-scale breakthroughs to market.

REFERENCES

Ahmad, R., Sardar, M. (2015). Enzyme immobilization: An overview on nanoparticles as immobilization matrix. *Biochemistry & Analytical Biochemistry* 4(2). https://doi.org /10.4172/2161-1009.1000178

Alshammari, A., Kalevaru, V.N., Martin, A. (2016). Metal nanoparticles as emerging green catalysts. In: M.L. Larramendy, S. Soloneski (eds.), *Green Nanotechnology - Overview and Further Prospects*, IntechOpen, London. https://doi.org/10.5772/63314

An, S., Joshi, B.N., Lee, M.W., Kim, N.Y., Yoon, S.S. (2014). Electrospun graphene-ZnO nanofiber mats for photocatalysis applications. *Applied Surface Science* 294, 24–28. https://doi.org/10.1016/j.apsusc.2013.12.159

Arco, J.D., Pérez, E., Naitow, H., Matsuura, Y., Kunishima, N., Fernández-Lucas, J. (2019). Structural and functional characterization of thermostable biocatalysts for the synthesis of 6-aminopurine nucleoside-5″-monophospate analogues. *Bioresource Technology* 276, 244–252. https://doi.org/10.1016/j.biortech.2018.12.120

Basumatary, S. (2013). Transesterification with heterogeneous catalyst in production of biodiesel: A review. *Journal of Chemical and Pharmaceutical Research* 5(1), 1–7.

Betancor, L., Luckarift, H.R. (2008). Bioinspired enzyme encapsulation for biocatalysis. *Trends in Biotechnology* 26(10), 566–572. https://doi.org/10.1016/J.TIBTECH.2008.06.009

Bolina, I.C.A., Salviano, A.B., Tardioli, P.W., Cren, É.C., Mendes, A.A. (2018). Preparation of ion-exchange supports via activation of epoxy-SiO2 with glycine to immobilize microbial lipase – Use of biocatalysts in hydrolysis and esterification reactions. *International Journal of Biological Macromolecules* 120(B), 2354–2365. https://doi.org/10.1016/j.ijbiomac.2018.08.190

Chang, F., Zhou, Q., Pan, H., Liu, X., Zhang, H., Xue, W., Yang, S. (2014). Solid mixed-metal-oxide catalysts for biodiesel production: A review. *Energy Technology* 2(11), 865–873. https://doi.org/10.1002/ente.201402089

Chen, C., Xie, L., Wang, Y. (2019). Recent advances in the synthesis and applications of anisotropic carbon and silica-based nanoparticles. *Nano Research* 12(6), 1267–1278. https://doi.org/10.1007/s12274-019-2324-9

Deng, X., He, T., Li, J., Duan, H.L., Zhang, Z.Q. (2020). Enhanced biochemical characteristics of β-glucosidase via adsorption and cross-linked enzyme aggregate for rapid cellobiose hydrolysis. *Bioprocess & Biosystems Engineering* 43(12), 2209–2217. https://doi.org/10.1007/s00449-020-02406-5

Devi, A., Singh, A., Bajar, S., Owamah, H.I. (2021). Nanomaterial in liquid biofuel production: Applications and current status. *Environmental Sustainability* 4(2), 343–353. https://doi.org/10.1007/s42398-021-00193-7

Gkantzou, E., Chatzikonstantinou, A.V., Fotiadou, R., Giannakopoulou, A., Patila, M., Stamatis, H. (2021). Trends in the development of innovative nanobiocatalysts and their application in biocatalytic transformations. *Biotechnology Advances* 51, 107738. https://doi.org/10.1016/j.biotechadv.2021.107738

Galanakis, C.M. (2012). Recovery of high added-value components from food wastes: Conventional, emerging technologies and commercialized applications. *Trends in Food Science & Technology* 26(2), 68–87. https://doi.org/10.1016/j.tifs.2012.03.003

Harun, R., Danquah, K., Forde, G.M. (2010). Microalgal biomass as a fermentation feedstock for bioethanol production. *Journal of Chemical Technology & Biotechnology* 85, 199–203. https://doi.org/10.1002/jctb.2287

Huang, L., Zhou, Y., Guo, X., Chen, Z. (2015). Simultaneous removal of 2,4-dichlorophenol and Pb(II) from aqueous solution using organoclays: Isotherm, kinetics and mechanism. *Journal of Industrial & Engineering Chemistry* 22, 280–287. https://doi.org/10.1016/j.jiec.2014.07.021

Hunt, A., Farmer, T. (2015). Elemental sustainability for catalysis. https://pubs.rsc.org/en/content/chapterhtml/2015/bk9781782626381-00001?isbn=978-1-78262-638-1&sercode=bk

Iravani, S., Varma, R. (2020). Nano- and bio-catalysis in green chemistry: Recent trends and future prospects. *Phytochem and BioSub Journal* 14, 8–12.

Karpagam, R., Jawaharraj, K., Gnanam, R. (2021). Review on integrated biofuel production from microalgal biomass through the outset of transesterification route: A cascade approach for sustainable bioenergy. *Science of the Total Environment* 766, 144236. https://doi.org/10.1016/j.scitotenv.2020.144236

Karpagam, R., Rani, K., Ashokkumar, B., Ganesh Moorthy, I., Dhakshinamoorthy, A., Varalakshmi, P. (2020). Green energy from *Coelastrella* sp. M-60: Bio-nanoparticles mediated whole biomass transesterification for biodiesel production. *Fuel* 279 (September 2019), 118490. https://doi.org/10.1016/j.fuel.2020.118490

Keighron, J.D., Keating, C.D. (2010). Enzyme: Nanoparticle bioconjugates with two sequential enzymes: Stoichiometry and activity of malate dehydrogenase and citrate synthase on Au nanoparticles. *Langmuir* 26(24), 18992–19000. https://doi.org/10.1021/la1040882

Klaochanpong, N., Puttanlek, C., Rungsardthong, V., Puncha-arnon, S., Uttapap, D. (2015). Physicochemical and structural properties of debranched waxy rice, waxy corn and waxy potato starches. *Food Hydrocolloids* 45, 218–226. https://doi.org/10.1016/j.foodhyd.2014.11.010

Lin, J., Fan, L., Miao, R., Le, X., Chen, S., Zhou, X. (2015). Enhancing catalytic performance of laccase via immobilization on chitosan/CeO_2 microspheres. *International Journal of Biological Macromolecules* 78, 1–8. https://doi.org/10.1016/j.ijbiomac.2015.03.033

Liu, Y., Ji, X., He, Z. (2019). Organic–inorganic nanoflowers: from design strategy to biomedical applications. *Nanoscale* 11(37), 17179–17194. https://doi.org/10.1039/C9NR05446D

Lv, Z., Yu, Q., Wang, Z., Liu, R. (2019). Immobilization and performance of penicillin G acylase on magnetic Ni0.7 Co0.3Fe2O4@SiO2-CHO nanocomposites. *Journal of Microbiology & Biotechnology* 29(6), 913–922.

Madhavan, A., Arun, K.B., Binod, P., Sirohi, R., Tarafdar, A., Reshmy, R., Kumar Awasthi, M., Sindhu, R. (2021). Design of novel enzyme biocatalysts for industrial bioprocess: Harnessing the power of protein engineering, high throughput screening and synthetic biology. *Bioresource Technology* 325, 124617. https://doi.org/10.1016/j.biortech.2020.124617

Mauro, A, Maria, E.F., Vittorio, P., Giuliana, I. (2017). ZnO for application in photocatalysis: From thin films to nanostructures. *Materials Science in Semiconductor Processing* 69, 44–51. https://doi.org/10.1016/j.mssp.2017.03.029

Misson, M., Zhang, H., Jin, B. (2015). Nanobiocatalyst advancements and bioprocessing applications. *Journal of the Royal Society. Interface* 12(102). https://doi.org/10.1098/RSIF.2014.0891

Mitchell, D.T., Lee, S.B., Trofin, L., Li, N., Nevanen, T.K., Söderlund, H., Martin, C.R. (2002). Smart nanotubes for bioseparations and biocatalysis. *Journal of the American Chemical Society* 124(40), 11864–11865. https://doi.org/10.1021/JA027247B

Nguyen, V.D., Styevkó, G., Madaras, E., Haktanirlar, G., Tran, A.T.M., Bujna, E., Dam, M.S., Nguyen, Q.D. (2019). Immobilization of β-galactosidase on chitosan-coated magnetic nanoparticles and its application for synthesis of lactulose-based galactooligosaccharides. *Process Biochemistry* 84, 30–38. https://doi.org/10.1016/j.procbio.2019.05.021

Pascariu, P., Homocianu, M. (2019). ZnO-based ceramic nanofibers: Preparation, properties and applications. *Ceramics International* 45(9), 11158–11173. https://doi.org/10.1016/j.ceramint.2019.03.113

Raina, S., Roy, A., Bharadvaja, N. (2020). Degradation of dyes using biologically synthesized silver and copper nanoparticles. *Environmental Nanotechnology, Monitoring & Management* 13, 100278. https://doi.org/10.1016/j.enmm.2019.100278

Roy, A., Elzaki, A., Tirth, V., Kajoak, S., Osman, H., Algahtani, A., Islam, S., Faizo, N.L., Khandaker, M.U., Islam, M.N., Emran, T.B., Bilal, M.. (2021). Biological synthesis of nanocatalysts and their applications. *Catalysts* 11(12), 1494. https://doi.org/10.3390/catal11121494

Sadeghzadeh, S., Ghobadi Nejad, Z., Ghasemi, S., Khafaji, M., Borghei, S.M. (2020). Removal of bisphenol A in aqueous solution using magnetic cross-linked laccase aggregates from Trameteshirsuta. *Bioresource Technology* 306, 123169. https://doi.org/10.1016/j.biortech.2020.123169

Schwaminger, S.P., Fehn, S., Steegmüller, T., Rauwolf, S., Löwe, H., Pflüger-Grau, K., Berensmeier, S. (2021). Immobilization of PETase enzymes on magnetic iron oxide nanoparticles for the decomposition of microplastic PET. *Nanoscale Advances* 3(15), 4395–4399. https://doi.org/10.1039/d1na00243k

Seelajaroen, H., Bakandritsos, A., Otyepka, M., Zbořil, R., Sariciftci, N.S. (2020). Immobilized enzymes on graphene as nanobiocatalyst. *ACS Applied Materials & Interfaces* 12(1), 250–259. https://doi.org/10.1021/ACSAMI.9B17777

Selvaraj, R., Praveenkumar, R., Moorthy, IG. (2019). A comprehensive review of biodiesel production methods from various feedstocks. *Biofuels* 10, 325–333. https://doi.org/10.1080/17597269.2016.1204584

Singhvi, M., Kim, B.S. (2020). Current developments in lignocellulosic biomass conversion into biofuels using Nanobiotechology approach. *Energies* 13(20), 5300. https://doi.org/10.3390/en13205300

Thangaraj, B., Solomon, P.R., Muniyandi, B., Ranganathan, S., Lin, L. (2019). Catalysis in biodiesel production — A review. *Clean Energy* 3(1), 2–23. https://doi.org/10.1093/ce/zky020

Verma, M.L., Barrow, C.J., Puri, M. (2013). Nanobiotechnology as a novel paradigm for enzyme immobilisation and stabilisation with potential applications in biodiesel production. *Applied Microbiology & Biotechnology* 97(1), 23–39.

Verma, M.L., Kumar, S., Das, A., Randhawa, J.S., Chamundeeswari, M. (2020). Chitin and chitosan-based support materials for enzyme immobilization and biotechnological applications. *Environmental Chemistry Letters* 18(2), 315–323. https://doi.org/10.1007/S10311-019-00942-5

Wang, F.Q., Xie, H., Chen, W., Wang, E.T., Du, F.G., Song, A.D. (2013). Biological pretreatment of corn stover with ligninolytic enzyme for high efficient enzymatic hydrolysis. *Bioresource Technology* 144, 572–578. https://doi.org/10.1016/j.biortech.2013.07.012

Yang, Y., Yu, C. (2019). Advances in silica-based nanoparticles for targeted cancer therapy. *Nanomedicine* 12(2), 317–332. https://doi.org/10.1016/j.nano.2015.10.018

Yang, Y., Zhang, M., Song, H., Yu, C. (2020). Silica-based nanoparticles for biomedical applications: From nanocarriers to biomodulators. *Accounts of Chemical Research* 53(8), 1545–1556. https://doi.org/10.1021/ACS.ACCOUNTS.0C00280

Yavaşer, R., Karagözler, A.A. (2021). Laccase immobilized polyacrylamide-alginate cryogel: A candidate for treatment of effluents. *Process Biochemistry* 101, 137–146. https://doi.org/10.1016/j.procbio.2020.11.021

Zdarta, J., Jankowska, K., Bachosz, K., Degórska, O., Kaźmierczak, K., Nguyen, L.N., Nghiem, L.D., Jesionowski, T. (2021). Enhanced wastewater treatment by immobilized enzymes. *Current Pollution Reports* 7(2), 167–179. https://doi.org/10.1007/s40726-021-00183-7

6 Biomimetic Concepts and Their Applications in Green Energy

Y.M.S.M Yapa and Jayani J. Wewalwela

CONTENTS

6.1   INTRODUCTION

Life on earth has evolved over an estimated 3.8 billion years. Many of nature's challenges have been solved, leading to lasting solutions providing maximum performance with minimal resources. The inventions of nature have always inspired and contributed to human achievement through effective algorithms, methods, materials, processes, structures, tools, mechanisms and systems (Fratzl, 2007). Nature has given efficient solutions to animals, plants and insects for their survival such as self-repair, self-cleaning, super hydrophobicity, drag reduction, dry adhesion and adaptive growth (Eadie & Ghosh, 2011), energy conversion and conservation, high adhesion, reversible adhesion, aerodynamic lift, materials and fibres with high mechanical strength, biological self-assembly, antireflection, structural coloration, thermal insulation, self-healing and sensory-aid mechanisms (Bhushan, 2009). These solutions

DOI: 10.1201/9781003316374-6

of nature have led humans to synthesise outstanding outcomes; for example, the fishing net came from the idea of a spider web, the structure of honeycomb inspired many applications in airplanes, etc. (Bello et al., n.d). Such biologically inspired adaptations and derivations are known as 'biomimicry' (Vincent et al., 2006). The word biomimetics is derived from the Greek word bio-mimesis (Bhushan, 2009). The word biomimetics was first coined by an academic and inventor, Otto Schmitt, in 1957, to describe the transferring of ideas and concepts from biology to technology (Bhushan, 2009; Bello et al., n.d).

In fact, biological structures are a huge source of inspiration for many fields, including mechanical engineering, architecture, material science, and aerodynamics (Fratzl, 2007). This involves biologists, chemists, physicists, material scientists, and engineers working in a variety of fields using biological structures, concepts, and principles of natural objects (Bhushan, 2009). When considering the survival of humanity and other resources of the world along with the development of technologies, study and research on sustainable energy production, conversion and storage is an essential and significant field. Therefore, the engineering of nanostructures and device-architectures is very significant in energy conversion and storage flat forms (Kim, 2020).

Nevertheless, the science of bio template and biomimetic materials has taken a significant role in the field of sustainable energy production due to the current problems associated with the environment such as greenhouse gas emissions and climate change (Faggal, 2013; Varshabi et al., 2022) and due to the changes in the climate patterns and composition of the atmospheric composition because of the continuous usage of fossil fuels and other natural resources on earth in past decades which has led the world to a unhealthy condition (Wahl, 2006).

6.2 OVERVIEW OF BIOMIMICRY

6.2.1 THE HISTORY OF BIOMIMICRY

The history of biomimicry goes back thousands of years to 500 BC. The sketches of a flying machine drawn by Leonardo Da Vinci (1452–1519) after the study of birds and their anatomy are considered one of the early examples of biomimicry (Radwan & Osama, 2016a). An American biophysicist and polymath Otto Schmitt developed the concept of 'biomimetics' in the 1950s. In his doctoral research he attempted to produce a physical device called a Schmitt trigger through studying replicates of the biological system of nerve propagation in squid. Then, by 1957, he had seen it as the opposite, disregarding the conventional understanding of biophysics but yet being crucial and vital. Later, he would refer to it as 'biomimetics' (Vincent et al., 2006). Otto Schmitt also said that

> Biophysics is not so much a subject matter as it is a point of view. It is an approach to problems of biological science utilizing the theory and technology of the physical sciences. Conversely, biophysics is also a biologist's approach to problems of physical science and engineering, although this aspect has largely been neglected.
>
> **(Vincent, 2009)**

In 1960, a similar word, bionics, was coined by Jack Steele of the US Air Force at a meeting at Wright-Patterson Air Force Base in Dayton, Ohio, and he defined bionics as 'the science of systems which have some function copied from nature, or which represent characteristics of natural systems or their analogues' (Vincent, 2009). Also, in a later meeting in 1963, Otto Schmitt stated that

> Let us consider what bionics has come to mean operationally and what it or some word like it (I prefer biomimetics) ought to mean in order to make good use of the technical skills of scientists specializing, or rather, I should say, specializing into this area of research. Presumably our common interest is in examining biological phenomenology in the hope of gaining insight and inspiration for developing physical or composite biophysical systems in the image of life.
>
> **(Vincent, 2009; Vincent et al., 2006)**

Nevertheless, in 1974, the word biomimetics was published in *Webster's Dictionary*, an English language dictionary, with the definition:

> The study of the formation, structure, or function of biologically produced substances and materials (as enzymes or silk) and biological mechanisms and processes (as protein synthesis or photosynthesis) especially for the purpose of synthesizing similar products by artificial mechanisms which mimic natural ones.
>
> **(Vincent, 2009)**

However, the term biomimicry first appeared in 1982. At present, there are many researchers who have defined biomimicry (Radwan & Osama, 2016a).

6.2.2 LEVELS OF BIOMIMICRY

There are three levels that biomimicry can work on. They are:

1. *The organism level*
   At this level, a particular whole organism or a part of the whole organism is mimicked.
2. *The behaviour level*
   On the behaviour level, the mimicking of an organism's interactions with and adaptations to its environment is considered.
3. *The ecosystem level*
   This level involves mimicking how the many components of the environment work together. The whole ecosystem and how it helps the organisms to work together are considered. This is considered the hardest level among the three levels of biomimicry.

6.2.3 BIOMIMICRY APPROACHES

The approach of biomimicry especially helps designers and architects who need to apply biomimicry concepts in buildings to enhance the quality of the building and

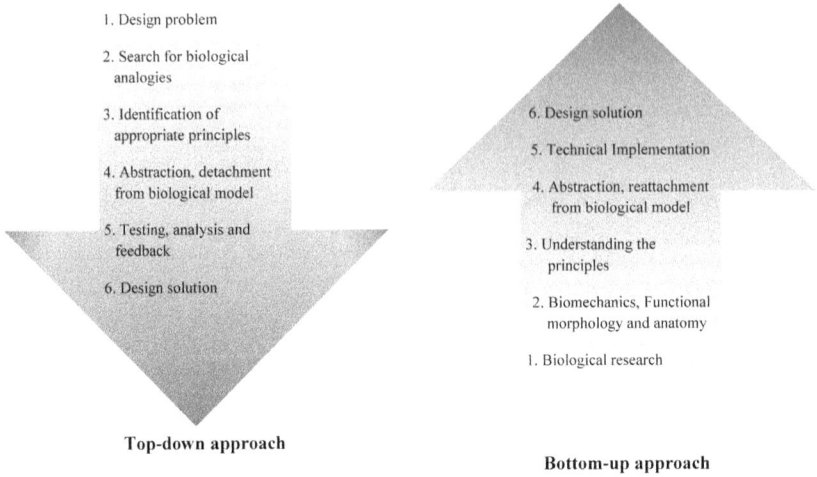

FIGURE 6.1   Biomimicry: top-down and bottom-up approaches.

its surrounding environment. As a design process there are two types of approaches, shown in Figure 6.1 (Pathak, 2019).

1. *Top-down approach*

    This has different names with the same meaning: 'problem-based approach', 'design looking to biology', 'problem driven biologically inspires design' and 'challenge to biology'. The top-down approach focuses on identifying the design issue and investigating how organisms and ecosystems have solved similar issues in the past (Pathak, 2019).

2. *Bottom-up approach*

    Like the previous approach this also is called by many names such as 'solution-based approach', 'solution-driven biologically inspired design', 'biology to design' and 'biology influencing design'. This refers to applying the organism's solutions and previous knowledge to the design problem we already have (Pathak, 2019).

6.3   APPLICATIONS OF THE BIOMIMETIC CONCEPTS IN THE SYNTHETIC WORLD

Although the words bio templating and biomimetics are relatively new, people of many centuries ago have looked to nature for inspiration and the development of various devices, structures and materials (Bhushan, 2009; Vincent et al., 2006). Since 1980, neural networks in information technology and artificial intelligence have been inspired by the human brain (Bhushan, 2009).

There are many sources of inspiration from organisms such as bacteria, aquatic and land animals and also from plants and other objects and structures which are

TABLE 6.1

Organisms, Objects of Nature and Their Inspiration Functions in the Synthetic Designs

Organisms/Their Objects	Inspiration Function
Bacteria and its flagella	Flagella motor
Plants and presence of spines, hair, waxes, fibres, wood	Hydrophobicity, adhesive effect, self-cleaning, drag reduction, optical appearance, superhydrophobic, motion, porous structure
Spider and cobwebs	Self-arrangement
Shark skin	Self-cleaning, low fluid drag, superoleophobicity
Tooth and bones	Toughness, strength
Fish	Polarization of light, underwater movements
Moth eye	Structural coloration, antireflective surface
Birds	Locomotion, aerodynamic lift, light coloration
Insects	Reversible adhesion in dry and wet environment
Sea shells	Mechanical strength

made by organisms (Bhushan, 2009). Table 6.1 indicates the organisms, objects of the nature and their inspiring functions in synthetic designs.

In fact, in the synthetic world there are many man-made structurers which were inspired by nature. Some of them include the design of the Sydney Opera House in Australia, the Egg chair in Denmark and the Shinkansen bullet train in Japan (Ulhoi, 2021). Also, some of nature's creations, behaviours and conditions have inspired success in business and other industries too. One of the best examples is the Velcro company that produced hook and loop fasteners for the first time. The concept for the hook and loop was invented by a Swiss engineer George de Mestral through a close inspection of the seeds from the genus *Arctium* commonly known as Burdock (Ivanic et al., 2015; Bello, 2013). Giovanni Borelli has taken ideas from the swimming motions of animals with applications in submarine technology (Ivanic et al., 2015).

The lotus effect also has been used as inspiration due to the self-cleaning property of the lotus leaves and due to their super hydrophobicity, drag reduction in fluid flow and low adhesion (Bhushan, 2009; Ivanic et al., 2015). Ivanic et al. (2015) also mentioned that many nanotech engineers have developed treatments, tiles, textiles, paints, coatings and other surfaces that can stay dry and self-clean with the lotus effect. The hydrophobic surface of the lotus leaf has been an inspiration in manufacturing exterior coatings and paints, and all these reduce the water waste when cleaning surfaces (Faggal, 2013).

Faggal (2013) has mentioned that the ability of cells to regenerate and form scabs when wounded has been applied in the production of self-healing cement, plastics and ceramics. Chaurasia and Srivastava (2020) have described two interesting biomimetic-inspired applications in the world. One of them is artificial skin made from spider silk, and the other is 'Gecko Vision' which is used in cameras and lenses inspired by the gecko lizard (Table 6.2).

TABLE 6.2

Some Existing Biomimetic Materials and Their Applications in the Synthetic World

Product	Natural Source of Inspiration	Function and Character	Reference
Tec Eco eco-cement	Plants, algae and sea snails	Traps atmospheric CO_2 and provides extra strength and rigidity	Oguntona and Aigbavboa, 2016
Dye-sensitized solar cells and panels	Artificial photosynthesis of plants	Exceptional stability and produce electricity efficiently with low cost	Oguntona and Aigbavboa, 2016
Sustainable carpet and floor finish	Chaos of a blanket of fallen leaves and a bed of river stones	High flexibility, faster installation, easy repairs and reduced waste	Oguntona and Aigbavboa, 2016
Sharklet	Shark skin with small spines	Restrains bacterial development through physical surface modification	Oguntona and Aigbavboa, 2016
Lotusan paint	Lotus plant (*Nelumbo nucifera*)	Self-cleaning property and reducing needs for chemical detergents and labour cost	Oguntona and Aigbavboa, 2016
Solar collectors	Butterfly wings	Absorb light more efficiently than conventional dye-sensitized cells	Oguntona and Aigbavboa, 2016
Wind turbines	Whale flipper	Increasing efficiency by improvement in lift and reduction in drag	Faggal, 2013
Robotic arm	Elephant's trunk	Freedom of movement	Shahda et al., 2014

6.3.1 APPLICATIONS OF BIOMIMICRY IN BUILDING STRUCTURES

Considering the biomimetic concepts that have been used in the past, they are still being using at present in the architecture field, especially in building structures. The evidence for bio-inspired structures in human civilization goes back to ancient Egyptian civilization. Their old temples and the columns of the temples were said to be inspired by lotus plants. Also, in the Greek and Roman ages plants and trees were considered to be taken as inspiration for ornamented structural columns. Years later, in the 12th century, bio-inspired structures appeared in churches. Then late in the 19th century and 20th century many architectural designs were found which were constructed according to concepts inspired by nature.

Comparing man-made systems and biological and natural systems, man-made systems are simple and they are resistant to change in structure, while natural systems are more complex and always adapting to changes in conditions. Man-made systems are mono cultured and centralized with one goal; meanwhile, biological systems are always diverse, distributed and act as a whole system. Most human-made systems produce much waste, but natural or biological systems do not produce any waste and they are always eco-friendly to the world (Table 6.3).

TABLE 6.3
Applications of Biomimetic Concepts in Architecture

Name of the Building	Inspiration from Nature	Advantageous Characteristics	Reference
Council House 2 (CH2) in Melbourne	Termite mounds and mammalian skin	1. Increasing energy efficiency 2. Zero greenhouse gases emissions	Faggal, 2013; Zari, 2007
Eastgate Centre in Zimbabwe	Termite mounds	1. Energy efficiency in thermal control	Faggal, 2013; Pathak, 2019
Hydrological Center project in Namibia	Namibian beetle's ability to capture fog	1. Picking water from fog on the building's roof to provide all the building's water needs	Faggal, 2013; Shahda et al., 2014
Bird's Nest Stadium in China	Bird nest	1. Reducing pollution 2. Rationalization of energy	Radwan and Osama, 2016
Eiffel Tower	Thigh bone	1. Effective ventilation 2. Withstands bending and shearing effects due to wind	Radwan and Osama, 2016
Beijing National Aquatic Centre in China	Soap bubbles	1. Collecting solar energy that heats swimming pools 2. Temperature regulation 3. Rationalization of energy 4. Reducing artificial lighting	Radwan and Osama, 2016
Esplanade Theatre, Marina Bay in Singapore	Durian fruit	1. Increasing energy efficiency 2. Capturing solar energy 3. Reducing artificial lighting	Radwan and Osama, 2016; Faggal, 2013
Ministry of Municipal Affairs and Agricultural building in Doha, Qatar	Cactus plant	1. Increasing energy efficiency 2. Shading the surface of the building	Faggal, 2013

6.4   SUSTAINABLE GREEN ENERGY

With the increasing population, current energy demand also is increasing greatly, therefore leading to many global problems, particularly global warming with respect to climate change due to the emission of enormous amounts of greenhouse gases (Ahmad & Zhang, 2020; Arroyo & Miguel, 2020). This is because the worldwide usage of energy has increased from 8,588.9 million tons in 1995 to 13,147.3 million tons in 2015 (Ahmad & Zhang, 2020). Greenhouse gas emissions from energy production and consumption are almost 70% of global emissions, and energy is considered the largest source of greenhouse gas emissions in the world (Graaf & Colgan, 2016).

Arroyo and Miguel (2020) mention that global economic stability is strongly dependent on the price of energy sources. Furthermore, 1.6 billion people in the world still don't have access to basic energy services and therefore, they are far away from economic development due to the prevailing energy poverty. Therefore, to ensure the health and protection of the environment, the energy sector has developed more innovations (Graaf & Colgan, 2016), as energy and power are the most essential requirements of every nation in the world at present (Ahmad & Zhang, 2020).

According to Arroyo and Miguel (2020) the past decade has seen a growth in the usage of renewable energy technologies with solar photovoltaic and wind power. Particularly, this is because of the new technologies and their development.

There are two sources of energy in the world. These are renewable energy and non-renewable energy sources (Kalyani et al., 2015).

Renewable energy sources are obtained from nature such as solar, wind, hydropower, geothermal energy, biomass and tidal energy (Kalyani et al., 2015; Strielkowski et al., 2021). On the other hand, the non-renewable energy sources are not eco-friendly and they cause serious negative impacts to the world. The non-renewable energy sources are oil, natural gas, coal and nuclear energy (Kalyani et al., 2015).

The renewable energy sources are considered alternative sources, because most industrialized countries do not depend on renewable energy sources as their main source of energy and most of the energy need is supplied through non-renewable sources such as fossil fuels or nuclear power.

The energy produced from renewable sources such as sunlight, wind, rain, tides, plants, algae and geothermal heat, with less impact on the environment, is known as green energy (Kalyani et al., 2015).

This green energy should be sustainable in order to ensure the continuous survival of living beings, as sustainable energy is energy that meets the needs of the present without jeopardizing the ability of future generations to meet their own (Taghizadeh, 2014).

6.4.1 Types of Green Energies

6.4.1.1 Solar Energy

The sun is the energy source which provides energy to all the living beings on earth. It has enormous potential for providing clean, safe and reliable power (Alrikabi, 2014; Kalyani et al., 2015).

The sun is a source of energy which produces about 10,000 times the total energy that the earth can produce in the 21st century (Kalyani et al., 2015).

Alrikabi (2014) mentioned that the solar energy which falls on the earth is more than 200 times the total annual commercial energy consumption by humans. There are two forms of solar energy. These are passive solar and active solar (Alrikabi, 2014).

Direct use or indirect use of energy from the sun is passive solar energy. Active solar energy is using the sun's electromagnetic radiation in generating electrical energy (Alrikabi, 2014).

6.4.1.2 Hydro Energy

Hydropower generates electricity via capturing the energy of falling water. The turbines can convert the kinetic energy of falling water into a usable form. Hydropower can provide 19% of the world's electricity needs approximately. Nevertheless, this power is used globally in dams both at small scale and large scale (Kalyani et al., 2015).

6.4.1.3 Geothermal Energy

Geothermal energy is a renewable energy source as the heat is continuously produced inside the earth. Therefore, it is known to be clean and sustainable (Alrikabi, 2014). Geothermal energy is used indirectly to generate electricity or in the heating of houses and water or in space heating. It is also used in aquaculture, industrial processes and laundries with the use of many technologies such as dry steam, binary cycle steam, flash steam and hot dry rock and resources at high temperatures (Kalyani et al., 2015).

6.4.1.4 Wind Energy

Wind energy is a renewable energy source and also it is a clean energy as it does not produce any pollution or harmful gases like greenhouse gases to the surrounding environment (Kalyani et al., 2015).

Wind energy is produced by the movement of atmospheric air through turbines. Therefore, the location of turbines is a vital factor in producing energy through the wind (Alrikabi, 2014).

6.5 BIOMIMETIC MATERIALS USAGE IN SUSTAINABLE GREEN ENERGY

Biomimicry has become a new approach to endowing energy efficiency in many fields such as transportation, the car industry, electronics, the clothing industry and also in building designs (Faggal, 2013). Biomimetic concepts in building designs are popular now as they can reduce energy consumption in buildings. In fact, it is very important to reduce the energy usage of buildings as around 40% of the energy consumption and CO_2 emissions is used in buildings.

The building of the Ministry of Municipal Affairs in Doha was designed with inspiration from the cactus plant and its ability to shade itself as well as to avoid the release of moisture through its spines and thorn-like structures. This building, therefore, has been designed to save energy consumption while being highly energy efficient (Faggal, 2013).

Son et al. (2022) mentioned that one of the biomimetic applications in buildings is the biomimetic window system that has been innovated so as to bring more daylight to the basement floor of buildings. This has been proven to reduce the energy usage of the building. It is a new way of capturing and transmitting solar heat and light via biomimetic applications. Furthermore, this concept can be applied to the interior non-window spaces of academic and other related types of buildings. According to Radwan and Osama (2016) there are many findings about building skins as a tool for energy management in the building industry. They described the building skin as the boundary through which interactions occur between buildings and the environment and it is where most energy and material exchange happen. This building skin is similar to natural skin, and it regulates the organs of the building such as the mechanical, electrical and plumbing systems and ensures the energy efficiency of the interior design.

At present, many people are using photovoltaic solar cells as an alternative to electricity as those solar cells are a renewable and green energy production method. In fact, non-uniformly red rooftop solar cells require a structural colour that is independent of both the viewing angle and the direction of the incident light. Such a non-iridescent red-rejection filter could possibly be fabricated by upscaling the linear dimensions of biomimetic filters nanoimprinted to reproduce the morpho blue. Therefore, naturally inspired structural colour has been used in the roof solar cells with biomimetic applications in order to increase their energy efficiency (Lenau et al., 2019). A study carried out by Machín et al. (2021) has found possibilities for the development of high-performance photocatalysts developing efficient catalysts through biomimetic catalysts.

The lithium-sulphur (Li-S) battery is one of the most promising battery types with a stationary energy storage system that has longer charge times and overall battery life. A novel ant-nest electrode structure for Li-S batteries (named CNT-nest-S), which is designed to imitate nature's own ant-nest structure, has taken advantage of ant-nest's abundant storage and highly efficient transport system (Ai et al., n.d.).

6.6 CONCLUSION

Although the biomimetic and bio-templating materials have positive effects on sustainable green energy and its applications, still, it is a new approach for some countries in the world. Also, fully designing or producing a biomimetic concept of a living being or of a natural ecosystem in any environment is challenging with even with today's technologies as nature is moving ahead of our technologies. To reliably find more and more sustainable applications of nature, there should be collaboration between many fields, researchers, studies, innovations and also many new technologies. Then, the world will be preserved with less impacts on the people, animals, plants and the whole environment; this can be done with the application of biomimetic concepts in all fields and industries globally.

REFERENCES

Ahmad, T., & Zhang, D. (2020). A critical review of comparative global historical energy consumption and future demand: The story told so far. *Energy Reports*, 6, 1973–1991. https://doi.org/10.1016/j.egyr.2020.07.020

Ai, G., Dai, Y., Mao, W., Zhao, H., Fu, Y., Song, X., En, Y., Battaglia, V. S., Srinivasan, V., & Liu, G. (2016). Biomimetic ant-nest electrode structures for high sulfur ratio lithium-sulfur batteries. *Nano Letters*, 16(9), 5365–5372.

Alrikabi, N. Kh. M. A. (2014). Renewable energy types. *Journal of Clean Energy Technologies*, 2, 61–64. https://doi.org/10.7763/jocet.2014.v2.92

Arroyo, F. R. M., & Miguel, L. J. (2020). The role of renewable energies for the sustainable energy governance and environmental policies for the mitigation of climate change in ecuador. *Energies*, 13(15). https://doi.org/10.3390/en13153883

Bhushan, B. (2009). Biomimetics: Lessons from nature - An overview. *Philosophical Transactions of the Royal Society A: Mathematical, Physical and Engineering Sciences*, 367(1893), 1445–1486. https://doi.org/10.1098/rsta.2009.0011

Chaurasia, M., & Srivastava, S. (2020). Biomimicry and its applications - A review. *International Journal of Engineering Applied Sciences and Technology*, 04(12), 545–549. https://doi.org/10.33564/ijeast.2020.v04i12.098

Eadie, L., & Ghosh, T. K. (2011). Biomimicry in textiles: Past, present and potential. An overview. *Journal of the Royal Society Interface*, 8(59), 761–775. https://doi.org/10.1098/rsif.2010.0487

Faggal, A. A. (2013). Biomimetic energy conservation techniques & its applications in buildings. 10.13140/RG.2.2.15652.07046

Fratzl, P. (2007). Biomimetic materials research: What can we really learn from nature's structural materials? *Journal of the Royal Society Interface*, 4(15), 637–642. https://doi.org/10.1098/rsif.2007.0218

Kalyani, V. L., Dudy, M. K., & Pareek, S. (2015). Green energy: The need of the world. *International Journal of Computer Science and Information Security*, 2(5), 18–26.

Kim, J. K. (2020). Novel materials for sustainable energy conversion and storage. *Materials*, 13(11). https://doi.org/10.3390/ma13112475

Lenau, T. A., Ahmad, F., & Lakhtakia, A. (2019). Towards biomimetic red solar cells. In *Proceedings of SPIE 10965, Bioinspiration, Biomimetics, and Bioreplication IX*, 109650E (13 March 2019)13. https://doi.org/10.1117/12.2513259

Machín, A., Soto-Vázquez, L., Colón-Cruz, C., Valentín-Cruz, C. A., Claudio-Serrano, G. J., Fontánez, K., Resto, E., Petrescu, F. I., Morant, C., & Márquez, F. (2021). Photocatalytic activity of silver-based biomimetics composites. *Biomimetics*. https://doi.org/10.3390/biomimetics

Oguntona, O., & Aigbavboa, C. (2016). Promoting biomimetic materials for a sustainable construction industry. *Bioinspired, Biomimetic and Nanobiomaterials*, 6, 1–1. https://doi.org/10.1680/jbibn.16.00014

Pathak, S. (2019). Biomimicry: (Innovation inspired by nature). International Journal of New Technology and Research. https://doi.org/10.31871/IJNTR.5.6.17

Radwan, G., & Osama, N. (2016). Biomimicry, an approach, for energy effecient building skin design. *Procedia Environmental Sciences*, 34, 178–189. https://doi.org/10.1016/j.proenv.2016.04.017

Shahda, M., Elmokadem, A., & Elhafeez, M. A. (2014). Biomimicry levels as an approach to the architectural sustainability. *Port Said Engineering Research Journal*, 18, 117–125.

Solomon Bello, O., Adesina Adegoke, K., & Oyeladun Oyewole, R. (n.d.). Biomimetic materials in our world: A review. *IOSR Journal of Applied Chemistry*, 5(3), 22–35. www.iosrjournals.org

Strielkowski, W., Civín, L., Tarkhanova, E., Tvaronavičienė, M., & Petrenko, Y. (2021). Renewable energy in the sustainable development of electrical power sector: A review. *Energies* 14(24). https://doi.org/10.3390/en14248240

Taghizadeh, M. M. (2014). Review on renewable energy, sustainable energy and clean energies. https://www.researchgate.net/publication/264829312

Ulhøi, J. P. (2021). From innovation-as-usual towards unusual innovation: Using nature as an inspiration. *Journal of Innovation and Entrepreneurship*, 10(1). https://doi.org/10.1186/s13731-020-00138-0

van de Graaf, T., & Colgan, J. (2016). Global energy governance: A review and research agenda. *Palgrave Communications*, 2, 15047. https://doi.org/10.1057/palcomms.2015.47

Varshabi, N., Selçuk, S. A., & Avinç, G. M. (2022). Biomimicry for energy-efficient building design: A bibliometric analysis. *Biomimetics*, 7(1). https://doi.org/10.3390/biomimetics7010021

Vincent, J. F. V. (2009). Biomimetics - A review. *Proceedings of the Institution of Mechanical Engineers, Part H: Journal of Engineering in Medicine*, 223(8), 919–939. https://doi.org/10.1243/09544119JEIM561

Vincent, J. F. V., Bogatyreva, O. A., Bogatyrev, N. R., Bowyer, A., & Pahl, A. K. (2006). Biomimetics: Its practice and theory. *Journal of the Royal Society Interface*, 3(9), 471–482. https://doi.org/10.1098/rsif.2006.0127

Wahl, D. C. (2006). Bionics vs. biomimicry: From control of nature to sustainable participation in nature. *WIT Transactions on Ecology and the Environment*, 87, 289–298. https://doi.org/10.2495/DN060281

Zari, M. P. (2007). Biomimetic approaches to architectural design for increased sustainability. In *Proceedings of thethe SB07 NZ Sustainable Building Conference*, Auckland, New Zealand, 14–16 November 2007, p. 10.

7 Fabrication of Hybrid Electrode Materials for Supercapacitors and Energy Storage Devices

Salman Khan

CONTENTS

7.1 INTRODUCTION

Due to the obvious dramatic increase in pollution and global warming, as well as other geopolitical problems throughout the world, the urgency of discovering a renewable fuel source that may end the use of fossil fuels is growing by the day. Storage technology development is just as vital as creating alternative energy sources, which is why high power and high energy density storage systems have received a lot of attention recently. Electrochemical energy is unquestionably one of the most essential components of the clean renewable energy spectrum. The electrochemical energy transformation concept is used by a variety of technologies, including fuel cells, batteries, and supercapacitors. The diminishing supply of fossil fuels, the occurrence of hazy weather, and the development of hybrid electric cars all push researchers to investigate sustainable energy storage systems (ESSs) to meet our everyday needs and the needs of industrial manufacturing. Supercapacitors achieve higher capacitance thanks to the use of electrode materials with a larger surface area and very thin dielectrics, which distinguishes them from

DOI: 10.1201/9781003316374-7

normal capacitors. Supercapacitors are a potential technology for manufacturing improved energy storage devices since they have higher power capacities and a longer cycle life than batteries (Abdel Maksoud et al., 2021). Supercapacitors can handle high power rates, which is impressive when compared to batteries, but their inability to hold a similar amount of charge as batteries (3 to 30 times lower) is their most significant limitation (González et al., 2016). That's why supercapacitors are employed in circumstances where a huge amount of energy storage isn't required but just high-power bursts. Supercapacitors can be used in battery-based energy storage devices to decouple the energy and power properties of the storage system. This might increase the system's size and longevity while also meeting the demands for electricity and energy (Figure 7.1).

An electrolyte, two electrodes, and a separator that electrically isolates the two electrodes are the components of supercapacitors. These electrodes are the most important and basic components of supercapacitors (Pope, 2013). According to Iro et al. (2016), supercapacitors' beneficial qualities are mostly based on their capacity to supplement the power of batteries during emergency power supply and in electric vehicle power systems. Supercapacitors have been described as having a wide range of applications in fuel cell cars, low-emission hybrid vehicles, electric vehicles, forklifts, power quality upgrades, and load cranes (Miller and Simon, 2008; Cai et al., 2016). Various nanomaterials, including conductive polymers, electrolytes, transition metal carbides, transition metal dichalcogenides, nitrides, and hydroxides, have been used in the fabrication of supercapacitors employing printing technology (Figure 7.2). Magnetic metal oxide nanoparticles are a popular form of inorganic solid because they are inexpensive and simple to make in large

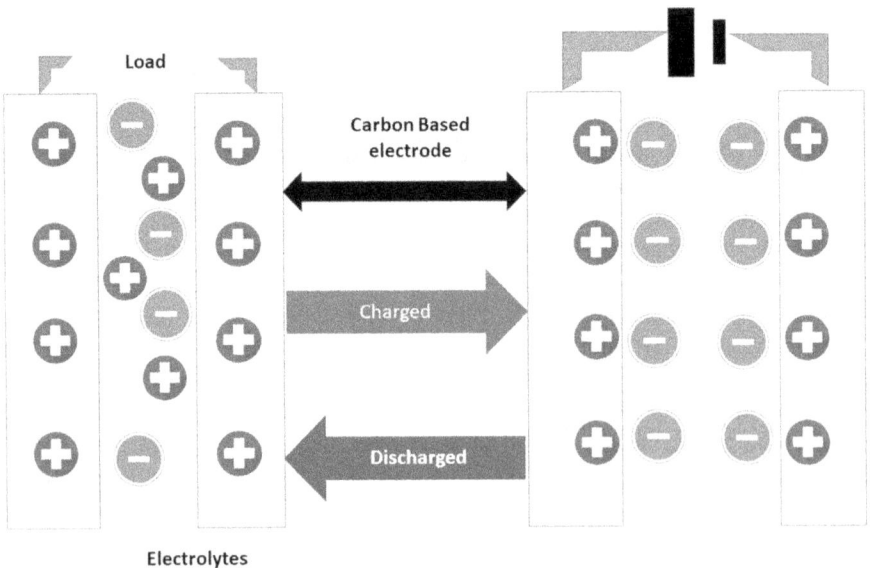

FIGURE 7.1 Common structure of a supercapacitor.

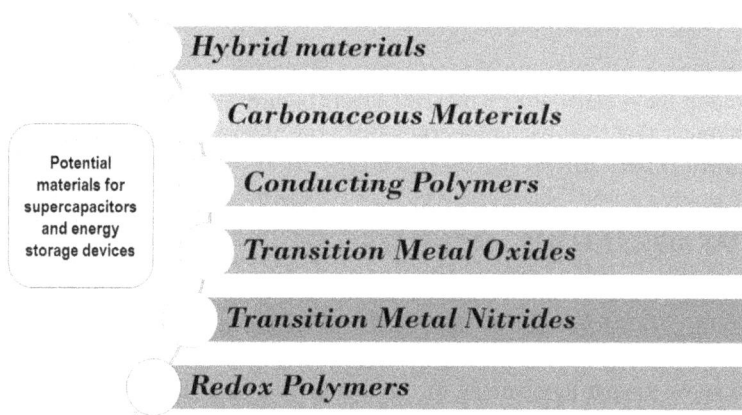

FIGURE 7.2   A list of potential materials that may be used for supercapacitors and energy storage devices.

quantities (Masala and Seshadri, 2004). Spinel ferrites and inorganic perovskite oxides outperform other magnetic materials as electrode materials in supercapacitor applications (Sharma et al., 2019; Kouchachvili et al., 2018; Wang et al., 2021; Afif et al., 2019; Miller et al., 2018; Raj et al., 2020; Xu et al., 2015; Brousse et al., 2017; Balasubramaniam et al., 2020).

According to new data, spinel ferrites of various elements are now being used in the construction of supercapacitor energy storage devices. Spinel ferrite nanoparticles have a high energy density, durability, and capacitance retention, as well as high power and long-term effectiveness. Perovskite oxides are multifunctional nanomaterials that have attracted a lot of interest for their prospective uses, and they are often used to make anion-intercalation supercapacitors. The valence state of the B-site element, surface area, and internal resistance all has a big impact on these nanomaterials. More crucially, there is a scarcity of research on the energy and power densities of perovskite oxides (Nan et al., 2019). The performance of a supercapacitor is determined by the manufacturing technique and components such as electrodes, current collector, binder, separator, and electrolyte. When it comes to the efficiency of supercapacitors, electrode materials are one of the most critical factors. They determine the capacitance by resolving the charge storage in the device. When it comes to energy density and stability, porous carbon, conductive polymers, metal hydroxide, and metal oxides, which are some of the materials usually employed for the electrodes in supercapacitors, have certain limitations (Miller et al., 2018). The development of robust, high-efficiency electrodes at a reasonable cost has been a major focus of supercapacitor research. Metal-organic frameworks, graphene-based materials, ceramic materials, activated carbon-based materials, and other modern electrode materials utilised in supercapacitors are covered in this study. This study seeks to give readers a fundamental understanding of the various electrode materials used in supercapacitors, as well as the research that will be needed in the future to increase their performance.

7.2 HYBRID MATERIALS

Hybrid materials are made up of two (or more) materials, space, and composites that are linked at the molecular or nanoscale level (Figure 7.3). They are put together in such a manner that they have properties that no single substance can provide on its own (Sarasini et al., 2017).

7.3 CARBONACEOUS MATERIALS

Carbon-based materials are the most often utilised material for diverse applications in supercapacitors, as previously said, due to their great availability and strong manufacturing procedures in the industry, which result in lower costs. Carbon-based electrodes are quite prominent among the applications. From 1D to 3D structures, they may be made in numerous forms such as fibres, nanotubes, and foams (Figure 7.4). The specific capacitance is usually proportional to the electrode surface area made of carbon; however, this is not always the case. When compared to an electrode with a high specific area, some forms of carbon will have a greater specific capacitance while having a lower surface area (Xie et al., 2016; Béguin et al., 2014).

A single atom thick sheet of graphene is made up of sp2 connected carbon atoms in a poly aromatic honeycomb crystal lattice (Yang et al., 2019; Wu et al., 2012).

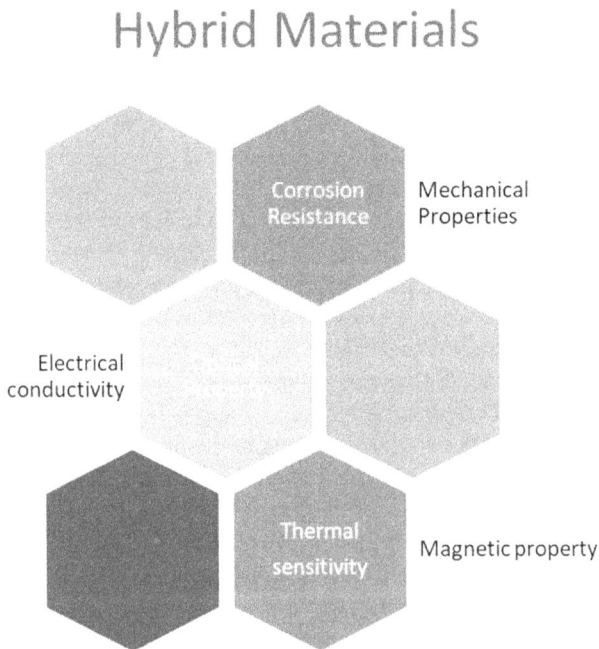

FIGURE 7.3   The general significance of hybrid material in various fields.

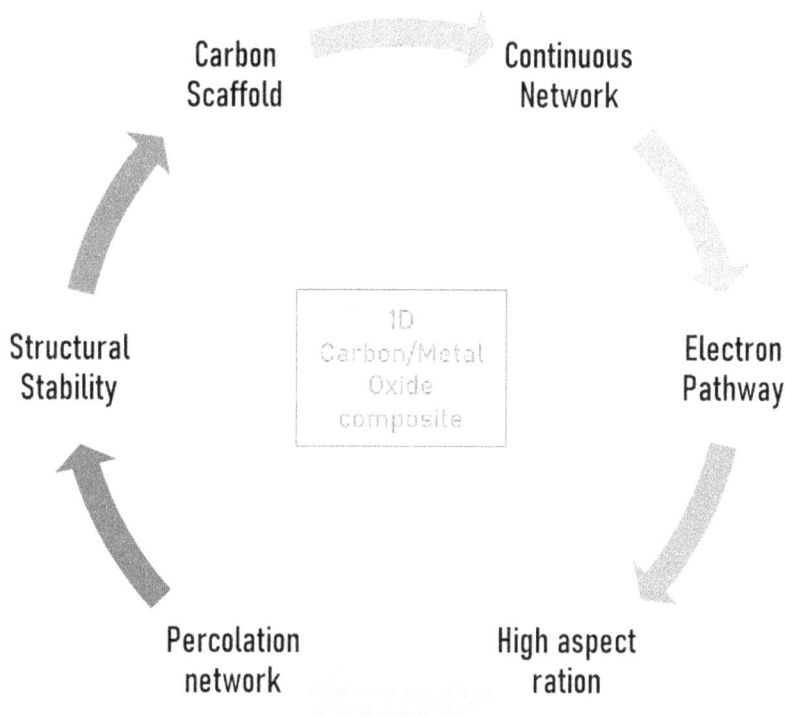

FIGURE 7.4 Overview of one-dimensional (1D) carbon/metal-oxide composite for energy storage device.

They're ideal for high-performance energy storage systems because of their superior physiochemical characteristics, capacity, and cycle capability. The high surface area, superior electrical conductivity, good flexibility, and favourable thermal and chemical stability are all characteristics of this material (Geim, 2009). The biggest downside of this material is permanent capacity loss due to re-stacking of graphene sheets. Because of this issue, the coulombic efficiency is also lowered. The re-stacking is caused by the van der Waals contact between the sheets, which reduces the surface area and hence the energy density (Yang et al., 2019). The goal of mixing graphene with metal oxide is to solve the challenges that each material has on its own. The compatibility and chemical functionality of graphene/metal oxide composites will be strengthened by graphene, allowing for easy processing of metal oxides, while metal oxide will give better capacity. Many different graphene-based composites are being examined and researched across the world for their potential use in supercapacitors. These include graphene and conductive polymers (e.g., polyaniline (Zhang et al., 2010), polypyrrole (Zhao et al., 2013), polythiophene (Alabadi et al., 2016)), as well as graphene/metal oxides (e.g., Mn_3O_4 (Chen et al., 2017), Co_3O_4 (Liao et al., 2015), SnO_2 (Velmurugan et al., 2016)), graphene/metal nitrides (e.g., VN (Balamurugan et al., 2016), Ni_3N (Yu et al., 2015)), graphene/sulphides (e.g.,

FeS$_2$ (Venkateshalu et al., 2018), MoS$_2$ (Zhou et al., 2017)), graphene/hydroxides (e.g., MnNi-LDH (Lee et al., 2018), NiCo-LDH (Qin et al., 2018)), and graphene/MXenes (Couly et al., 2018; Fan et al., 2018). In addition to these characteristics, the supercapacitor's minimal self-discharge demonstrated its suitability for use in future hybrid electric cars (Manoharan et al., 2021). In the first section, a quick description of activated carbon was provided. When it comes to implementing supercapacitors on a bigger scale, cost has always been a huge stumbling block. Even though there are several high-performance electrode materials available, they are still much too pricey for practical usage (Zhang et al., 2016). In the case of supercapacitors, one of the key research focuses has been on cost-effective materials. Multiple studies have looked at the possibility of activating and using carbon from agricultural waste as electrode materials in SCs, such as cassava peels (Ni et al., 2017) or apricot seeds (Augustyn et al., 2014). One of the key study topics has been the manufacture of electrodes for supercapacitors utilising renewable resource-based materials because of the lower environmental impact, low cost, and availability in nature (Figure 7.5). This material is made from plant parts such as shaddock peel, bamboo, and petals, as well as animal raw materials such as silk, crab shell, and honeycomb, and metabolic products such as starch and cellulose. All of these materials are readily available in our daily lives, which makes them appealing electrode materials (Park et al., 2019; Herou et al., 2019; Leguizamon et al., 2015; Zhang et al., 2009; Fic et al., 2018; Rolison et al., 2009; Yu et al., 2015; Gao et al., 2019; Lv et al., 2018; Tang et al., 2019; Guo et al., 2017; Zhu et al., 2019).

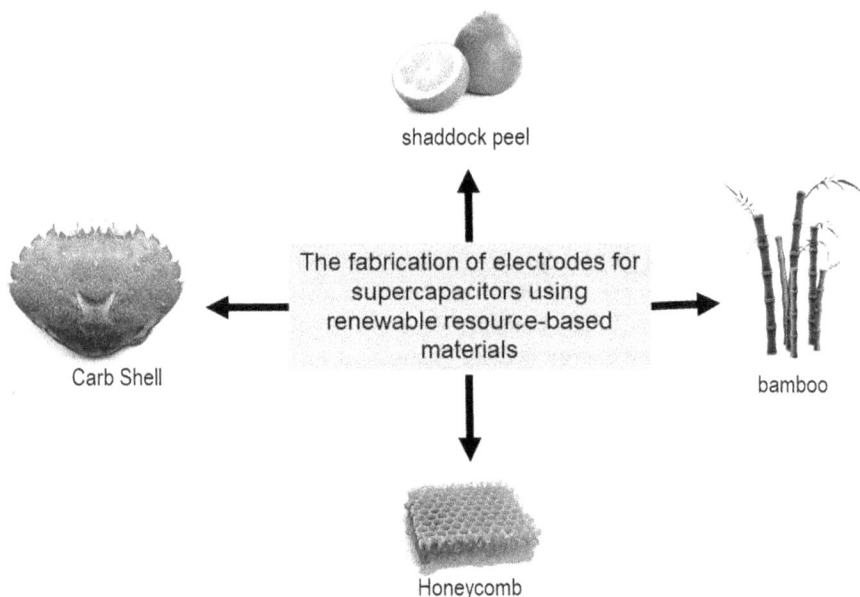

FIGURE 7.5 The fabrication of electrodes for supercapacitors using renewable resource-based materials.

Carbon nanotubes (CNTs) are one of the most promising materials for electrochemical energy storage applications because of their increased electrical conductivity and active surface area due to their one-dimensional structure. CNT electrodes in supercapacitors are made up of an interconnected mat of carbon nanotubes. The electrodes are made up of a linked network of open and accessible mesopores. This aids in the formation of a continuous distribution, which in turn aids in the more effective utilisation of the particular surface area (Pandolfo et al., 2006).

7.4 CONDUCTING POLYMERS

Conducting polymers (CPs) can enhance the specific capacitance of supercapacitors by enabling faradaic redox processes. They're most commonly found in composite materials used to make supercapacitor electrodes. The ions from the electrolyte move into and out of the polymer during the redox process, resulting in increased capacitance and lower cyclability (Ates, 2016). Polyaniline (PANI) is an example of this type of polymer that has been studied extensively. It is an appealing material because of its low cost, superior conductivity, and considerably easier production procedures.

7.5 TRANSITION METAL OXIDES

Although it has a greater specific capacitance and strong cycle stability, real-time applications are difficult due to its higher cost and environmental toxicity (Figure 7.6). Transition metal oxides were chosen for supercapacitors because of their excellent capacitive performance, low cost, and environmental friendliness. Their charge storage mechanism behaves like a pseudo capacitor. The key properties of TMO are its high intrinsic stability and challenging and difficult valence, which allows electrons and ions to be intercalated into the lattice of metallic elements (Jing et al., 2020; Jing

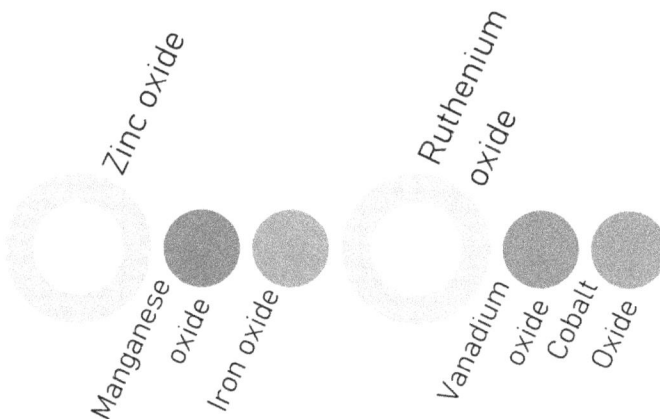

FIGURE 7.6 Different types of transition metal oxides.

et al., 2021; Li et al., 2017). MnO_2, NiCo, and other well-known materials have been investigated for use as electrode materials in supercapacitors.

7.6 TRANSITION METAL NITRIDES

Transition metal nitrides have attracted interest for supercapacitor applications because of their remarkable chemical bonding and intrinsic structures, which allow them to display superior conductivity and physiochemical characteristics. When it comes to using metal oxides and conducting polymers as pseudocapacitive materials, poor cycle stability, and low conductivity are important roadblocks. Metal nitrides emerge as a possible option in this situation (Figure 7.7). The material's characteristics are determined by the metal nitrogen bond formation. Group 1 and group 2-based nitrides display excellent ionic conductivity, but group 3 and 4-based nitrides are stiff and resistant to high temperatures due to their inherent covalent bonding (Chen, 1996; Balogun et al., 2014; Gage et al., 2016; Alexander et al., 2010; Liu et al., 1998; Choi et al., 2006; Cheng et al., 2011; Jiang et al., 2009; Sun et al., 2017; Nagae et al., 2001).

7.7 REDOX POLYMERS

Redox polymers are also an interesting material that might be employed as electrodes in the development of a long-lasting, low-cost supercapacitor. They are an appealing material because of their mechanical elasticity, greater processability, low cost, molecular diversity, and strong electrochemical activity (Figure 7.8). These characteristics make them a better contender for supercapacitor applications than inorganic materials such as metal oxides (Larcher and Tarascon, 2014). The reversible redox reaction of electrode materials in supercapacitors depends on this material not just to generate pseudocapacitance but also to accumulate energy at the electrical double layer. As a result, the supercapacitor has a substantially higher specific capacitance and energy density than electric double layer capacitors (EDLCs) (Esser et al., 2021). Regardless of the potential of redox polymers for the development of high-performance supercapacitors, one important stumbling block to overcome is

Transition metal nitrides

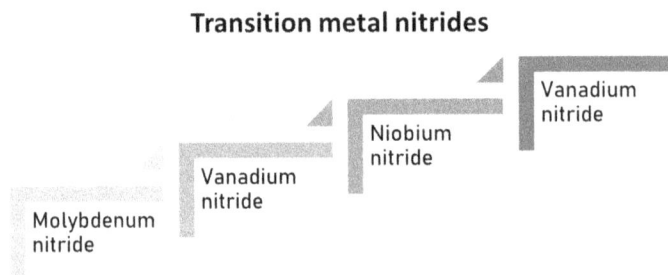

Vanadium nitride

Niobium nitride

Vanadium nitride

Molybdenum nitride

FIGURE 7.7   Types of transition metal nitrides.

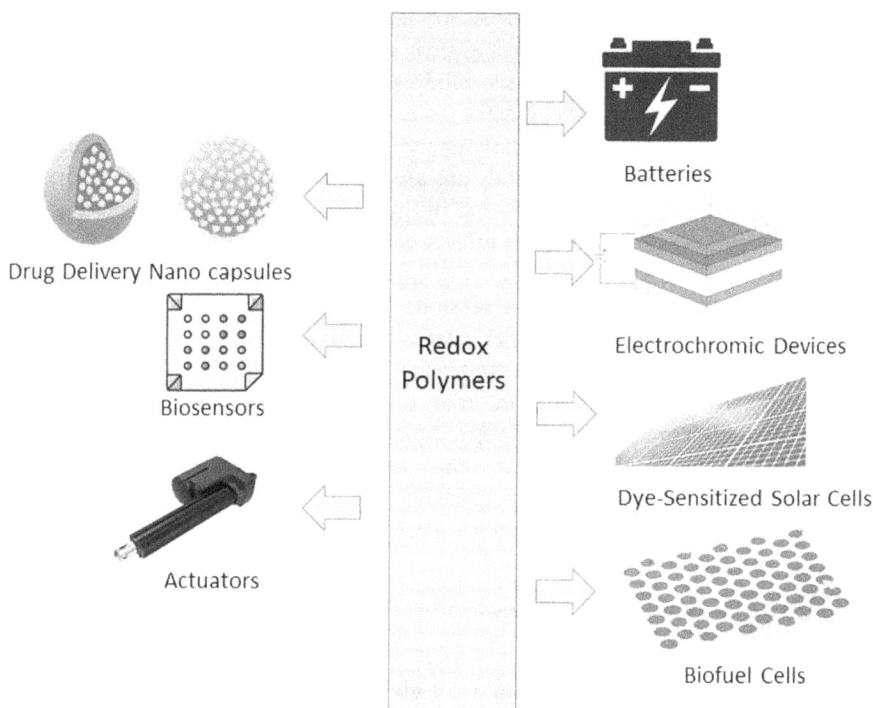

FIGURE 7.8 Application of redox polymers in the various aspects of bionanotechnology.

their low electrical conductance. Because of the limited electrical conductivity, active areas may be neglected, resulting in poor processability. Several researchers have been working to circumvent this problem, and some promising investigations have shown that the low electrical conductivity may be solved via clever molecular design (Meng et al., 2017; Zhang et al., 2016; Zhang et al., 2015; Zhang et al., 2021).

7.8 CONCLUSIONS

The demand for electric hybrid and electric vehicles has grown in response to growing pollution and other environmental issues in the automobile industry. The electrochemical energy storage technology, despite substantial improvements in recent decades, is lagging very far behind, and this is the HEV and EV's biggest disadvantage when compared to traditional fuel-based cars. This is where the use of supercapacitors in conjunction with a battery system becomes useful. This coupled system, known as a hybrid capacitor, has lately gotten a lot of attention for its promising outcomes in energy storage technology. When a battery is unable to give energy quickly, the supercapacitors could do so for a short time, and when a continuous energy flow is necessary, the battery can once again offer the required energy. Many manufacturers have already begun to increase their investments in supercapacitor technology as demand in the automotive and other sectors grows. Supercapacitors have become a potential option in the energy market as their energy density has increased dramatically over the years. Despite extensive study, high-performance supercapacitors are still in the early phases of development. To maximise the speed of supercapacitors, further research is needed, particularly in energy storage methods and suitable electrode design. The electrode materials are among the important issues in supercapacitors that need further research, and in this study, we briefly examined several electrode materials, described their effectiveness in supercapacitors, and evaluated their benefits and drawbacks.

However, rising energy needs necessitate more research into electrode materials for use in high-performance supercapacitors with high capacitance, higher cycle stability, and exceptional rate. The study should also concentrate on the production parameters and material qualities in order to synthesise the best electrode material that will aid in the improvement of supercapacitors and meeting energy needs.

This research clearly demonstrates that supercapacitors have become an established energy storage solution, since recent advancements in electrode materials have led to high performance, and they will play an essential part in the development of energy storage systems.

ACKNOWLEDGEMENTS

The author is grateful to the book editor for giving him the opportunity to contribute.

REFERENCES

Abdel Maksoud, M.I.A., Fahim, R.A., Shalan, A.E., AbdElkodous, M., Olojede, S.O., Osman, A.I., Farrell, C., Al-Muhtaseb, A.H., Awed, A.S., Ashour, A.H., Rooney, D.W. (2021). *Advanced Materials and Technologies for Supercapacitors Used in Energy Conversion and Storage: A Review.* Springer International Publishing, Berlin/Heidelberg, Germany, Volume 19.

Afif, A., Rahman, S.M., Tasfiah Azad, A., Zaini, J., Islam, M.A., Azad, A.K. (2019). Advanced materials and technologies for hybrid supercapacitors for energy storage—A review. *J. Energy Storage* 25, 100852.

Alabadi, A., Razzaque, S., Dong, Z., Wang, W., Tan, B. (2016). Graphene oxide-polythiophene derivative hybrid nanosheet for enhancing performance of supercapacitor. *J. Power Sources* 306, 241–247.

Alexander, A.M., Hargreaves, J.S.J. (2010). Alternative catalytic materials: Carbides, nitrides, phosphides and amorphous boron alloys. *Chem. Soc. Rev.* 39(11), 4388–4401.

Ates, M. (2016). Graphene and its nanocomposites used as an active materials for supercapacitors. *J. Solid State Electrochem.*, 20(6), 1509–1526.

Augustyn, V., Simon, P., Dunn, B. (2014). Pseudocapacitive oxide materials for high-rate electrochemical energy storage. *Energy Environ. Sci.* 7(5), 1597–1614.

Balamurugan, J., Karthikeyan, G., Thanh, T.D., Kim, N.H., Lee, J.H. (2016). Facile synthesis of vanadium nitride/nitrogen-doped graphene composite as stable high performance anode materials for supercapacitors. *J. Power Sources* 308, 149–157.

Balasubramaniam, S., Mohanty, A., Balasingam, S.K., Kim, S.J., Ramadoss, A. (2020). Comprehensive insight into the mechanism, material selection and performance evaluation of Supercapatteries. *Nano Micro Lett.* 2(1), 85.

Balogun, M.S., Yu, M., Li, C., Zhai, T., Liu, Y., Lu, X., Tong, Y. (2014). Facile synthesis of titanium nitride nanowires on carbon fabric for flexible and high-rate lithium ion batteries. *J. Mater. Chem. A* 2(28), 10825–10829.

Béguin, F., Presser, V., Balducci, A., Frackowiak, E. (2014). Carbons and electrolytes for advanced supercapacitors. *Adv. Mater.* 26(14), 2219–2251.

Brousse, T., Bélanger, D., Chiba, K., Egashira, M., Favier, F., Long, J., Miller, J.R., Morita, M., Naoi, K., Simon, P. (2017). Materials for electrochemical capacitors. In: *Springer Handbook of Electrochemical Energy*. Ed: Cornelia Breitkopf, Karen Swider-Lyons, Springer, Berlin/Heidelberg, Germany, pp. 495–561.

Cai, W., Lai, T., Lai, J., Xie, H., Ouyang, L., Ye, J., Yu, C. (2016). Transition metal sulfides grown on graphene fibersfor wearable asymmetric supercapacitors with high volumetriccapacitance and high energy density. *Sci. Rep.* 6, 26890.

Chen, J.G. (1996). Carbide and nitride overlayers on early transition metal surfaces: Preparation, characterization, and reactivities. *Chem. Rev.* 96(4), 1477–1498.

Chen, S., Wang, L., Huang, M., Kang, L., Lei, Z., Xu, H., Shi, F., Liu, Z.H. (2017). Reduced graphene oxide/Mn3O4 nanocrystals hybrid fiber for flexible all-solid-state supercapacitor with excellent volumetric energy density. *Electrochim. Acta* 242, 10–18.

Cheng, F., He, C., Shua, D., Chen, H., Zhang, J., Tang, S., Finlow, D.E. (2011). Preparation of nanocrystalline VN by the melamine reduction of V2O5 xerogel and its supercapacitive behavior. *Mater. Chem. Phys.* 131(1–2), 268–273.

Choi, D., Blomgren, G.E., Kumta, P.N. (2006). Fast and reversible surface redox reaction in nanocrystalline vanadium nitride supercapacitors. *Adv. Mater.* 18(9), 1178–1182.

Couly, C., Alhabeb, M., Van Aken, K.L., Kurra, N., Gomes, L., Navarro-Suárez, A.M., Anasori, B., Alshareef, H.N., Gogotsi, Y. (2018). Asymmetric flexible MXene-reduced graphene oxide micro-supercapacitor. *Adv. Electron. Mater.* 4(1), 1700339.

Esser, B., Dolhem, F., Becuwe, M., Poizot, P., Vlad, A., Brandell, D. (2021). A perspective on organic electrode materials and technologiesfor next generation batteries. *J. Power Sources* 482, 228814.

Fan, Z., Wang, Y., Xie, Z., Wang, D., Yuan, Y., Kang, H., Su, B., Cheng, Z., Liu, Y. (2018). ModifiedMXene/Holey graphene films for advanced supercapacitor electrodes with superior energy storage. *Adv. Sci.* 5(10), 1800750.

Fic, K., Platek, A., Piwek, J., Frackowiak, E. (2018). Sustainable materials for electrochemical capacitors. *Mater. Today* 21(4), 437–454.

Gage, S.H., Trewyn, B.G., Ciobanu, C.V., Pylypenko, S., Richards, R.M. (2016). Synthetic advancements and catalytic applications of nickel nitride. *Catal. Sci. Technol.* 6(12), 4059–4076.

Gao, S., Tang, G., Hua, D., Xiong, R., Han, J., Jiang, S., Zhang, Q., Huang, C. (2019). Stimuli-responsive bio-based polymeric systems and their applications. *J. Mater. Chem. B* 7(5), 709–729.

Geim, A.K. (2009). Graphene: Status and prospects. *Science* 324(5934), 1530–1534.

González, A., Goikolea, E., Barrena, J.A., Mysyk, R. (2016). Review on supercapacitors: Technologies and materials. *Renew. Sustain. Energy Rev.* 58, 1189–1206.

Guo, Z., Tang, G., Zhou, Y., Shuwu, L., Hou, H., Chen, Z., Chen, J., Hu, C., Wang, F., De Smedt, S.C., Xiong, R., Huang, C. (2017). Fabrication of sustained-release CA-PU coaxial electrospun fiber membranes for plant grafting application. *Carbohydr. Polym.* 169, 198–205.

Herou, S., Ribadeneyra, M.C., Madhu, R., Araullo-Peters, V., Jensen, A., Schlee, P., Titirici, M. (2019). Ordered mesoporous carbons from lignin: A new class of biobased electrodes for supercapacitors. *Green Chem.* 21(3), 550–559.

Iro, Z.S. (2016). A brief review on electrode materials for supercapacitor. *Int. J. Electrochem. Sci.* 11, 10628–10643.

Jiang, Q.W., Li, G.R., Gao, X.P. (2009). Highly ordered TiN nanotube arrays as counter electrodes for dye-sensitized solar cells. *Chem. Commun.*, 44, 6720–6722.

Jing, C., Dong, B., Zhang, Y. (2020). Chemical modifications of layered double hydroxides in the supercapacitor. *Energy Environ. Mater.* 3(3), 346–379.

Jing, C., Liu, X.D., Li, K., Liu, X., Dong, B., Dong, F., Zhang, Y. (2021). The pseudocapacitance mechanism of graphene/CoAl LDH and its derivatives: Are all the modifications beneficial? *J. Energy Chem.* 52, 218–227.

Jing, C., Song, X., Li, K., Zhang, Y., Liu, X., Dong, B., Dong, F., Zhao, S., Yao, H., Zhang, Y. (2020). Optimizing the rate capability of nickel cobalt phosphide nanowires on graphene oxide by the outer/inter-component synergistic effects. *J. Mater. Chem.* A 8(4), 1697–1708.

Kouchachvili, L., Yaïci, W., Entchev, E. (2018). Hybrid battery/supercapacitor energy storage system for the electric vehicles. *J. Power Sources* 374, 237–248.

Larcher, D., Tarascon, J. (2014). Towards greener and more sustainable batteries for electrical energy storage. *Nat. Publ. Gr.* 7, 19–29.

Lee, I., Jeong, G.H., An, S., Kim, S.W., Yoon, S. (2018). Facile synthesis of 3D MnNi-layered double hydroxides (LDH)/graphene composites from directly graphites for pseudocapacitor and their electrochemical analysis. *Appl. Surf. Sci.* 429, 196–202.

Leguizamon, S., Díaz-Orellana, K.P., Velez, J., Thies, M.C., Roberts, M.E. (2015). High charge-capacity polymer electrodes comprising alkali lignin from the Kraft process. *J. Mater. Chem.* A 3(21), 11330–11339.

Li, X., Du, D., Zhang, Y., Xing, W., Xue, Q., Yan, Z. (2017). Layered double hydroxides toward high-performance supercapacitors. *J. Mater. Chem.* A 5(30), 15460–15485.

Liao, Q., Li, N., Jin, S., Yang, G., Wang, C. (2015). All-solid-state symmetric supercapacitor based on Co3O4 nanoparticles on vertically aligned graphene. *ACS Nano* 9(5), 5310–5317.

Liu, T.C., Pell, W.G., Conway, B.E., Roberson, S.L. (1998). Behavior of molybdenum nitrides as materials for electrochemical capacitors: Comparison with ruthenium oxide. *J. Electrochem. Soc.* 145(6), 1882–1888.

Lv, D., Wang, R., Tang, G., Mou, Z., Lei, J., Han, J., De Smedt, S., Xiong, R., Huang, C. (2019). Ecofriendly electrospun membranes loaded with visible-light-responding nanoparticles for multifunctional usages: Highly efficient air filtration, dye scavenging, and bactericidal activity. *ACS Appl. Mater. Interfaces* 11(13), 12880–12889.

Manoharan, S., Krishnamoorthy, K., Sathyaseelan, A., Kim, S.J. (2021). High-power graphene supercapacitors for the effective storage of regenerative energy during the braking and deceleration process in electric vehicles. *Mater. Chem. Front.* 5(16), 6200–6211.

Masala, O., Seshadri, R. (2004). Synthesis routes for large volumes of nanoparticles. *Annu. Rev. Mater. Res.* 34(1), 41–81.

Meng, Q., Cai, K., Chen, Y., Chen, L. (2017). Research progress on conducting polymer based supercapacitor electrode materials. *Nanoenergy* 36, 268–285.

Miller, E.E., Hua, Y., Tezel, F.H. (2018). Materials for energy storage: Review of electrode materials and methods of increasing capacitance for supercapacitors. *J. Energy Storage* 20, 30–40.

Miller, J.R., Simon, P. (2008). Electrochemical capacitors for energy management. *Science* 321(5889), 651–652.

Nagae, M., Yoshio, T., Takemoto, Y., Takada, J. (2001). Microstructure of a molybdenum nitride layer formed by nitriding molybdenum metal. *J. Am. Ceram. Soc.* 84(5), 1175–1177.

Nan, H.S., Hu, X., Tian, H. (2019). Recent advances in perovskite oxides for anionintercalationsupercapacitor: A review. *Mater. Sci. Semicond. Process.* 94, 35–50.

Ni, L., Zhang, W., Wu, Z., Sun, C., Cai, Y., Yang, G., Chen, M., Piao, Y., Diao, G. (2017). Supramolecular assembled three-dimensionalgraphene hybrids: Synthesis and applications in supercapacitors. *Appl. Surf. Sci.* 396, 412–420.

Novoselov, K.S., Geim, A.K., Morozov, S.V., Jiang, D., Katsnelson, M.I., Grigorieva, I.V., Dubonos, S.V., Firsov, A.A. (2005). Twodimensionalgas of massless Dirac fermions in graphene. *Nature* 438(7065), 197–200.

Novoselov, K.S., Geim, A.K., Morozov, S.V., Jiang, D., Zhang, Y., Dubonos, S.V., Grigorieva, I.V., Firsov, A.A. (2004). Electric field in atomically thin carbon films. *Science* 306(5696), 666–669.

Pandolfo, A.G., Hollenkamp, A.F. (2006). Carbon properties and their role in supercapacitors. *J. Power Sources* 157(1), 11–27.

Park, J.H., Rana, H.H., Lee, J.Y., Park, H.S. (2019). Renewable flexible supercapacitors based on all-lignin-based hydrogel electrolytes and nanofiber electrodes. *J. Mater. Chem. A* 7(28), 16962–16968.

Pope, M.A. (2013). Supercapacitor electrodes produced through evaporative consolidation of graphene oxide-water-ionic liquidgels. *J. Electrochem. Soc.* 160, A1653–A1660.

Qin, K., Wang, L., Wen, S., Diao, L., Liu, P., Li, J., Ma, L., Shi, C., Zhong, C., Hu, W., Liu, E., Zhao, N. (2018). Designed synthesis of NiCo-LDH and derived sulfide on heteroatom-doped edge-enriched 3D rivet graphene films for high-performance asymmetric supercapacitor and efficient OER. *J. Mater. Chem. A* 6(17), 8109–8119.

Raj, B., Padhy, A.K., Basu, S., Mohapatra, M. (2020). Review—Futuristic direction for R&D challenges to develop 2D advanced materials based supercapacitors. *J. Electrochem. Soc.* 167(13), 136501.

Rolison, D.R., Long, J.W., Lytle, J.C., Fischer, A.E., Rhodes, C.P., Mc Evoy, T.M., Bourg, M.E., Lubers, A.M. (2009). Multifunctional 3D nanoarchitectures for energy storage and conversion. *Chem. Soc. Rev.* 38(1), 226–252.

Sharma, K., Arora, A., Tripathi, S.K. (2019). Review of supercapacitors: Materials and devices. *J. Energy Storage* 21, 801–825.

Sun, P., Lin, R., Wang, Z., Qiu, M., Chai, Z., Zhang, B., Meng, H., Tan, S., Zhao, C., Mai, W. (2017). Rational design of carbon shell endowsTiN@C nanotube based fiber supercapacitors with significantly enhanced mechanical stability and electrochemical performance. *Nano Energy* 31, 432–440.

Tang, G., Xiong, R., Lv, D., Xu, R.X., Braeckmans, K., Huang, C., De Smedt, S.C. (2019). Gas-shearing fabrication of multicompartmental microspheres: A one-step and oil-free approach. *Adv. Sci.* 6(9), 1–9.

Velmurugan, V., Srinivasarao, U., Ramachandran, R., Saranya, M., Grace, A.N. (2016). Synthesis of tin oxide/graphene (SnO2/G) nanocomposite and its electrochemical properties for supercapacitor applications. *Mater. Res. Bull.* 84, 145–151.

Venkateshalu, S., Goban Kumar, P., Kollu, P., Jeong, S.K., Grace, A.N. (2018). Solvothermal synthesis and electrochemical properties of phase pure pyrite FeS2 for supercapacitor applications. *Electrochim. Acta* 290, 378–389.

Wang, C., Song, Z., Shi, P., Lv, L., Wan, H., Tao, L., Zhang, J., Wang, H., Wang, H. (2021). High-rate transition metal-based cathode materials for battery-supercapacitor hybrid devices. *Nanoscale Adv.* 3(18), 5222–5239.

Wu, Z.S., Zhou, G., Yin, L.C., Ren, W., Li, F., Cheng, H.M. (2012). Graphene/metal oxide composite electrode materials for energy storage. *Nano Energy* 1(1), 107–131.

Xie, L., Sun, G., Su, F., Guo, X., Kong, Q., Li, X., Huang, X., Wan, L., Song, W., Li, K., Lv, C., Chen, C. (2016). Hierarchical porous carbon microtubesderived from willow catkins for supercapacitor applications. *J. Mater. Chem. A* 4(5), 1637–1646.

Xu, Y., Shi, G., Duan, X. (2015). Self-assembled three-dimensional graphene macrostructures: Synthesis and applications in supercapacitors. *Acc. Chem. Res.* 48(6), 1666–1675.

Yang, Z., Tian, J., Yin, Z., Cui, C., Qian, W., Wei, F. (2019). Carbon nanotube- and graphene-based nanomaterials and applications in high-voltage supercapacitor: A review. *Carbon N. Y.* 141, 467–480.

Yu, Y., Gao, W., Shen, Z., Zheng, Q., Wu, H., Wang, X., Song, W., Ding, K. (2015). A novel Ni3N/graphene nanocomposite as supercapacitor electrode material with high capacitance and energy density. *J. Mater. Chem. A* 3(32), 16633–16641.

Yu, Z., Tetard, L., Zhai, L., Thomas, J. (2015). Supercapacitor electrode materials: Nanostructures from 0 to 3 dimensions. *Energy Environ. Sci.* 8(3), 702–730.

Zhang, K., Zhang, L.L., Zhao, X.S., Wu, J. (2010). Graphene/polyaniline nanofiber composites as supercapacitor electrodes. *Chem. Mater.* 22(4), 1392–1401.

Zhang, X., Xiao, Z., Liu, X., Mei, P., Yang, Y. (2021). Redox-active polymers as organic electrode materials for sustainable supercapacitors. *Renew. Sustain. Energy Rev.* 147, 111247.

Zhang, Y., Cheng, T., Wang, Y., Lai, W., Pang, H., Huang, W. (2016). A simple approach to boost capacitance: Flexible supercapacitors based on manganese oxides@MOFs via chemically induced in situ self-transformation. *Adv. Mater.* 28(26), 5242–5258.

Zhang, Y., Feng, H., Wu, X., Wang, L., Zhang, A., Xia, T., Dong, H., Li, X., Zhang, L. (2009). Progress of electrochemical capacitor electrode materials: A review. *Int. J. Hydr. Energy* 34(11), 4889–4899.

Zhang, Y., Wen, G., Gao, P., Bi, S., Tang, X., Wang, D. (2016). High-performance supercapacitor of macroscopic graphene hydrogels bypartial reduction and nitrogen doping of graphene oxide. *Electrochim. Acta* 221, 167–176.

Zhang, Y.Z., Wang, Y., Cheng, T., Lai, W.Y., Pang, H., Huang, W. (2015). Flexible supercapacitors based on paper substrates: A new paradigm for low-cost energy storage. *Chem. Soc. Rev.* 44(15), 5181–5199.

Zhao, Y., Liu, J., Hu, Y., Cheng, H., Hu, C., Jiang, C., Jiang, L., Cao, A., Qu, L. (2013). Highly compression-tolerant supercapacitor based on polypyrrole-mediated graphene foam electrodes. *Adv. Mater.* 25(4), 591–595.

Zhou, R., Han, C., Wang, X. (2017). Hierarchical MoS2-coated three-dimensional graphene network for enhanced supercapacitor performances. *J. Power Sources* 352, 99–110.

Zhu, M., Xiong, R., Huang, C. (2019). Bio-based and photocrosslinkedelectrospun antibacterial nanofibrous membranes for air filtration. *Carbohydr. Polym.* 205, 55–62.

8 Tailored Bio-Nanomaterial Electrodes for Bioelectricity Generation

Md. Merajul Islam

CONTENTS

8.1 INTRODUCTION

Bio-nanomaterials with biological organism (mostly enzymes or related biomolecules) on electrode surfaces have piqued the interest of researchers all over the world who are working on biofuel cells, sensors/biosensors, and bioprocessing based on biomolecules as catalysts (Walcarius et al., 2013; Minteer et al., 2012). The biological fuel cell (Shukla et al., 2004; Calabrese et al., 2004; Bullen et al., 2006; Cooney et al., 2008; Heller et al., 2004; Willner et al., 2009) is a type of fuel cell in which a biocatalyst is used to convert chemical reactions directly into electrical energy. The nature of the electrodes (cathode and anode) in any fuel cell has a considerable impact on the generation of electricity. Biocatalysts play a crucial part in

DOI: 10.1201/9781003316374-8

bioelectricity generation using microbial fuel cells (MFC) or enzyme-based biofuel cells, where the cathode and/or anode can be made up of biocatalysts. Microbial fuel cells are those in which the biocatalyst is a living cell, whereas enzymatic biofuel cells are those in which the biocatalyst is made up of subcellular biological components (Walcarius et al., 2013).

The MFC is the first biofuel cell to use microbes to oxidize the fuel at the anode (generally organic matter), after the invention of the enzymatic fuel cell in 1960, which uses mediators and oxidoreductase enzymes to catalyze the oxidation of biomolecules at the anode (Yahiro et al., 1964). Microbes/enzymes-based cathodes have also been used in a variety of applications in recent decades, including wastewater treatment, underwater electricity, fuel generation, oxygen or peroxide reduction in water, and so on (Balat et al., 2010; Du et al., 2007; Logan, 2010; Nevin et al., 2010; Rabaey et al., 2010; Brunel et al., 2007; Pizzariello et al., 2002).

8.2 CARBON-BASED BIO-NANOMATERIALS AS ELECTROCATALYSTS

Sustainable energy conversion technologies and sophisticated biorelated applications rely heavily on catalysis (Hu et al., 2019; Friend et al., 2017; Kuchler et al., 2016; Zhao et al., 2016). Pt, Rd, Ru, Ir, Pd, and other novel metals or their compounds (usually oxides) are utilized as catalysts, but their scarcity on the planet and high cost limit their use (Dai et al., 2015). The development of plentiful and cost-effective materials is a pressing necessity, and research into such materials is taking place all across the world. To replace innovative metal-based catalysts, carbon and carbon-based compounds are the only earth-abundant, ecofriendly, cost-effective, and biocompatible materials (Dai et al., 2015; Dai, 2013; Liu et al., 2017). The incorporation of numerous co-existing active sites and multiple catalytic functions is conceivable in carbon-based materials, which is challenging to do with innovative metal-based catalysts (Hu et al., 2019). Carbon-based electrodes are thus a versatile material for bio-nanomaterial-based applications, such as bioelectricity generation.

Based on the configurations of the carbon atoms in carbon-based materials, they are classified as diamond, graphite, and amorphous carbon (Dai et al., 2015). C_{60} or fullerenes, carbon nanotubes (CNTs), and various types of graphene such as nanosheets, quantum dots, and nanoribbon have all recently been investigated, opening up a new vista in carbon material research and technology (Dai et al., 2015). These individual carbon units also serve as building blocks for three-dimensional (3D) entities with increased porousness, a large effective surface area, a variety of active sites, high ion diffusion and conductivity, and better mechanical strength as well (Hu et al., 2019; Dai et al., 2015; Shin et al., 2012; Du et al., 2011).

8.3 PROPERTIES OF TAILORED BIO-NANOMATERIAL ELECTRODES

Over time, new materials will be discovered, paving the way for improved biofuel cell performance. Using bio-nanomaterial customized electrodes, a considerable boost in biofuel cell performance can be seen. The improvement of charge-transfer from the biocatalyst to the electrode surface, the development of immobilized patterns of biocatalysts, materials with enhanced active surface area and electronic conductivity, facile mass transport, and reducing the high overpotentials commonly observed for various bioelectrochemical systems are the main inspirations for developing such electrode morphologies in biofuel cells (Walcarius et al., 2013; Minteer et al., 2012). At the nanoscale level, there are two main ways to change the properties of electrode surfaces: first, changing the inherent properties of the nanomaterials used, and second, tuning the architecture of the electrode interface (where electron transfer and related phenomena occur), or a combination of the two (Walcarius et al., 2013). As a result, several nanoscale adjustments ensure that the biofuel cell's overall performance improves. Various forms of carbon-based nanomaterials such as nanoparticles, nanotubes, graphene-based materials, especially designed functional polymers, and bio-functionalized electrodes are among the nanomaterials to be studied in depth.

8.3.1   BIO-NANOMATERIALS THAT IMPROVE THE ELECTRON TRANSFER

Electron transport is one of the fundamental characteristics that determine how much bioelectricity can be generated using a biological fuel cell. The functioning of a biological fuel cell, or any electrochemical device in general, is considered to be improved if the electron transport is better. In a biological fuel cell, electron transfer occurs via two mechanisms: direct electron transfer (DET) and mediated electron transfer (MET) (Minteer et al., 2012; Cooney et al., 2008). In DET, the electron transfers directly from the electrode surface to the biocatalyst or vice-versa, while in MET there is a shuttling of electrons between the biocatalysts and the electrode surface through the diffusible redox species (Figure 8.1) (Cooney et al., 2008).

In this method, mediator species interact directly with the biocatalyst, converting it to either an oxidized or reduced state before transporting electrons to or from the electrode surface. The mediator species must be stable and exhibit selectivity for

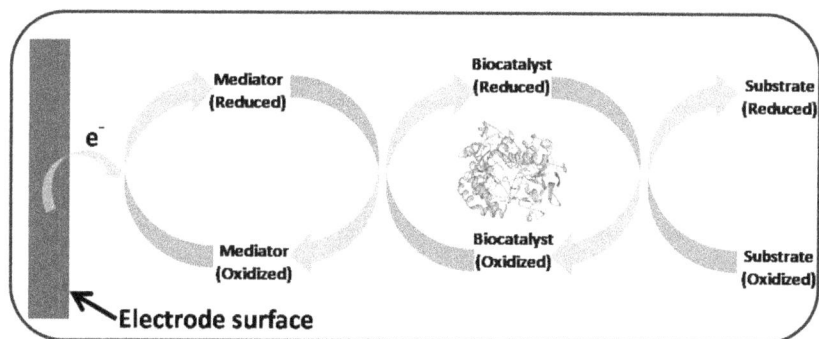

FIGURE 8.1   Schematic of a mediated electron transfer-based bioelectrode (Cooney et al., 2008).

both oxidized and reduced versions of the species. If the mediator's redox chemistry is reversible and has a modest overpotential, MET can be an efficient process (Cooney et al., 2008). MET has a number of drawbacks, including (Walcarius et al., 2013; Cooney et al., 2008): (i) lability: most mediators are labile in nature, limiting the time span of the species and affecting their selectivity; (ii) leaching: cells experience leaching as a result of continuous operation, which reduces the mediator's efficiency; (iii) overpotential: due to the presence of co-factors in biomolecules which act as a mediator and cause the increase in the half-cell overpotential; and (iv) Crossover: One of the most common difficulties with all fuel cells is crossover. To minimize the crossover problem, the fuel cell's two half-cells must be properly spaced. The above-mentioned constraints are more noticeable in microbial fuel cells than enzymatic fuel cells.

The advent of such DET-capable biocatalysts has resulted in a paradigm change in biofuel cell technology. DET eliminates most of the constraints of MET by allowing electrons to tunnel between the electrode and the biocatalyst without the use of a shuttle species (Cooney et al., 2008). Designing DET-capable bioelectrodes, on the other hand, is difficult. The redox centers of the enzyme must be close enough to the bioelectrode to facilitate the DET in enzymatic fuel cells. The non-conducting shell of the enzyme structure must not shield the redox center since this inhibits electron transmission. It's difficult to get the enzyme structure to be oriented in this way. However, the current innovation and development of new sophisticated nanomaterials provide the path for the creation of such a bio-nano interface that allows for precise control over the redox center in order to establish a favorable DET orientation (Walcarius et al., 2013). With optimal orientation of the redox center and bioelectrode, such a bio-nano interface provides increased electron transfer with lower charge-transfer resistance between biocatalyst and electrode, facilitating charge-transfer (Bard et al., 2000). As a result, proper protein (or biocatalyst) orientation combined with enhanced bioelectrode architecture can significantly improve the efficiency of DET operations, resulting in a more efficient biological fuel cell technology (Arechederra et al., 2009).

Several functionalized bioelectrode nanomaterials with excellent conductivity have recently been produced (Kim et al., 2006; Gorby et al., 2006). The foundation for developing a bioelectrode with suitable architecture for microbial and/or enzymatic biofuel cells, on the other hand, is completely different. Electrodes with a macroporous structure (or morphology with increased effective surface area) are desirable in microbial fuel cells because they facilitate the adhesion of a large population of cells or microorganisms. Microorganisms, on the other hand, can create nanowires (known as pilli) that provide a better environment for electron transfer via DET between the microbes and the electrode surface (Gorby et al., 2006). Electrodes with correct protein connections and alignment with the redox center at the nanoscale level are desired in enzymatic fuel cells. Redox biocatalysts have an intimate link with the material at the nanoscale, which reduces the electron tunneling distance and improves charge-transfer (Minteer et al., 2012).

Carbon black, CNTs, and graphene or graphene-based materials are some of the most common carbon-based electrode materials which generate a favorable bio-nano interface with improved charge-transfer efficiency, desirable attributes of biofuel cells (Minteer et al., 2012).

8.3.1.1 Carbon Black Nanomaterials (CBN)

Those nanomaterials are suitable for the electrodes used in biological fuel cells which demonstrate high surface area, porosity, and conductivity because these properties are favorable for a biological interface (Minteer et al., 2012). These properties of CBN make it suitable to be used in enzyme-functionalized electrodes. Due to hydrophilic-hydrophobic interactions in biomolecules, proteins in particular adsorb onto CBN which facilitates DET between the electrode and biocatalyst if the redox centers of the enzymes are oriented preferentially (Tominaga et al., 2006). Several examples are available in this regard. Ma et al. (Ma et al., 2007) reported hemoglobin on CBN in powdered form which is properly immobilized with diameters of 30–100 nm. He demonstrated the direct redox reaction of the heme with iron through voltammetry. Kano and co-workers (Kontani et al., 2009) at Kyoto University developed a biocatalyst (cuperous oxidase (CuO) + Ketjen black) that uses oxygen in the air as a terminal electron acceptor and generates a small cathodic current (Kontani et al., 2009). The Ketjen black incorporated electrodes increase the current density (~4 mA/cm^2) for cuprous oxidase on highly ordered pyrolytic graphite electrodes (HOPGE) to ~20 mA/cm^2. Similarly, when Vulcan® XC-72, a CBN, is added to PQQ-dependent alcohol dehydrogenase or PQQ-dependent glucose dehydrogenase fuel cells, an increase of five- to ten-fold in power density is observed (Arechederra et al., 2009).

CBN can be used as composites and emulsion with Teflon® as well which is used to fabricate a kind of material matrix with well-suited hydrophilic and hydrophobic interactions. This creates tri-phase "electrolyte-carbon-air" interface and acts as a gas diffusion electrode (GDE) (Bidault et al., 2009; Sheintenberg et al., 1982). This composite or emulsion of CBN with Teflon provides controllable attachment on several traditional electrode surfaces as well which can be used in biological fuel cells (Bidault et al., 2009). The interplay between CBN, GDE, and the Ketjen black-based architectures further enhances the capability as an electrode surface manipulated into a biological fuel cell (Gupta et al., 2011a; Shleev et al., 2010).

8.3.1.2 Carbon Nanotubes (CNTs)

CNTs are one of the most suitable architectures to combine with a bio-nano interface. Their inherent properties provide enhanced conductivity at the bio-nano interface which is used in several applications such as in sensors, bio-electronics, and fuel cell technology (Gong et al., 2005; Hu et al., 2009; Jia et al., 2008). As we know, there are two types of CNTs viz. multi-walled carbon nanotubes (MWNTs) and single-walled carbon nanotubes (SWNTs). These carbon nanotubes have unique chemical and physical properties and possess direct charge/electronic interplay in the presence of the redox centers of the biocatalysts. Figure 8.2 shows the simple two-dimensional interactions of carbon nanotubes with the redox centers of the enzymes (Minteer et al., 2012).

The desired functional groups which act as redox centers such as carboxyl, amine, hydroxyl, etc., can be incorporated into the CNTs, providing additional sites to the

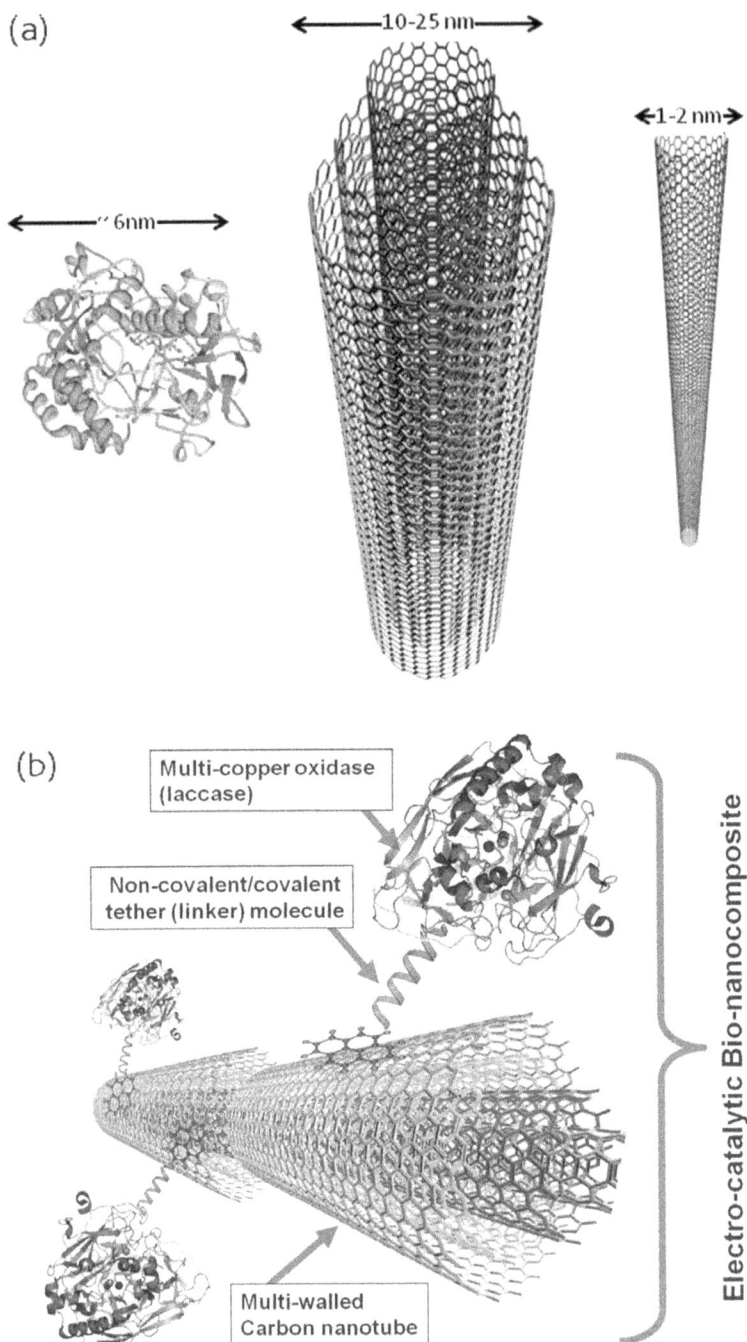

FIGURE 8.2   (a) A glucose oxidase molecule with typical multiwall and single wall carbon nanotubes. (b) Diagram of a bionanocomposite made with phenol oxidase and multiwalled carbon nanotubes (Minteer et al., 2012).

biocatalyst and increasing the conductivity. Generally, in MWCNTs, there are weak interactions between the redox centers of the biocatalyst and nanotubes due to the presence of multiple walls. But curved MWNTs are considered to be a flat surface and provide stronger interactions and attachment with the enzymes. This lets different parts of enzymes be put on MWNTs and makes it easier for ionic, van der Waals, hydrophilic-hydrophobic, and hydrophilic–hydrophobic interactions, as well as covalent interactions, between the tailored and untailored nanotubes (Jia et al., 2008). Considering the above dimensions of MWNTs, SWNTs offer better interactions and close proximity with the redox centers of the biocatalyst. Thus, further reducing the tunneling distance of electrons in the middle of the electrode and catalyst consequently increases the conductivity remarkably.

In the recent literature, there are numerous cases of CNT and/or CNT-based biocatalyst materials which are particularly beneficial for biological fuel cells (Arechederra et al., 2009; Higgins et al., 2011a; 2011b). CNTs are used by several researchers to enhance the effective surface area of bioelectrodes, conductivity, and porosity, and to promote the DET. For example, due to the physisorption of a redox protein on CNT, proteins clinging to the hydrophobic CNT surface mostly by van der Waals forces may take place. However, due to non-covalent interactions, DET, through chemical oxidation which produces functional groups such as carboxylic acid groups on the CNT surface, is one of the most frequent ways of functionalizing CNTs. Carbodiimide chemistry can then activate the carboxyl groups, forming an unstable ester that reacts with available nitrogen-based groups such as amino groups of the protein to produce covalent amide connections. The covalent bond promotes electronic connection and DET by stabilizing contacts and reducing the tunneling of electrons between the protein and exterior of the CNT. Vaze et al., using SWNT-based nanomaterials as electrodes and glucose oxidase (GOx) as a bioelectrocatalyst, achieved the expected half-cell redox potential for FAD/FADH2 (0.455V versus SCE) in which current densities are dependent on glucose concentration (Vaze et al., 2009). Later, Vashist et al. confirmed the DET between the enzyme and the electrode while the enzyme's catalytic function was preserved (Vashist et al., 2011). However, there are certain drawbacks to this method, as oxidation produces toxins and flaws in the CNTs that reduce the material's conductivity. Furthermore, the short covalent bond may cause steric restrictions on the structure of the protein that reduce its catalytic activity (Luo et al., 2006). CNTs have been coupled with metal colloids in experimental materials in order to make use of the features of the metal nanoparticles of each material for improved electrocatalysis (Wang et al., 2011).

8.3.1.3 Graphene Nanomaterials

Graphene is one of the most studied materials nowadays, and worldwide research is going on to make it technologically viable. Although it is still unclear whether it is a feasible material for making electrodes, preliminary research suggests that it has all the characteristics to be an electrochemically potential material, though it is less explored. Graphene, like CNTs, can have its surface functionalized in both covalent and non-covalent ways. If done appropriately, the functionalization does not affect the enhanced conductivity of virgin graphene (Malig et al., 2011). Thus, the combination of biocatalysts and graphene is gaining attention in the literature. Simple physisorption between GOx and a graphene-glassy carbon electrode is used and shows catalytic activity with the flavin adenosine nucleotide (FAD) co-factor at −454 mV versus SCE in one study (Wu et al., 2010). Chitosan, a biopolymer, is used as the dispersion nanomaterial, which includes graphene also to make thin film electrodes easier to manufacture. A biocatalyst replica, such as GOx, is used to illustrate the benefits of the hybrid nanocomposite. Chitosan-graphene thin film adsorbed on GOx facilitates DET, and its amperometric response to the glucose concentrations was 37.93 $\mu A\ mM^{-1}cm^{-2}$ compared to 7.36 $\mu A\ mM^{-1}cm^{-2}$ observed for chitosan/MWNT. When GOx was immobilized on graphene alone, the detection sensitivity increased two-fold using chitosan architecture (Kang et al., 2009). It has been reported elsewhere that the same method is used to blend chitosan with cytochrome C and graphene on a glassy-carbon electrode. Thus, DET was reported for nitric oxide reduction biocatalytically (Wu et al., 2010). Alongside a comparison of bioelectrodes based on graphene with SWNT-based, an oxygen reduction reaction is achieved when GOx is utilized as the anode and billirubin oxidase as the cathode. The manufactured graphene fuel cell has a SWNT design, double the current density, and three times the power density of an equivalent SWNT fuel cell (Liu et al., 2010). Parallel research has recently discovered that graphene is useful in the construction of MFC. Li and colleagues have reduced graphene to facilitate DET using the bacteria Shewanella sp. (Wang et al., 2011).

8.3.2 Materials That Improve the Immobilization and Stability of Biocatalysts

The effective alignment and proper contact between the electrode and biocatalyst surface are the essence of effectively using biomolecules in biofuel cells, as stated above. *Ex-situ* enzymes typically have a short lifespan, especially when the physiological conditions thrust the biomolecules towards the optimal activity. If the stability of the biocatalyst used is poor, this reduces power density and their lifetimes as well. Applications such as in bioelectronics where enzymes are dissolved in a solution at room temperature result in low activity, and such materials are required for catalysis at the electrode surface which can transport the electrons efficiently (Kim et al., 2006). Subsequently, maintaining the integrity of the enzyme is critical to its efficiency, which is commonly accomplished using several methods of enzyme immobilization (Osman et al., 2011). Immobilization allows biomolecules to be anchored in a way that preserves their inherent tertiary structure. Enzyme stabilization can also improve mass transfer kinetics and increase selectivity. Methods such as physical adsorption, for which electrostatic binding is responsible, provide conducting polymer matrices and/or covalent extension to functionalized polymers which are the most usual strategies of immobilization (Moehlenbrock et al., 2011). Although electrocatalytic activity can be maintained by physical adsorption, the power density is frequently low due to insufficient biocatalyst loading, and/or leaching is an issue that limits durability. Immobilization techniques based on covalent interaction, on the other hand, have better electrocatalytic properties but can sometimes obstruct protein conformation (Noll et al., 2011). Furthermore, anchoring functional groups on the enzyme should be avoided for catalysis; otherwise, inactivation of the enzyme will result.

Modifications are made to enzymes immobilized in hydrophobic pockets in micellar polymers. Nafion and Chitosan are two examples of materials that have been proven to be effective at stabilizing biocatalysts at the surface of electrodes and lengthening their useful lifetimes over time (Besic et al., 2011; Kim et al., 2011a; Tan et al., 2009) (Figure 8.3).

Enzymes immobilized in hydrophobic pockets, micellar polymers, are modified. One example is Nafion® and chitosan, which have been found to successfully stabilize biocatalysts at the surface of electrodes and enhance their operating lives of over few years (Besic et al., 2011).

Encapsulation during silica sol-gel production can stabilize a broad range of redox catalysts (Lim et al., 2007; Sarma et al., 2009; Wang, 1999). To increase the conductivity of a silica-based matrix, the co-immobilization of the conductive material, like CNTs, is used. Lysozyme, for example, is a cationic protein that catalyzes and guides the synthesis of silica onto a conductive electrode made of carbon paper. When CNTs with GOx are added to the reaction mixture, they produce a composite that is enclosed as silica forms with good catalytic activity (Ivnitski et al., 2008). Within the silica matrix, the CNTs form nanowires and construct an electrical connection between the biocatalyst and the electrode. This also increases the effective surface area of the electrode to a significant value. A biocatalyst can also catalyze the reduction of metal salt to generate particular metal-based structures, viz. gold nanoparticles. The reduction of gold (III) chloride by GOx results in the creation of size-controllable gold nanoparticles where the protein will be encapsulated in the metal structure. GOx/gold composite shows all the catalytic activity of protein with DET for the FAD co-factor and the electrode. This is evidenced by a voltammetric peak at 0.44V vs Ag|AgCl, with catalytic current observed in response to glucose, which increases non-linearly from 5 mM to 25 mM concentration of glucose (Luckarift et al., 2010). The co-factor in GOx enzyme is buried deep inside the structure, resulting in difficulty in direct interaction between the enzyme and electrode. These constraints can be eliminated using physical adhering of the FAD co-factor to the electrode surface. After this step, a suitable enzyme with no co-factor, known as

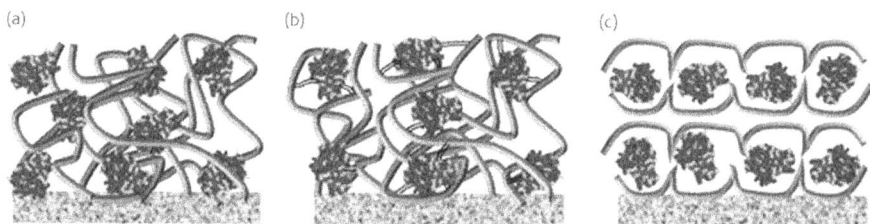

FIGURE 8.3 Enzymes that have been immobilized on an electrode surface: (a) adsorption to a polymer through physical means, (b) the black and white tethers demonstrate covalent bonding to a polymer, and (c) polymer micelles encapsulation (Minteer et al., 2012).

apoenzyme, is added. This ensures the proper attachment of FAD which guarantees that the biocatalyst has made contact with the electrode surface. The co-factor harboring is usually done using gold nanoparticles and/or CNTs which act as an electron bridge (Willner et al., 2006; Xiao et al., 2003). As an example, FAD is coupled with SWNT which is utilized to place the apoenzyme of GOx; this ensures better conductivity with the length of the CNTs (Fernando et al., 2004). Similarly, Ivnitski et al. showed that GOx is synergized with the CNT, and a DET is observed between the active site of the enzyme and MWNT produced on a Toray® carbon electrode (Ivnitski et al., 2006). A variety of CNTs as conductive material such as CNT paper, carbon nanofiber, and CNT in the form of gels have been reported. All these materials show a remarkable improvement in electron-transfer characteristics for the electrode catalysts (Yu et al., 2011; Hussein et al., 2011a; Ramasamy et al., 2010; Kim et al., 2011a). On CNT paper, for example, reducing the amount of oxygen at the cathode is done through simple physical adsorption (Hussein et al., 2011b). However, a bifunctional cross-linking agent such as 1-pyrenebutanoic acid and/or succinimidyl ester (PBSE) can be added that interacts with CNT and ensures a favorable orientation (Ramasamy et al., 2010). With PBSE laccase tethering, constant reduction currents and possible losses of 0.1 V are guaranteed. Buckygels (CNT gels), on the other hand, can be integrated with ionic liquids and CNTs form a composite material. This composite material with NAD(P)H electrocatalysts (such as methylene green) can be attached to favor the regeneration of the enzyme co-factor NAD(P)$^+$ at adequate overpotentials (Yu et al., 2011).

The emerging Os- or Ru-based complexes used as redox hydrogels are some of the most remarkable materials to be used to efficiently co-immobilize material and have significant applications in biosensors and biological fuel cell technology. This technical orientation contributed to the demonstration of biofuel cells' applications as implantable gadgets and their implications in diabetes care (Heller et al., 2010; Heller et al., 2006). Long-term stability is further achieved by attaching the hydrogel to the electrode's surface through amine and/or carboxylate groups (Boland et al., 2009a; 2009b). Both anodic and cathodic electrodes have been made with osmium-based redox hydrogels. Although hydrogels are often thought to be brittle, enzymes in hydrogel matrices have been found to have lives of more than 14 days. Microbial bioelectrocatalysis has also been done with redox hydrogels (Timur et al., 2007). Due to the self-immobilizer ability of microorganisms whose nanowires, called pilli, communicate with the electrode directly, these types of strategies of immobilization with electron mediation are less pronounced in microbial fuel cells (Gorby et al., 2006).

8.3.2.1 Some Immobilization Strategies

Immobilization techniques for immobilizing and/or stabilizing living entities at electrode surfaces come in a variety of forms. Adsorbing the biomolecule is the simplest technique, for example an enzyme, organelle, or microbe as a biocatalyst or antibodies or aptamers for recognition elements. However, the above plan of action does not expose the biomolecule to harsh chemical environments. The denaturation of some biomolecules at the electrode surfaces can take place; thus, the present methodology is useful for disposable or single-use applications only. Other problems such as leaching of the biomolecule from the electrode surface can take place; therefore, it is a temporary immobilization which suffers from stabilization problems.

Another typical way is to use cross-linking or self-assembled monolayer methods to attach the biocatalyst directly to the electrode surface (Fischback et al., 2012; Takeshi et al., 2005; Kim et al., 2011b; Bardea et al., 1997; Katz et al., 2001; Willner et al., 1998). Although the above two procedures are sometimes mixed together when talking about immobilization, actually the two are entirely different. In the cross-linking method several strategies are adopted, for example, on the cross-linker strength, the ability of the enzyme to be cross-linked to the electrode surface or the ability to be cross-linked into clusters or nanoparticles on the surface of the electrode. Wang and Kim invented a method in which the bio-functionalization of electrodes by the cross-linking of the biocatalyst into clusters or nanoparticles on the surface of the electrode is achieved (Kim et al., 2011b; Kwon et al., 2010; Kim et al., 2007b). The biocatalyst or enzyme loading becomes high enough using the above method; sometimes, a decrease in enzyme activity is observed due to cross-linking but it can be still increased if optimized properly. This approach commonly affects the enzyme's activity, but it considerably improves stability and reduces biomolecule leaking from the electrode surface. Producing a self-assembled monolayer of protein on the electrode surface is the other surface approach to increase the enzyme loading. Willner and Katz reported a remarkable method, in which redox molecules are linked to the electrode surface and the enzyme to the redox molecules, which is supposed to be one of the most useful strategies (Bardea et al., 1997; Katz et al., 2001). Although sometimes denaturation of the enzyme on these types of electrode surface may happen, it lowers the leaching and can show several effects on the enzyme function as well. This approach, on the other hand, is intriguing since it can be expanded and amalgamated with the nanomaterials mentioned above to create complicated electrode architectures.

Encapsulation and/or entrapment of the biological organism in a polymer matrix are the other two prevalent approaches. The biomolecule/biocatalyst is incorporated quite differently in these two techniques. The biomolecule is injected with the monomer and cross-linked into the polymer matrix directly on the electrode surface, resulting in entrapment (Meredith et al., 2011; Merchant et al., 2009). This method often prevents enzyme leaching because of the cross-linking and the fact that the enzyme is present during the difficult chemistry of polymer cross-linking, although it may negatively affect enzyme function. On the other hand, encapsulation permits a biomolecule to be implanted in a polymer matrix without chemically adhering the enzyme to it. Both strategies have their own benefits and drawbacks. An inert polymer (Klotzbach et al., 2008; Cooney et al., 2008; Moehlenbrock et al., 2012), an ion conducting polymer (Moore et al., 2004), a redox polymer (Meredith et al., 2011; Merchant et al., 2009; Barriere et al., 2006; Barton et al., 2001; Gallaway et al., 2009; Kavanagh et al., 2009; Mano et al., 2002; Stoica et al., 2009; Tasca et al., 2008), or a conducting polymer can be used (Gerard et al., 2002; Shin et al., 2009; Schuhmann 1995; Trojanowicz et al., 1995) for this purpose. The polymer can also be used in the form of composites containing nanoparticles or nanotubes (Ivnitski et al., 2008;

Higgins et al., 2011; Lau et al., 2008), or a composite of polymers (for example, Naon has been used in composites of poly (3,4- ethylenedioxythiophene)-polystyrenesulfo nic acid (PEDOT:PSS) to improve the stability of PEDOT:PSS and prevent swelling and the loss of immobilized biomolecules such as ascorbate oxidase (Wen et al., 2012)). Because the biomolecule is present in some situations during polymer synthesis and not in others, there are changes in enzyme activity in some cases and not in others. If adjusted, this approach can prevent enzyme leaching and ensure long-term stability for the bio-functionalized electrode.

8.3.3 Materials for Increased Conductivity and Surface Area of Electrodes

Electrode conductivity, components of the electrode used, and ionic conductivity of the electrolyte used are the critical factors in the functioning of biofuel cells. The ionic conductivity is generally divided into two categories: the ionic conductivity of the electrolyte solutions and the ionic conductivity of the polymer electrolyte membrane (PEM) used which separates the catholyte and anolyte. Low conductivity results in high ohmic losses in biological fuel cells, hence degrading the performance. Early in developing traditional metal-catalyzed fuel cells, researchers designed fuel cells specifically to minimize the distance between electrodes. These strategies have been equally crucial to biological fuel cells over the last decade. The original H-cell setup has transitioned toward membrane-free systems and membrane electrode assembly (MEA)-style biofuel cells. Figure 8.4 depicts several arrangements of biofuel cells used, from H-cells to membrane-free electrochemical cells to MEA-style fuel cells.

H-cells usually have a minimum distance of 1 cm between the anode and the cathode (although many cell designs have spaces greater than 10 cm). The bulk of

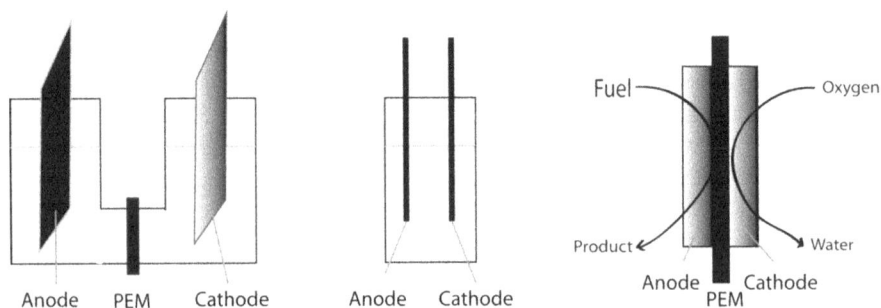

Anode   PEM   Cathode   Anode   Cathode   Anode   Cathode
                                                                   PEM

FIGURE 8.4 The transition from the original biological fuel cell design (left figure, also known as an H-cell) where the two electrodes are submerged in two different solutions separated by a polymer electrolyte membrane (PEM), to the membrane-less biological fuel cell (center figure) where the two electrodes are submerged in the same solution with no separator/ PEM, to the membrane electrode assembly (MEA) design (right figure) where the anode and cathode are submerged in the same solution with the PEM (Minteer et al., 2012).

that distance is filled with a low conductivity electrolyte solution (such as a buffer of biological nature). The membrane-free technique allows for a smaller electrode spacing (typically less than 5 mm) and a low conductivity electrolyte solution to fill the gap (Mano et al., 2004). MEA-style fuel cell designs, on the other hand, often have a cathode and anode spacing of less than 1 mm, with the entire gap filled with a polymer electrolyte membrane (Bhatnagar et al., 2011; Hudak et al., 2005).

Worldwide, researchers have found several polymer electrolyte membranes to improve ionic conductivity from a materials standpoint. The most common polymer electrolyte membrane (PEM) used in biofuel cells is Nafion. However, this PEM is not ideal for most biofuel cells because a neutral pH is required to operate the biofuel cells. Generally, potassium or sodium buffers are used with higher resistance at near-neutral pH than in the usually acidic environment of traditional fuel cells. As a result, recent research has focused on the creation of alternate cation exchange membranes (such as Ultrex) (Logan et al., 2006) and alkaline exchange membranes (Kim et al., 2007). Bipolar membranes for microbial fuel cells have also been studied (Ter Heijne et al., 2006). A PEM with good conductivity at near-neutral pH, or that can manage pH fluctuations of the electrolyte or electrodes, is yet to be developed. The lack of a suitable PEM with high ionic conductivity at neutral pH and that can efficiently handle pH variations of the electrolyte or electrodes is a critical difficulty that will need to be addressed in the next phase of biological fuel cell development.

In the absence of a standardized fuel cell design, there is no practical way to compare variances in the conductivity of different electrochemical cells. The enzymatic biological fuel cell field rarely determines cell Ohmic resistances, but microbial fuel cell systems regularly report this parameter. However, there is no standard method for comparing this type of performance data. Some studies, for example, provide bioelectrochemical system Ohmic resistance per cubic meter, whereas others show Ohmic resistance per square meter. The resistance is a function of the thickness of the membrane and the properties of the individual membrane. Sleutels et al. started addressing this limitation by directly analyzing the differences between cation and anion exchange membranes (Sleutels et al., 2009). An internal resistance of 192 mΩ/m^2 was reported for a biofuel cell structure with an anion exchange membrane versus 435 mΩ/m^2 for a comparable cation exchange membrane-based cell (Sleutels et al., 2009).

The electrode, which uses as a current collector, must have a large surface area to measure the conductivity. Electrodes with a large surface area are the second critical issue associated with conductivity. High surface area materials are required to load higher amounts of the biocatalyst, such as proteins, organelles, and living cells with limited volumetric catalytic activity. The aim was to maintain the conductivity while expanding the effective surface area. The purpose of these highly effective surface area materials is to have a large surface area to volume or mass ratio. Graphite, glassy carbon, and associated vitreous carbon were the most common early biological fuel cell materials (RVC). Mesoporous carbon (Kim et al., 2006), carbon foams, CNT paper, and CNT gels possess higher surface area and are suitable for biofuel cell applications.

8.3.4 Materials That Enhance Facile Mass Transport

Biofuel cells reported in the literature are mainly "bio-batteries"; they are made of two types, that is, catalytic bio-electrodes immersed in a suitable solution of the fuel and/or fuel cells that include the fuel as a part of their design. As a result, the reactive layer does not receive a continuous fuel supply. Actual "biological fuel cells" have just recently begun to appear, necessitating improved mass transport to and from biocatalysts (Higgins et al., 2011; Rincon et al., 2011a; Rincon et al., 2011b). The necessity of matching the transport qualities at the appropriate scale naturally accelerates the invention of materials for biological fuel cell applications. The fluid flow must be accommodated at a macro-scale, requiring appropriate solutions with large void volumes (ideally greater than 0.6), pore diameters in the range of 10–100 μm to 1 mm, and minimal unevenness of the porous media. This design scale accommodates convective flow rates of less than 1 cm³/s. These materials have high electrical conductivity. They are typically believed to enhance the mechanical stability or rigidity required for including them as a structural component in biological fuel cell designs. Various varieties of carbon such as graphite, felts, and carbon papers like Toray paper, one of the classic sources, are among the most extensively used materials. Metal foams and RVC have also been presented as materials of choice, especially for 3D architecture and cylindrical flow-through electrodes working in a plug-flow regime (Higgins et al., 2011; Rincon et al., 2011a; Rincon et al., 2011b).

These macroporous materials, on the other hand, do not have enough surface area to immobilize biocatalysts. Their inherent surface area is often less than 10 m²/g (and less than 1 m²/g in some cases). This reality, together with the above-mentioned functional improvement of biocatalyst interactions with nanomaterials, necessitates the incorporation of such materials with microporous or nanoscale, high-surface area materials. The direct transfer of nanomaterials onto exposed pore structured substrates, such as CNT produced on Toray paper, is one example of such integration (Ivnitski et al., 2008; Barton et al., 2007). This macro-nano composite structure allows for high enzyme loading due to the increase in the effective surface area, and promotes the nanomaterial/biocatalyst interactions that are required. Such integration is complex, and much research is needed to focus on developing hierarchically structured materials with all three porosity scales present. The macro-scale porosity enables convective flow and fuel delivery. The meso-scale porous architecture is designed for the integration of material properties, and the nano-materials such as CNTs or gold nanoparticles, as in the case of GDE (Gupta et al., 2011b), to enhance the progress of gaseous species (Ivnitski et al., 2008; Luckarift et al., 2010; Lau et al. 2011). The interconnectivity of such a composite matrix is usually provided by the mesoporous component, which also ensures the matrix's electrical conductivity (Lee et al., 2006; Wen et al., 2009). Figure 8.5 shows all three degrees of porosity/structure and an SEM microphotograph of one such composite bioelectrode. For example, RVC is used as a conductive, macroporous matrix, and a conductive polymer composite of CNT/chitosan forms on the wall of its "foam-like" structure. The porous structure is optimized for CNT content using a freeze-drying process and demonstrates substantial conductivity. Furthermore, the "surface" of the CNT

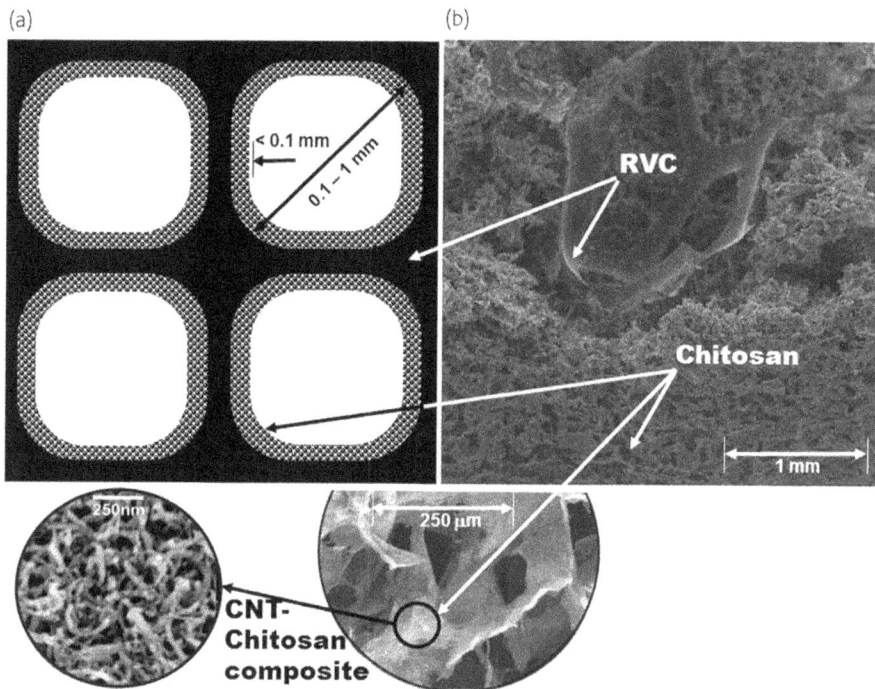

FIGURE 8.5 A flow-through electrode material for a biological fuel cell, with macropores in the reticulated vitreous carbon (RVC), micropores from freeze drying chitosan, and nanopores from the chitosan/CNT composite, is shown in this diagram (Minteer et al., 2012).

is exposed for biocatalyst immobilization. When utilized as enzyme anodes with immobilized oxidases or dehydrogenases or with microorganisms inhabiting their inner space in a microbial fuel cell as anodes, such gradually constructed electrodes are advantageous.

When constructing a porous electrode for flow, a variety of essential parameters must be considered, such as hydrophobicity/hydrophilicity of the high effective surface area material to ensure wetting and minimize idle sites and the material's diffusional transport qualities. Improved biological fuel cell performance can be achieved by paying strict attention to these aspects.

8.4 POWER DENSITY AND BIOCATALYST LOADING

Power density is defined as power generation per surface area of the electrode, or per weight or volume of the cell. Power density is one of the immediate difficulties in producing efficient biofuel cells. For a high output current density, a large enzyme loading is necessary. The enzyme loading was only 1.7×10^{-12} mol/cm^2 (0.27 µg/cm^2) when GOx was randomly packed as a monolayer over a smooth surface, which was determined by the physical size of the biocatalyst. The maximum current density was observed to be only approximately 0.2 mA/cm^2 assuming all enzyme molecules are as active as in aqueous solutions with an average turnover number of 600 s^{-1} (Willner et al., 1996). GOx is assumed to be one of the most efficient redox enzymes. The theoretical current density for other enzymes with lower specific activity should be much lower. Because enzymatic biofuel cells usually operate at lower voltages than 1 V, this provides a power density of less than ~0.2 mW/cm^2 through the above current density, theoretically. Many attempts have been made to improve power density by enhancing enzyme loading in various methods. Multiple-layer enzyme assemblies can be tested to improve enzyme loading in biofuel cells. The power density of newly created biofuel cells was roughly 1~2 orders of magnitude higher when compared to the performance of biofuel cells reported about two decades ago (Schroder et al., 2003; Niessen et al., 2004a; Niessen et al., 2004b).

8.5 CONCLUSION

In the last decade, significant advances in materials engineering have contributed to significant increases in enzymatic and microbial biological fuel cell performance. However, further innovation is required to realize the promise of biological fuel cells fully. The inclusion of high surface area materials to increase biocatalyst loading and DET, materials for improved enzyme immobilization and stabilization, and the fabrication of hierarchical material architectures for advanced electrode design are just a few materials engineering breakthroughs. Research is needed in the design of materials to improve the bio-nano interface, so that biocatalysts can be used more effectively, and in the fabrication of structures that facilitate DET so that high current density electrodes with long-term stability may be created. Second, a paradigm shift in creating ion exchange membrane materials is required to build membrane materials that are specially intended for biological fuel cells rather than the standard fuel cell in usually highly acidic or highly alkaline settings.

REFERENCES

Arechederra, R., & Minteer, S. D. (2009). In *Nanomaterials for Energy Storage Applications, Vol 1*, ed. H. S. Nalwa. American Scientific Publishers, Stevenson Ranch, CA, p. 287.

Balat, M., & Kirtay, E. (2010). Major technical barriers to a "hydrogen economy". *Energ. Sources A*, *32*(9), 863–876.

Bard, A. J., & Faulker, L. R. (2000). *Electrochemical Methods: Fundamentals and Applications* (2nd ed.). John Wiley & Sons, Inc., New York.

Bardea, A., Katz, E., Bückmann, A. F., & Willner, I. (1997). NAD$^+$-dependent enzyme electrodes: Electrical contact of cofactor-dependent enzymes and electrodes. *J. Am. Chem. Soc.*, *119*(39), 9114–9119.

Barrière, F., Kavanagh, P., & Leech, D. (2006). A laccase–glucose oxidase biofuel cell prototype operating in a physiological buffer. *Electrochim. Acta*, *51*(24), 5187–5192.

Barton, S. C., Kim, H. H., Binyamin, G., Zhang, Y., & Heller, A. (2001). Electroreduction of O2 to water on the "wired" laccase cathode. *J. Phys. Chem. B*, *105*(47), 11917–11921.

Barton, S. C., Sun, Y., Chandra, B., White, S., & Hone, J. (2007). Mediated enzyme electrodes with combined micro-and nanoscale supports. *Electrochem. Solid State Lett.*, *10*(5), B96.

Besic, S., & Minteer, S. D. (2011). Micellar polymer encapsulation of enzymes. In: Minteer, S. (ed) *Enzyme Stabilization and Immobilization*. Humana Press, Totowa, NJ, pp. 113–131.

Bhatnagar, D., Xu, S., Fischer, C., Arechederra, R. L., & Minteer, S. D. (2011). Mitochondrial biofuel cells: Expanding fuel diversity to amino acids. *Phys. Chem. Chem. Phys.*, *13*(1), 86–92.

Bidault, F., Brett, D. J. L., Middleton, P. H., & Brandon, N. P. (2009). Review of gas diffusion cathodes for alkaline fuel cells. *J. Power Sources*, *187*(1), 39–48.

Boland, S., Foster, K., & Leech, D. (2009a). A stability comparison of redox-active layers produced by chemical coupling of an osmium redox complex to pre-functionalized gold and carbon electrodes. *Electrochim. Acta*, *54*(7), 1986–1991.

Boland, S., Jenkins, P., Kavanagh, P., & Leech, D. (2009b). Biocatalytic fuel cells: A comparison of surface pre-treatments for anchoring biocatalytic redox films on electrode surfaces. *J. Electroanal. Chem.*, *626*(1–2), 111–115.

Brunel, L., Denele, J., Servat, K., Kokoh, K. B., Jolivalt, C., Innocent, C., & Tingry, S. (2007). Oxygen transport through laccase biocathodes for a membrane-less glucose/O_2 biofuel cell. *Electrochem. Commun.*, *9*(2), 331–336.

Bullen, R. A., Arnot, T. C., Lakeman, J. B., & Walsh, F. C. (2006). Biofuel cells and their development. *Biosens. Bioelectron.*, *21*(11), 2015–2045.

Calabrese Barton, S., Gallaway, J., & Atanassov, P. (2004). Enzymatic biofuel cells for implantable and microscale devices. *Chem. Rev.*, *104*(10), 4867–4886.

Cooney, M. J., Lau, C., Windmeisser, M., Liaw, B. Y., Klotzbach, T., & Minteer, S. D. (2008). Design of chitosan gel pore structure: Towards enzyme catalyzed flow-through electrodes. *J. Mater. Chem.*, *18*(6), 667–674.

Cooney, M. J., Svoboda, V., Lau, C., Martin, G., & Minteer, S. D. (2008). Enzyme catalysed biofuel cells. *Energy Environ. Sci.*, *1*(3), 320–337.

Dai, L. (2013). Functionalization of graphene for efficient energy conversion and storage. *Acc. Chem. Res.*, *46*(1), 31–42.

Dai, L., Xue, Y., Qu, L., Choi, H. J., & Baek, J. B. (2015). Metal-free catalysts for oxygen reduction reaction. *Chem. Rev.*, *115*(11), 4823–4892.

Du, F., Yu, D., Dai, L., Ganguli, S., Varshney, V., & Roy, A. K. (2011). Preparation of tunable 3D pillared carbon nanotube–graphene networks for high-performance capacitance. *Chem. Mater.*, *23*(21), 4810–4816.

Du, Z., Li, H., & Gu, T. (2007). A state of the art review on microbial fuel cells: A promising technology for wastewater treatment and bioenergy. *Biotechnol. Adv.*, *25*(5), 464–482.

Fernando, P., Yossi, W., & Itamar, W. (2004). Long-range electrical contacting of redox enzymes by SWCNT connectors. *Angew. Chem. Int. Ed.*, *43*(16), 2113.

Fischback, M., Kwon, K. Y., Lee, I., Shin, S. J., Park, H. G., Kim, B. C., ... & Ha, S. (2012). Enzyme precipitate coatings of glucose oxidase onto carbon paper for biofuel cell applications. *Biotechnol. Bioeng.*, *109*(2), 318–324.

Friend, C. M., & Xu, B. (2017). Heterogeneous catalysis: A central science for a sustainable future. *Acc. Chem. Res.*, *50*(3), 517–521.

Gallaway, J. W., & Barton, S. A. C. (2009). Effect of redox polymer synthesis on the performance of a mediated laccase oxygen cathode. *J. Electroanal. Chem.*, *626*(1–2), 149–155.

Gerard, M., Chaubey, A., & Malhotra, B. D. (2002). Application of conducting polymers to biosensors. *Biosens. Bioelectron.*, *17*(5), 345–359.

Gong, K., Yan, Y., Zhang, M., Su, L., Xiong, S., & Mao, L. (2005). Electrochemistry and electroanalytical applications of carbon nanotubes: A review. *Anal. Sci.*, *21*(12), 1383–1393.

Gorby, Y. A., Yanina, S., McLean, J. S., Rosso, K. M., Moyles, D., Dohnalkova, A., ... & Fredrickson, J. K. (2006). Electrically conductive bacterial nanowires produced by

Shewanella oneidensis strain MR-1 and other microorganisms. *Proc. Natl. Acad. Sci. U. S. A.*, *103*(30), 11358–11363.

Gupta, G., Lau, C., Branch, B., Rajendran, V., Ivnitski, D., & Atanassov, P. (2011a). Direct bio-electrocatalysis by multi-copper oxidases: Gas-diffusion laccase-catalyzed cathodes for biofuel cells. *Electrochim. Acta*, *56*(28), 10767–10771.

Gupta, G., Lau, C., Rajendran, V., Colon, F., Branch, B., Ivnitski, D., & Atanassov, P. (2011b). Direct electron transfer catalyzed by bilirubin oxidase for air breathing gas-diffusion electrodes. *Electrochem. Commun.*, *13*(3), 247–249.

Heller, A. (2004). Miniature biofuel cells. *Phys. Chem. Chem. Phys.*, *6*(2), 209–216.

Heller, A. (2006). Potentially implantable miniature batteries. *Anal. Bioanal. Chem.*, *385*(3), 469–473.

Heller, A., & Feldman, B. (2010). Electrochemistry in diabetes management. *Acc. Chem. Res.*, *43*(7), 963–973.

Higgins, S. R., Foerster, D., Cheung, A., Lau, C., Bretschger, O., Minteer, S. D., ... & Cooney, M. J. (2011a). Fabrication of macroporous chitosan scaffolds doped with carbon nanotubes and their characterization in microbial fuel cell operation. *Enzyme Microb. Technol.*, *48*(6–7), 458–465.

Higgins, S. R., Lau, C., Atanassov, P., Minteer, S. D., & Cooney, M. J. (2011b). Hybrid biofuel cell: Microbial fuel cell with an enzymatic air-breathing cathode. *ACS Catal.*, *1*(9), 994–997.

Hu, C., & Hu, S. (2009). Carbon nanotube-based electrochemical sensors: Principles and applications in biomedical systems. *J. Sens.*, *2009*. *Article ID 187615, 40 pages.* doi:10.1155/2009/187615

Hu, C., Qu, J., Xiao, Y., Zhao, S., Chen, H., & Dai, L. (2019). Carbon nanomaterials for energy and biorelated catalysis: Recent advances and looking forward. *ACS Cent. Sci.*, *5*(3), 389–408.

Hudak, N. S., & Barton, S. C. (2005). Mediated biocatalytic cathode for direct methanol membrane-electrode assemblies. *J. Electrochem. Soc.*, *152*(5), A876.

Hussein, L., Feng, Y. J., Alonso-Vante, N., Urban, G., & Krüger, M. (2011a). Functionalized-carbon nanotube supported electrocatalysts and buckypaper-based biocathodes for glucose fuel cell applications. *Electrochim. Acta*, *56*(22), 7659–7665.

Hussein, L., Rubenwolf, S., Von Stetten, F., Urban, G., Zengerle, R., Krueger, M., & Kerzenmacher, S. (2011b). A highly efficient buckypaper-based electrode material for mediatorless laccase-catalyzed dioxygen reduction. *Biosens. Bioelectron.*, *26*(10), 4133–4138.

Ivnitski, D., Artyushkova, K., Rincón, R. A., Atanassov, P., Luckarift, H. R., & Johnson, G. R. (2008). Entrapment of enzymes and carbon nanotubes in biologically synthesized silica: Glucose oxidase-catalyzed direct electron transfer. *Small*, *4*(3), 357–364.

Ivnitski, D., Branch, B., Atanassov, P., & Apblett, C. (2006). Glucose oxidase anode for biofuel cell based on direct electron transfer. *Electrochem. Commun.*, *8*(8), 1204–1210.

Jia, H., Zhao, X., Kim, J., & Wang, P. (2008). Carbon nanotube composite electrodes for biofuel cells. In *Biomolecular Catalysis: Nanoscale Science and Technology* (pp. 273–288). (ACS Symposium Series; Vol. 986). American Chemical Society. https://doi.org/10.1021/bk-2008-0986.ch018.

Kang, X., Wang, J., Wu, H., Aksay, I. A., Liu, J., & Lin, Y. (2009). Glucose oxidase–graphene–chitosan modified electrode for direct electrochemistry and glucose sensing. *Biosens. Bioelectron.*, *25*(4), 901–905.

Katz, E., Bückmann, A. F., & Willner, I. (2001). Self-powered enzyme-based biosensors. *J. Am. Chem. Soc.*, *123*(43), 10752–10753.

Kavanagh, P., Boland, S., Jenkins, P., & Leech, D. (2009). Performance of a glucose/O2 enzymatic biofuel cell containing a mediated Melanocarpus albomyces laccase cathode in a physiological buffer. *Fuel Cells*, *9*(1), 79–84.

Kim, B. C., Zhao, X., Ahn, H. K., Kim, J. H., Lee, H. J., Kim, K. W., ... & Kim, J. (2011b). Highly stable enzyme precipitate coatings and their electrochemical applications. *Biosens. Bioelectron.*, 26(5), 1980–1986.

Kim, H., Lee, I., Kwon, Y., Kim, B. C., Ha, S., Lee, J. H., & Kim, J. (2011a). Immobilization of glucose oxidase into polyaniline nanofiber matrix for biofuel cell applications. *Biosens. Bioelectron.*, 26(9), 3908–3913.

Kim, J., Jia, H., & Wang, P. (2006). Challenges in biocatalysis for enzyme-based biofuel cells. *Biotechnol. Adv.*, 24(3), 296–308.

Kim, J. R., Cheng, S., Oh, S. E., & Logan, B. E. (2007a). Power generation using different cation, anion, and ultrafiltration membranes in microbial fuel cells. *Environ. Sci. Technol.*, 41(3), 1004–1009.

Kim, M. I., Kim, J., Lee, J., Jia, H., Na, H. B., Youn, J. K., ... & Chang, H. N. (2007b). Crosslinked enzyme aggregates in hierarchically-ordered mesoporous silica: A simple and effective method for enzyme stabilization. *Biotechnol. Bioeng.*, 96(2), 210–218.

Klotzbach, T. L., Watt, M., Ansari, Y., & Minteer, S. D. (2008). Improving the microenvironment for enzyme immobilization at electrodes by hydrophobically modifying chitosan and Nafion® polymers. *J. Membr. Sci.*, 311(1–2), 81–88.

Kontani, R., Tsujimura, S., & Kano, K. (2009). Air diffusion biocathode with CueO as electrocatalyst adsorbed on carbon particle-modified electrodes. *Bioelectrochemistry*, 76(1–2), 10–13.

Küchler, A., Yoshimoto, M., Luginbühl, S., Mavelli, F., & Walde, P. (2016). Enzymatic reactions in confined environments. *Nat. Nanotechnol.*, 11(5), 409–420.

Kwon, K. Y., Youn, J., Kim, J. H., Park, Y., Jeon, C., Kim, B. C., ... & Kim, J. (2010). Nanoscale enzyme reactors in mesoporous carbon for improved performance and lifetime of biosensors and biofuel cells. *Biosens. Bioelectron.*, 26(2), 655–660.

Lau, C., Cooney, M. J., & Atanassov, P. (2008). Conductive macroporous composite chitosan– carbon nanotube scaffolds. *Langmuir*, 24(13), 7004–7010.

Lee, J., et al. (2006). *Adv. Mater.*, 182, 073.

Lim, J., Cirigliano, N., Wang, J., & Dunn, B. (2007). Direct electron transfer in nanostructured sol–gel electrodes containing bilirubin oxidase. *Phys. Chem. Chem. Phys.*, 9(15), 1809–1814.

Liu, C., Alwarappan, S., Chen, Z., Kong, X., & Li, C. Z. (2010). Membraneless enzymatic biofuel cells based on graphene nanosheets. *Biosens. Bioelectron.*, 25(7), 1829–1833.

Liu, H., Zhang, L., Yan, M., & Yu, J. (2017). Carbon nanostructures in biology and medicine. *J. Mater. Chem. B*, 5(32), 6437–6450.

Logan, B. E. (2010). Scaling up microbial fuel cells and other bioelectrochemical systems. *Appl. Microbiol. Biotechnol.*, 85(6), 1665–1671.

Logan, B. E., Hamelers, B., Rozendal, R., Schröder, U., Keller, J., Freguia, S., ... & Rabaey, K. (2006). Microbial fuel cells: Methodology and technology. *Environ. Sci. Technol.*, 40(17), 5181–5192.

Luckarift, H. R., Ivnitski, D., Rincón, R., Atanassov, P., & Johnson, G. R. (2010). Glucose oxidase catalyzed self-assembly of bioelectroactive gold nanostructures. *Electroanal. Int. J. Devoted Fundam. Pract. Aspects Electroanal.*, 22(7–8), 784–792.

Luo, X., Killard, A. J., & Smyth, M. R. (2006). Reagentless glucose biosensor based on the direct electrochemistry of glucose oxidase on carbon nanotube-modified electrodes. *Electroanal. Int. J. Devoted Fundam. Pract. Aspects Electroanal.*, 18(11), 1131–1134.

Ma, G. X., Lu, T. H., & Xia, Y. Y. (2007). Direct electrochemistry and bioelectrocatalysis of hemoglobin immobilized on carbon black. *Bioelectrochemistry*, 71(2), 180–185.

Malig, J., Englert, J. M., Hirsch, A., & Guldi, D. M. (2011). Wet chemistry of graphene. *Electrochem. Soc. Interface*, 20(1), 53.

Mano, N., Kim, H. H., & Heller, A. (2002). On the relationship between the characteristics of bilirubin oxidases and O2 cathodes based on their "wiring". *J. Phys. Chem. B, 106*(34), 8842–8848.

Mano, N., Mao, F., & Heller, A. (2004). A miniature membrane-less biofuel cell operating at+ 0.60 V under physiological conditions. *Chem.Bio.Chem, 5*(12), 1703–1705.

Merchant, S. A., Tran, T. O., Meredith, M. T., Cline, T. C., Glatzhofer, D. T., & Schmidtke, D. W. (2009). High-sensitivity amperometric biosensors based on ferrocene-modified linear poly (ethylenimine). *Langmuir, 25*(13), 7736–7742.

Meredith, M. T., & Minteer, S. D. (2011). Inhibition and activation of glucose oxidase bio-anodes for use in a self-powered EDTA sensor. *Anal. Chem., 83*(13), 5436–5441.

Minteer, S. D., Atanassov, P., Luckarift, H. R., & Johnson, G. R. (2012). New materials for biological fuel cells. *Mater. Today, 15*(4), 166–173.

Moehlenbrock, M. J., & Minteer, S.D. (2011).Introduction to the Field of Enzyme Immobilization and Stabilization. In: Minteer, S. (eds) *Enzyme Stabilization and Immobilization. Methods in Molecular Biology*, vol 679. Humana Press, Totowa, NJ. https://doi.org/10.1007/978-1-60761-895-9_1.

Moehlenbrock, M. J., Meredith, M. T., & Minteer, S. D. (2012). Bioelectrocatalytic oxidation of glucose in CNT impregnated hydrogels: Advantages of synthetic enzymatic metabo-lon formation. *ACS Catal., 2*(1), 17–25.

Moore, C. M., Akers, N. L., Hill, A. D., Johnson, Z. C., & Minteer, S. D. (2004). Improving the environment for immobilized dehydrogenase enzymes by modifying nafion with tetraalkylammonium bromides. *Biomacromolecules, 5*(4), 1241–1247.

Nevin, K. P., Woodard, T. L., Franks, A. E., Summers, Z. M., & Lovley, D. R. (2010). Microbial electrosynthesis: Feeding microbes electricity to convert carbon dioxide and water to multicarbon extracellular organic compounds. *MBio, 1*(2), e00103-10.

Niessen, J., Schröder, U., Rosenbaum, M., & Scholz, F. (2004a). Fluorinated polyanilines as superior materials for electrocatalytic anodes in bacterial fuel cells. *Electrochem. Commun., 6*(6), 571–575.

Niessen, J., Schröder, U., & Scholz, F. (2004b). Exploiting complex carbohydrates for micro-bial electricity generation–a bacterial fuel cell operating on starch. *Electrochem. Commun., 6*(9), 955–958.

Nöll, T., & Nöll, G. (2011). Strategies for "wiring" redox-active proteins to electrodes and applications in biosensors, biofuel cells, and nanotechnology. *Chem. Soc. Rev., 40*(7), 3564–3576.

Osman, M. H., Shah, A. A., & Walsh, F. C. (2011). Recent progress and continuing challenges in bio-fuel cells. Part I: Enzymatic cells. *Biosens. Bioelectron., 26*(7), 3087–3102.

Pizzariello, A., Stred'ansky, M., & Miertuš, S. (2002). A glucose/hydrogen peroxide biofuel cell that uses oxidase and peroxidase as catalysts by composite bulk-modified bioelec-trodes based on a solid binding matrix. *Bioelectrochemistry, 56*(1–2), 99–105.

Rabaey, K., Johnstone, A., Wise, A., Read, S., & Rozendal, R. (2010). Microbial electrosyn-thesis: from electricity to biofuels and biochemicals. *BIOTECH INTERNATIONAL, 22*(June).

Ramasamy, R. P., Luckarift, H. R., Ivnitski, D. M., Atanassov, P. B., & Johnson, G. R. (2010). High electrocatalytic activity of tethered multicopper oxidase–carbon nanotube conju-gates. *Chem. Commun. (Camb), 46*(33), 6045–6047.

Rincón, R. A., Lau, C., Garcia, K. E., & Atanassov, P. (2011a). Flow-through 3D biofuel cell anode for NAD⁺-dependent enzymes. *Electrochim. Acta, 56*(5), 2503–2509.

Rincón, R. A., Lau, C., Luckarift, H. R., Garcia, K. E., Adkins, E., Johnson, G. R., & Atanassov, P. (2011b). Enzymatic fuel cells: Integrating flow-through anode and air-breathing cathode into a membrane-less biofuel cell design. *Biosens. Bioelectron., 27*(1), 132–136.

Sarma, A. K., Vatsyayan, P., Goswami, P., & Minteer, S. D. (2009). Recent advances in material science for developing enzyme electrodes. *Biosens. Bioelectron.*, *24*(8), 2313–2322.

Schroder, U., Niessen, J., & Scholz, F. (2003). A generation of microbial fuel cells with current outputs boosted by more than one order of magnitude. *Angew. Chem. Int. Ed. Engl.*, *42*(25), 2880–2883.

Schuhmann, W. (1995). Conducting polymer based amperometric enzyme electrodes. *Microchim. Acta*, *121*(1), 1–29.

Shin, K. M., Kim, S. I., So, I., & Kim, S. J. (2009). A conducting polymer/ferritin anode for biofuel cell applications. *Electrochim. Acta*, *54*(16), 3979–3983.

Shleev, S., Shumakovich, G., Morozova, O., & Yaropolov, A. (2010). Stable 'floating' air diffusion biocathode based on direct electron transfer reactions between carbon particles and high redox potential laccase. *Fuel Cells*, *10*(4), 726–733.

Shteinberg, G. V., Dribinsky, A. V., Kukushkina, I. A., Musilova, M., & Mrha, J. (1982). Influence of structure and hydrophobic properties on the characteristics of carbon—Air electrodes. *J. Power Sources*, *8*(1), 17–33.

Shukla, A. K., Suresh, P., Sheela, B., & Rajendran, A. J. C. S. (2004). Biological fuel cells and their applications. *Curr. Sci.*, *87*(4), 455–468.

Sihn, S., Varshney, V., Roy, A. K., & Farmer, B. L. (2012). Prediction of 3D elastic moduli and Poisson's ratios of pillared graphene nanostructures. *Carbon*, *50*(2), 603–611.

Sleutels, T. H., Hamelers, H. V., Rozendal, R. A., & Buisman, C. J. (2009). Ion transport resistance in microbial electrolysis cells with anion and cation exchange membranes. *Int. J. Hydrog. Energy*, *34*(9), 3612–3620.

Stoica, L., Dimcheva, N., Ackermann, Y., Karnicka, K., Guschin, D. A., Kulesza, P. J., & Schuhmann, W. (2009). Membrane-less biofuel cell based on cellobiose dehydrogenase (anode)/laccase (cathode) wired via specific os-redox polymers. *Fuel Cells*, *9*(1), 53–62.

Takeshi, H., Masaya, M., Hiroyuki, N., & Hideaki, M. (2005). Immobilization of enzymes on a microchannel surface through cross-linking polymerization. *Chem. Commun.*, *40*, 5062–5064.

Tan, Y., Deng, W., Ge, B., Xie, Q., Huang, J., & Yao, S. (2009). Biofuel cell and phenolic biosensor based on acid-resistant laccase–glutaraldehyde functionalized chitosan–multiwalled carbon nanotubes nanocomposite film. *Biosens. Bioelectron.*, *24*(7), 2225–2231.

Tasca, F., Gorton, L., Harreither, W., Haltrich, D., Ludwig, R., & Noll, G. (2008). Highly efficient and versatile anodes for biofuel cells based on cellobiose dehydrogenase from *Myriococcum thermophilum*. *J. Phys. Chem. C*, *112*(35), 13668–13673.

Ter Heijne, A., Hamelers, H. V., De Wilde, V., Rozendal, R. A., & Buisman, C. J. (2006). A bipolar membrane combined with ferric iron reduction as an efficient cathode system in microbial fuel cells. *Environ. Sci. Technol.*, *40*(17), 5200–5205.

Timur, S., Haghighi, B., Tkac, J., Pazarlıoğlu, N., Telefoncu, A., & Gorton, L. (2007). Electrical wiring of *Pseudomonas putida* and *Pseudomonas fluorescens* with osmium redox polymers. *Bioelectrochemistry*, *71*(1), 38–45.

Tominaga, M., Otani, M., Kishikawa, M., & Taniguchi, I. (2006). UV–ozone treatments improved carbon black surface for direct electron-transfer reactions with bilirubin oxidase under aerobic conditions. *Chem. Lett.*, *35*(10), 1174–1175.

Trojanowicz, M., Geschke, O., vel Krawczyk, T. K., & Cammann, K. (1995). Biosensors based on oxidases immobilized in various conducting polymers. *Sens. Actuators B*, *28*(3), 191–199.

Vashist, S. K., Zheng, D., Al-Rubeaan, K., Luong, J. H., & Sheu, F. S. (2011). Advances in carbon nanotube based electrochemical sensors for bioanalytical applications. *Biotechnol. Adv.*, *29*(2), 169–188.

Vaze, A., Hussain, N., Tang, C., Leech, D., & Rusling, J. (2009). Biocatalytic anode for glucose oxidation utilizing carbon nanotubes for direct electron transfer with glucose oxidase. *Electrochem. Commun.*, *11*(10), 2004–2007.

Walcarius, A., Minteer, S. D., Wang, J., Lin, Y., & Merkoci, A. (2013). Nanomaterials for biofunctionalized electrodes: Recent trends. *J. Mater. Chem. B*, *1*(38), 4878–4908.

Wang, G., Qian, F., Saltikov, C. W., Jiao, Y., & Li, Y. (2011). Microbial reduction of graphene oxide by Shewanella. *Nano Res.*, *4*(6), 563–570.

Wang, J. (1999). Sol–gel materials for electrochemical biosensors. *Anal. Chim. Acta*, *399*(1–2), 21–27.

Wang, Y., Yuan, R., Chaia, Y., Li, W., Zhuo, Y., Yuan, Y., & Li, J. (2011). Direct electron transfer: Electrochemical glucose biosensor based on hollow Pt nanosphere functionalized multiwall carbon nanotubes. *J. Mol. Catal. B Enzym.*, *71*(3–4), 146–151.

Wen, Y., Xu, J., Li, D., Liu, M., Kong, F., & He, H. (2012). A novel electrochemical biosensing platform based on poly (3, 4-ethylenedioxythiophene): Poly (styrenesulfonate) composites. *Synth. Met.*, *162*(13–14), 1308–1314.

Wen, Z., & Li, J. (2009). Hierarchically structured carbon nanocomposites as electrode materials for electrochemical energy storage, conversion and biosensor systems. *J. Mater. Chem.*, *19*(46), 8707–8713.

Willner, B., Katz, E., & Willner, I. (2006). Electrical contacting of redox proteins by nanotechnological means. *Curr. Opin. Biotechnol.*, *17*(6), 589–596.

Willner, I., Arad, G., & Katz, E. (1998). A biofuel cell based on pyrroloquinoline quinone and microperoxidase-11 monolayer-functionalized electrodes. *Bioelectrochem. Bioenerg.*, *44*(2), 209–214.

Willner, I., Heleg-Shabtai, V., Blonder, R., Katz, E., Tao, G., Bueckmann, A. F., & Heller, A. (1996). Electrical wiring of glucose oxidase by reconstitution of FAD-modified monolayers assembled onto au-electrodes. *J. Am. Chem. Soc.*, *118*(42), 10321–10322.

Willner, I., Yan, Y. M., Willner, B., & Tel-Vered, R. (2009). Integrated enzyme-based biofuel cells–A review. *Fuel Cells*, *9*(1), 7–24.

Wu, J. F., Xu, M. Q., & Zhao, G. C. (2010). Graphene-based modified electrode for the direct electron transfer of cytochrome c and biosensing. *Electrochem. Commun.*, *12*(1), 175–177.

Wu, P., Shao, Q., Hu, Y., Jin, J., Yin, Y., Zhang, H., & Cai, C. (2010). Direct electrochemistry of glucose oxidase assembled on graphene and application to glucose detection. *Electrochim. Acta*, *55*(28), 8606–8614.

Xiao, Y., Patolsky, F., Katz, E., Hainfeld, J. F., & Willner, I. (2003). "Plugging into enzymes": Nanowiring of redox enzymes by a gold nanoparticle. *Science*, *299*(5614), 1877–1881.

Yahiro, A. T., Lee, S. M., & Kimble, D. O. (1964). Bioelectrochemistry: I. Enzyme utilizing bio-fuel cell studies. *Biochim. Biophys. Acta Spec. Sect. Biophys. Subj.*, *88*(2), 375–383.

Yu, P., Zhou, H., Cheng, H., Qian, Q., & Mao, L. (2011). Rational design and one-step formation of multifunctional gel transducer for simple fabrication of integrated electrochemical biosensors. *Anal. Chem.*, *83*(14), 5715–5720.

Zhao, S., Wang, Y., Dong, J., He, C. T., Yin, H., An, P., Zhao, K., Zhang, X., Gao, C., Zhang, L. (2016). Ultrathin metal–organic framework nanosheets for electrocatalytic oxygen evolution. *Nat. Energy*, *1*(12), 16184.

9 Challenges and Future Scope of Bio-nanotechnology in Algal Biofuel and Life Cycle Analysis

Musa S.I., Isitua C.C., Nathan Moses, and Dioha I.J.

CONTENTS

9.1 INTRODUCTION

Energy is required for the socio-economic and cultural development of every complex society in the world. Over the years, the basic source of energy worldwide has been petroleum (Kumar et al., 2015). Petroleum is a complex mixture of hydrocarbons which occurs naturally through millions of years of decomposition, which is why it is also regarded as a fossil fuel (David et al., 2011). The decomposition of natural fossil materials and their transformation into petroleum usually occur under high stress (Earle, 2015). These fossil fuels are now used as the major source of energy, used in transportation, industry, mills, equipment, and many other areas of every-day life (David et al., 2011). Fossil fuels used for the generation of energy are coal, petroleum or gasoline and natural gas (Enzler, 2019). Recent studies and realities have shown that fossil fuels cannot meet the increasing energy demand of the world at large and

DOI: 10.1201/9781003316374-9

questions are raised about how long this non-renewable resource can continue to cater for world energy requirements. In addition, another study estimated that gasoline and natural gas may be depleted in approximately 40 years, while coal reserves may be depleted in 100 years. This finding clearly indicates that petroleum reserves will not last forever. Furthermore, another concern from environmental scientists is about the high rate of impact fossil fuels have on to the environment. It has been a serious challenge to find out how much fossil fuels can be burnt to reduce the negative impact on the environment (Earle, 2015). Different proposals by environmental research bodies have tried to investigate the amount of CO_2 emissions, but a reliable amount have not been documented (Enzler, 2019). For example, it has been proposed that CO_2 emissions should be less than 870 to 1240 gigatons to limit the global temperature increase to around 2°C (Jakob and Hilaire, 2015; Mcglade and Ekins, 2015). But even with these standards, fossil fuel usage has increased from 26,200 million barrels of oil equivalent as in 1965 to about 80,300 million barrels of oil equivalent by 2012 (Table 9.1). There is a projection that the consumption demand of different fossil-based fuels may increase by 30% to 60% by 2035.

Considering these rising concerns and realities, the world is seriously searching to find alternative sources of energy in order to supply the required energy to the increasing global pollution and reducing environmental impacts due to fossil fuel extraction and combustion. Global political leaders along with researchers at this point of time are considering works on alternative energies and determining what is scientifically possible, environmentally acceptable, and technologically promising (Dresselhaus and Thomas, 2001).

For this purpose, policies that discourage the use of fossil-based fuels in favor of alternative or renewable sources of energy has been considered important (Bacon and Silvana, 2004). An example of such energy sources is the renewable resources

TABLE 9.1

Data Showing 2019 Energy Consumption, a Case Study of the 10 Most Energy-Consumed Countries

S/N	Country	Energy Consumption (MTO)
1	China	3,164
2	USA	2,258
3	India	929
4	Russia	800
5	Japan	424
6	South Korea	307
7	Germany	301
8	Canada	301
9	Brazil	290
10	Iran	265

MTO = million ton of oil equivalent

through biomass-based fuel (Omar et al., 2014). These methods would provide a sustainable, greener energy without posing any direct damage to the environment that risks life here on earth. Different types of biofuel can be developed from biomass such as liquid fuel like ethanol (Isitua et al., 2018), methanol, and biodiesel and gaseous fuel (Dioha et al., 2013) like hydrogen and methane. The basic difference between biofuel and petroleum is the oxygen content. In addition, the environmental concerns with regards to petroleum have made biofuel a market competitive (Isitua et al., 2018).

For a decade, the hike in petroleum prices has affected the market consumption of the fuel. Such a hike was not seen previously. Biofuels have gotten a boost as a result. This is because biofuel is affordable and price changes only depends on feedstock value. The choice and availability of different types of feedstock for biofuel production make biofuel a sustainable fuel. International Energy Agency has indicated that there was a steep rise in the production of biofuel from 2000 to 2007, with the major producing nations being the USA (43%), Brazil (32%), and the European Union (15%) (Coyle, 2007). Because of the recent acceptability of biofuel as a sustainable source of energy, researchers have developed different strategies in the production of different types of biofuel, depending on different biomass employed (Rodionova et al., 2016).

9.2 BIOFUEL PRODUCTION

Biofuel can be seen as any fuel that is derived from biomass (Sheehan et al., 2002). Due to the non-environmentally friendly nature of petroleum fuel, plant and animal as well as algae biomass have been considered as the raw materials needed for the production of biofuel (Coyle, 2007). These raw materials were selected as the most important sources of biofuel because of their replenishable nature, unlike fossil fuels such as petroleum, coal, and natural gas (Sheehan et al., 2002). Historically, biofuel has been in use in heating and cooking since the 1800s, but recently the environmental impacts of petroleum has made the world consider commercializing biofuel (Boumesbah et al., 2015). Going by chronology and technological advancements, there are four main strategies in biofuel production (Coyle, 2007). These four strategies, aside from being dependent on time and technological advancements, are also classified based on the type of biofuel produced and the source of the raw materials used for production (Wilhelm et al., 2003).

9.2.1 Biofuel Production from Food Crops

The first method used in biofuel production employed the use of food crops. This method has been used since the 1800s. Food crops such as corn, sugar beets and sugar cane were used as biomass in the generation. In this process, biofuel is produced through ethanol from starch rich food. This biofuel was documented to have a high octane rating, which is the measure of the fuel burning capacity of engines; a major drawback for this biofuel was the high octane rating which made it non-feasible for combustion. This form of biofuel is usually either biodiesel or bioethanol, depending also on the type of raw materials used (Table 9.2). After the first strategy, researchers also attempted to employ another method in biofuel production.

The second method of biofuel production involve the use of non-food crops as raw materials (Dautzenberg et al., 2011). Industrial waste and forest biomass that has high cellulose levels were used for biofuel production (AndréA et al., 2010). In a study carried out in Nigeria, Isitua et al. (2018) successfully produced biofuel (bio-ethanol) from a low-cost cellulosic biomass, cassava (*Manihot esculenta*) residues posing problems of environmental pollution. Cellulose biomass seems to be more attractive because it is documented as the largest source of carbon available on earth (Zhang, 2009). Approximately, the overall chemical energy that is stored in plant biomass is about six to seven times the annual total human energy consumption (Zhang, 2009), and it does not compete with food industries (Andre et al., 2010). Another reason why cellulose received more consideration was because cellulose is easily fermentable because of its structure (Fischer et al., 2008). Biofuel production from cellulose containing biomass is usually done through the gasification, liquefaction, and hydrolysis of the biomass. In this case, the cellulose biomass is first disassembled to improve the isolation of cellulose from other constituents such as lignin and hemicelluloses. Then, cellulose macromolecules are depolymerized to enhance the chemical and biological conversion of cellulose to glucose and then converted to biofuel via biological treatments, and then the biofuel is purified (Zhang, 2009). In addition, cellulose biomass is deoxygenated because the presence of oxygen may influence the heat content of molecules and that can bring about high polarity. This high polarity can affect the ability of the newly produced biofuel to blend with previously produced biofuel (Fischer et al., 2008).

9.2.2 ALGAL BIOFUEL

Furthermore, the third and the most recent method of biofuel generation also doesn't involve food crops or industrial wastes; it involves the use of algae for the production of biofuel (Ramsurn and Gupta, 2013). This method helps in creating a cleaner environment in addition to the low emission of greenhouse gases.

TABLE 9.2
Different Raw Materials in Biofuel Production

Biofuels	Raw Material	Biofuel Yield (kg/ha)	Reference
Bioethanol	Corn	60–90	Johnson et al., 2003; Sheehan et al., 2002
	Sweet sorghum	553–790	McAloon et al., 2000
	Wheat	36–190	Wilhelm et al., 2003
Biodiesel	Rice husk	24–45	Wu et al., 2018
	Sunflower	40–70	Boumesbah et al., 2015
	Cotton stalk	17–23	Thirumurugaveera et al., 2018
	Soybean	27–41	JonVan and Gerhard, 2008

Algae are organisms that grow in aquatic environments and use light and carbon dioxide (CO_2) to create biomass. There are two classifications of algae: macro-algae and microalgae. Macro-algae, which are measured in inches, are the large, multi-cellular algae often seen growing in ponds. Algae are an extremely diverse group that contains many thousands of known species, and potentially hundreds of thousands. The great diversity of algal species provides a wide range of starting strains for fuel production. For example, in dry weight *Spirulina maxima* has 60–71% w/w of proteins, *Porphyridium cruentum* has 40–57% w/w of carbohydrates, and *Schizochytrium* species have 50–57 m% w/w of lipids (Tran et al., 2010). This confirms that different microalgae can be utilized as the preferred source of different biomass under certain conditions (Razaghifard, 2013). Presently, ethanol, alcohols, biodiesel, triglycerides, fatty acids, lipids, carbohydrates, cellulose, and the biomass of organisms are considered as the major biofuel sources. Chisti (2007) stated that it was possible for microalgae to grow quickly and contain high oil content when compared with terrestrial crops, which take a season to grow and only contain a maximum of about 5% dry weight of oil. Research into algae for the mass production of oil focuses mainly on microalgae (organisms capable of photosynthesis that are less than 0.4 mm in diameter, including the diatoms and cyanobacteria) as opposed to macroalgae, such as seaweed (Awudu and Zhang, 2012). The preference for microalgae has come about due largely to their less complex structure, fast growth rates, and high oil content (for some species). However, Lewis (2005) suggested that some research was being done using seaweeds for biofuels, probably due to the high availability of this resource.

Algae can be converted into various types of fuels, depending on the production technologies and the part of the cell used (Janda et al., 2012). The lipid, or oily part of the algae biomass, can be extracted and converted into biodiesel through a process similar to that used for any other vegetable oil (Figure 9.1) or converted in a refinery into "drop-in" replacements for petroleum-based fuels (Rodionova et al., 2016). These microalgae are an ideal biodiesel feedstock, which eventually could replace petroleum-based fuel due to several advantages, such as high oil content, high rates of production, less land required, etc. Currently, algal biodiesel production is still too expensive to be commercialized. Due to the static costs associated with oil extraction and biodiesel processing and the variability of algal biomass production, cost-saving efforts for algal oil production should focus on the production method of the oil-rich algae itself. Therefore, there is a need to enhance both algal biology (in terms of biomass yield and oil content) and culture-system engineering. In addition, Spolaore et al. (2006) explained that the aspects of microalgae used for producing various value-added products besides algal fuel, via an integrated biorefinery, is an appealing way to lower the cost of algal biofuel production. Indeed, microalgae contain a large percentage of oil, with the remaining parts consisting of large quantities of proteins, carbohydrates, and other nutrients.

Although both hydrogen- and sugar-based biofuel production could potentially impact the use of petroleum fuels, algae have the potential to produce a number of secondary metabolites that have characteristics much closer to existing petroleum fuels. The most promising of these are the terpenes, which offer a potential new

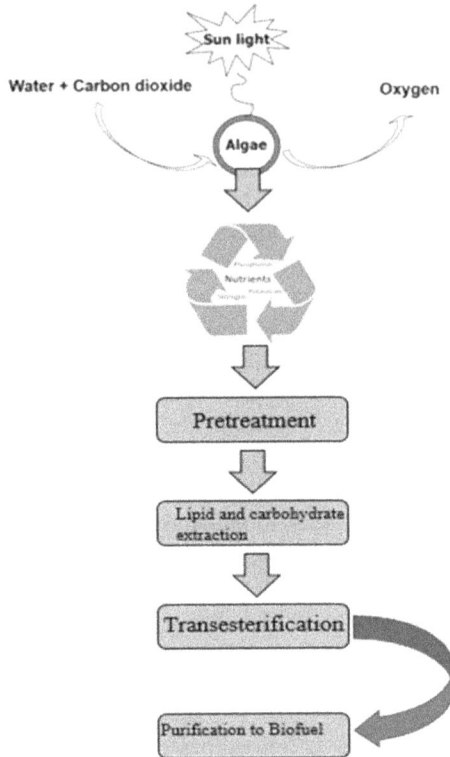

FIGURE 9.1 Schematic representation of algal biofuel production.

fuel source outside of fatty acids that is compatible with our existing fuel frame-work. Sialve et al. (2009) stated that terpenes are polymers of isoprene C_5H_8. If produced in high quantities, hemiterpenes (single isoprene units) and monoter-penes ($C_{10}H_{16}$) could function in current gasoline engines with only minor engine modifications. Longer isoprene chains could be used for biodiesel or cracked into biogasoline or jet fuels. To date, terpene accumulation for fuels has not been a large focus of algal biology, since many terpenes and terpene derivatives have properties that make them more valuable for other uses, such as flavor, fragrances, and antibiotics (Kirby et al., 2009).

Furthermore, studies have indicated that algae can synthesize about 20 times more fuel than soy and canola per hectare. Therefore, most researchers prefer the use of algal to food sources. Another factor is that food specialists argue that it is not sustainable to produce food for energy while people are hungry in society. Since the cost is very high, so much work is required on this aspect. Over the last decades much research have been conducted to understand the growth of algae in different environments and environmental conditions (Park et al., 2011). Such studies have shown that microalgae like *Scenedesmus obliquus*, *Neochlorisole abundans*, *Nanochloropsis* sp., and others have shown the most efficiency in fuel production and may become a major and important non-food crop in the coming generations. These species have several properties that researchers consider effective such as:

1. The high biomass level of the microalgae
2. The rapid growth rate and replication
3. The high CO_2 fixation capacity
4. The increased O_2 production
5. The ability to produce high lipid contents
6. The ability to produce high carbohydrate contents

Furthermore, the lipid content of some microalgae are high and can be used to produce higher concentrations of lipids. Moreover, the ability of microalgae to sequester CO_2 from the atmosphere boost its production in power plants so as to bring down the concentration of greenhouse gases. Much properties of the algal biofuel and general fuel are comparable (Gouveia and Oliveira, 2009). The steps involved in biofuel production from microalgae is as follows:

1. Ensuring the growth and replication of the microalgae
2. Evaluating the growth for assessment
3. Extraction of lipids from the microalgae
4. Characterization of the oil extracted from the microalgae using different methods

However, different methods are available for the cultivation of microalgae depending on the purpose of use. This means that the cultivation method chosen depends on how and where microalgae are being used, whether for biomass production to be utilized for different product or for the production of biofuel which requires high lipid content (Vonshak, 1986). The concentration of lipids in the microalgae largely depends on the nutritional media used in their cultivation. Studies have indicated that the accumulation of fatty acids occurs during the growth limiting step. On the development of such conditions, like growth limiting, the energy produced from photosynthesis is directed towards the accumulation of carbohydrates and the synthesis of lipids, also increasing nutrient absorption. Among the many available macronutrient, Poirreck et al. (1984) have suggested that limiting the nitrogen source has increased the lipid content in microalgae. It has shown an increase of up to 70% in lipid content and a decrease in protein content (Poirreck et al., 1983; Huerlimann et al., 2010; Cruz et al., 2018). Photobioreactors are widely used for the production of microalgae and are categorized into (1) open-air pond reactors (OPR) and (2) closed reactors (CR). In OPR, less amount of energy is utilized, while the CR produces more biomass than OPR. It works on the following principles:

1. Utilization of appropriate light sources
2. Increasing photon conversion efficiency
3. Maintenance of culture conditions

However, this technique has so far been used at lab scale and not at commercial level (Kothari et al., 2017). Oil extraction from microalgae produces many types of lipids which have proven to be beneficial for health. However, it has gained the research interest mainly due to biofuel production. Noteworthy advances have been made in oil purification strategies, but they still face many challenges in bringing down the cost of extraction at the industrial level (Mercer and Armenta, 2011). Neutral lipids present in the microalgae is meant for energy storage whereas polar ones are available for the bilipid layer. After microalgae has been cultivated, lipid extraction is carried out. The extraction process is an energy-intensive process. The most commonly used method is the solvent extraction method (Gong et al., 2017). The Folch method is the oldest method used for lipid extraction. In this method a combination of chloroform and methanol is used. A ratio of 2:1 (chloroform:methanol) is maintained. The homogenized cells are kept briefly in saline solution and the resultant mixture is kept until two distinct layers separate out. Lipids are found in the upper phase which is extracted. With some modifications this method is still in use; however it is less sensitive (Folch et al., 1957; Kumar et al., 2015). In the Bligh and Dyer method, like in the Folch method, a partition is created and lipids are extracted simultaneously and proteins are precipitated in the interphase. It differs from the above method in the solvent:solvent and solvent:tissue ratios. The lipids are then extracted from the chloroform (Bligh and Dyer, 1959). Many modifications have been made to the above method like the addition of 1M NaCl instead of water which inhibits the binding of acidic lipids to denatured lipids (Kumar et al., 2015). Superior solvent extraction method: the solvents used so far have environmental and health risks. Many substitute solvents have been tried by researchers which are less toxic. Among these, recent research has been reported which shows the use of 2-ethoxyethanol for superior lipid recovery. This is another accelerated solvent extraction which employs heat and pressure, reduces processing time and recovers solvent for future use (Kumar et al., 2015). Mechanical approach: when solvent process requires, the choice of solvent for the type of microalgae used, the mechanical method does not depend on the species of microalgae and causes less contamination. However, the heat generated during the process can cause some damage to end products, and therefore a cooling system is required. Dried biomass is pressed with tremendous pressure to squeeze out the oil from the algae. Bead beating is another method, in which microalgae is mixed with fine beads and is spun at high speed, causing the disruption of the cell and spillage of the oil (Demirbas, 2008; Lee et al., 1998). To characterize the microalgal oil, the extracted oil is subjected to fatty acid profile determination in which the oil is transesterified into fatty acid methyl ester (FAME). In this method the lipid sample is mixed with toluene and sulfuric acid and kept overnight at 55°C. After the due time NaCl is added along with hexane and the solution vortexed. Layers are separated, and the upper phase is taken for GLC analysis. In GLC, FAME is quantified. Through this, the amount of oil present in the microalgae is quantified (Kumar et al., 2011; Abubakar et al., 2012).

Several strategies have been considered in the production of microalgae, usually from various sources such as the open ponds or the controlled automatized bioreactors (CAB). However, both systems have proven unique advantages and disadvantages, which influences the choice of system to use. The increase in energy and operating costs ends up being less attractive. Thus, the choice of each system must be based on the intrinsic needs of each microorganism used, the climatic conditions of the region, and the costs involved in the system (Brennan and Owende, 2010). The closed systems have as a characteristic non-direct contact with atmospheric air, which allows the control of several cultivation conditions, such as temperature, light, pH, amount of nutrients, and cell density, thus reducing the risk of contamination (Bitog et al., 2011). This system is generally used to obtain products with higher added value, such as pharmaceuticals and food supplements. Closed photobioreactors have different configurations, and among them, we can mention flat panel, tubular, and spiral. The flat panel photobioreactor consists of two transparent glass plates arranged in a rectangular shape connected in a cascade that can be arranged vertically and in an inclined manner (Bitog et al., 2011). Presenting a higher contact surface, which allows for a higher brightness of the culture. However, it has as a disadvantage the possibility of biofilm formation on the internal surface of the glass plates (Chew et al., 2018). Acien et al., (2015) observed that the *Geitlerinema* sp. reached 47.7 gm^2 of dry biomass, in addition to the removal of 82%, 85%, and 100% of carbon, nitrogen, and phosphorus, respectively, when cultivated in 100 L flat panel photobioreactors under non-aseptic cultivation conditions. Tubular photobioreactors, on the other hand, consist of a series of connected transparent tubes, which can be arranged in an inclined, horizontal, or vertical manner (Demirlel et al., 2015).. Under natural conditions, microalgae are grown in open systems, with little or no control of growing conditions. The cultivation ponds are open, and thus in contact with atmospheric air, but they present a higher risk of external contamination. Generally, the ponds used are shallow to be more efficient in the use of light (Brennan and Owende, 2010). The open systems of microalgae production, generally of the raceways type, have been developed, aiming at higher production of algal biomass with lower operational cost. However, when this cultivation is carried out outdoors, it can alter the biochemical composition and the productivities of the algal cells, due to seasonal, temperature, and light variations, and contamination (Chew et al., 2018). Open systems are generally used to produce algal biomass with less added value and can be used in the feeding of aquatic organisms and the production of biofuels (He et al., 2016), among other uses. Adesanya et al. (2014) combined tubular photobioreactors of the airlift type in the first stage and raceways in a second stage for the cultivation of *Chlorella vulgaris*. In this study, the authors assessed the environmental impact of biodiesel produced using microalgae and fossil fuels. Thus, a reduction in global warming potential of 76% and a reduction in fossil energy requirements of 75% were observed.

Furthermore, the whole algal biomass can be used to produce different forms of biofuel such as biogas, liquid, and gaseous transportation fuel, kerosene, ethanol, aviation fuel, and biohydrogen following distinct techniques (Ikhajigbe and Musa, 2020). The biofuel conversion technologies for algal biomass can be separated into four basic categories as captured in Table 9.3.

TABLE 9.3
Different Methods of Producing Biofuel from Algae

Biofuel Technologies	Process	Product	References
Biochemical conversation	Fermentation	Ethanol, acetone, and butanol	Harun et al., 2011; Choi et al., 2010
	Aerobic digestion	CH_4, H_2	Sialve et al., 2009; Yang et al., 2011
	Photobiological H_2 production	H_2	Burgess and Fernandez-Velasco, 2007
Thermochemical conversion	Gasification	Fuel gas	Hirano et al., 1998
	Pyrolysis	Bio-oil, charcoal	Miao and Wu, 2004
	Liquefaction	Bio-oil	Minowa and Sawayama, 1999
Chemical reaction	Transesterification	Biodiesel	Huang et al., 2010
Direct combustion	Power generation	Heat and electricity	Bruhn et al., 2011

9.3   ALGAL LIFE CYCLE

Biofuel production using algae as a biomass source depends on the algal life cycle (Brennan and Owende, 2010). The algal life cycle may be asexual in the case of lower species, which reproduce through cell division or fragmentation. However, larger algae reproduce through spore formation. Some algae produce monospores (walled, non-flagellate, spherical cells) that are carried by water currents and upon germination produce a new organism, while some other algae produce non-motile spores called aplanospores. Others also produce zoospores, which lack true cell walls and bear one or more flagella. The flagella allow zoospores to swim to a favorable environment, whereas monospores and aplanospores have to rely on passive transport by water currents. Also, many of the microalgae strains are lipid-rich species, showing high growth rates and land use efficiency, when compared with traditional crops (Branco-Vieira et al., 2018). This is one of the major characteristics that allow algae to be used in biofuel production (Brennan and Owende, 2010). Microalgae have become a good alternative feedstock for biofuel production (Mata et al., 2010). Among microalgae, *Phaeodactylum tricornutum*, a marine diatom, can grow in freshwater. Branco-Vieira et al. (2018) state that *P. tricornutum* is one of the most thoroughly studied diatom species, is highly productive and environmentally adaptable, and can be used for biodiesel production but also has other applications through a biorefinery process. Chanj et al. (2017) explained that, for microalgae cultivation, either open or closed bioreactors can be used. However, there are expressive differences among these production systems, mostly related to energy and water use.

Sexual reproduction is characterized by the process of meiosis, in which progeny cells receive half of their genetic information from each parent cell. Sexual reproduction is usually regulated by environmental events such as temperature, salinity, inorganic nutrients (e.g., phosphorus, nitrogen, and magnesium), or day length. In many species the environmental factors become unfavorable to them. A sexually reproducing organism has two phases in its life cycle. In the first stage, each of the cell has a single set of chromosomes and is called haploid, while in the second stage each cell has two sets of chromosomes and is called diploid. In a mature algae, one haploid gamete fuses with another haploid gamete, leading to fertilization; the resulting combination, with two sets of chromosomes, is called a zygote. A diploid cell directly or indirectly undergoes a special reductive cell-division process (meiosis). Diploid cells in this stage are called sporophytes because they produce spores. During meiosis the chromosome number of a diploid sporophyte is halved, and the resulting daughter cells are haploid. At some time, immediately or later, haploid cells act directly as gametes.

The life cycles of sexually reproducing algae is different from asexually reproducing algae; in some, the dominant stage is the sporophyte, in others it is the gametophyte. Example, *Sargassum* (class Phaeophyceae) has a diploid (sporophyte) body, and the haploid phase is represented by gametes. *Ectocarpus* (class Phaeophyceae) has alternating diploid and haploid vegetative stages, whereas *Spirogyra* (class Charophyceae) has a haploid vegetative stage, and the zygote is the only diploid cell. In biofuel production, the algae are usually used at maturity, when cellulose and lipid bi layers are visible (Rodolfi et al., 2009).

Lllman et al. (2000) suggested in his work that the pure culture of *Chlorella vulgaris* can be achieved in an open raceway, facilities covering about 100 ha. Unlike many other microalgae species, *Chlorella* is known to react to nitrogen deprivation by accumulating lipids and carbohydrates but at the cost of a lower growth rate. As it is not evident which strategy will give better results, both options (normal and low N) will be evaluated. However, it is assumed that in both cases culture is carried out in one step, without using a specific facility dedicated to nitrogen deprivation or inoculum maintenance. Algae harvesting is achieved by continuous recirculation of culture ponds through a thickener; the flocculated stream is then dewatered. Oil extraction is subject to much discussion (Converti et al., 2009), and it is not clear now which technology is the more efficient. As a consequence, two options have been evaluated: either advanced drying followed by hexane extraction (similarly to soybeans), or direct extraction from the wet algal paste. Water collected at the thickener and dewatering unit is redirected to the pond. An oil extraction unit located in the facility extracts oil from the algal paste. The oil fraction is then shipped to an industrial transesterification facility where it is transformed into biodiesel.

Since a number of factors must be considered in choosing the best alternative for producing microalgae-based products, the use of life cycle assessment (LCA) is promising since it is a harmonized tool that supports the systematic assessment of environmental impacts, considering the entire life cycle (Teresa et al., 2012). LCA has been extensively applied to the assessment of biofuels (Collet et al., 2014), but a variety of theoretical assumptions are considered, making it difficult to compare results and different realities (Mata et al., 2013). Besides, most studies in the existing scientific literature make use of secondary data and design predictive scenarios to support their assessment (Bradley et al., 2015), which leads to unrealistic microalgae-based biofuel LCA studies. Such studies often disregard the economic reality of each location, as well as the availability of infrastructure and the considerable differences that exist in microalgae growth rates, mainly due to the climatic conditions of each site (Kugler et al., 2015).

9.4 CHALLENGES OF USING ALGAE AS BIOMASS FOR BIOFUEL PRODUCTION

Algae biofuels may provide a viable alternative to fossil fuels; however, this technology must overcome a number of hurdles before it can compete in the fuel market and be broadly deployed (Rodolfi et al., 2009). These challenges include strain identification and improvement, both in terms of oil productivity and crop protection, nutrient and resource allocation and use, and the production of co-products to improve the economics of the entire system.

The high growth rates, reasonable growth densities, and high oil contents have all been cited as reasons to invest significant capital to turn algae into biofuels. However, for algae to mature as an economically viable platform to offset petroleum and, consequently, mitigate CO_2 release, there is a need to develop solutions to problems such as how and where to grow these algae, how to improving oil extraction and fuel processing, how to improve the methods of oil extraction, and how to manipulate microalgae genes for improved biofuel production (Rodolfi et al., 2009).

Another aspect is how to balance the photosynthetic needs of the microalgae, especially the liberation of molecular oxygen during photosynthesis. This can irreversibly inhibit hydrogenases (Huesemann et al., 2009). Apart from this, the production cost of biohydrogen is not competitive enough to replace hydrogen production from fossil fuel (Hirano et al., 1998). Other factors that may influence the option of using microalgae as energy biomass for biofuel production include the different efficiencies of light utilization by phototrophs under different solar light intensities (Miao and Wu, 2004; Brennan and Owende, 2010), the high level of ambient oxygen that impairs the hydrogen production enzymes and the pathways in cells (Huang et al., 2010; Wang et al., 2008), and the rate of photosynthetic CO_2 assimilation necessary for an efficient accumulation of cell biomass and its further low rate of conversion to biohydrogen (Miao and Wu, 2004). Possible ways to enhance the microalgal lipid biosynthetic pathway are of immense significance for the optimal and improved production of biodiesel (Radakovits et al., 2010).

One of the challenges of using algal biofuel is the energy-intensive nature of the processes. Also, the techniques used in cell wall disruption and disassembling is also of significant importance since the mechanical methods available are energy consuming (sonification and hydrocavitation). Although freshwater is added to waste water for its neutralization, waste water still can cause pollution in the surrounding environment. Other challenges in biofuel production using algae include the energy-intensive mechanical method of algal cell wall disruption, and the problem of adding wastewater for neutralization which brings about pollution in the surrounding environment (Dyni, 2006). With increasing population and expanding economy, coupled with the possibilities of depleting petroleum hydrocarbon, there is an important need for biotechnologists to employ various means in improving algal biofuel production (Schindler et al., 2008).

9.5 BIONANOTECHNOLOGY IN IMPROVING ALGAE FOR BIOFUEL PRODUCTION

Considering the challenges observed in biofuel production using algal, scientists have made several proposals on possible strategies to improve algal biofuel production, especially to meet commercial requirements (Enzler, 2019). Nanomaterials with both magnetic and catalytic properties can be beneficial for algal biofuel production through bionanotechnology (Enzler, 2019). Bionanotechnology is the application of nanotechnology in solving biological problems (Ikhajiagbe and Musa, 2020). It involves the introduction of biological molecules into nano artifacts (Rodionova et al., 2016). One such nanoparticle (NP) is FeO_4 silica, which could be used as a catalyst for coupling process during algal biofuel production (Anoop et al., 2011). To this nanoparticle a base was attached, triazabicyclodecene; which could add functionality to the NP. Previous research has shown that this NP have the ability to attach to the microalgae and proved to be a good micro harvester. This construct was then added to the chloroform:methanol solvent in which lipid extraction is done. Since the NP acts as a catalyst, the lipid will easily be converted to biodiesel at a rapid rate (Tyagi, 2007) depending on the microalgae. From this it is evident that NPs can enhance biofuel production by providing surface area for the reaction to take place (Chiang et al., 2015). Furthermore, magnetic NP has been used in cultivating microalgae in photobioreactors; such an arrangement distributes nutrients uniformly throughout.

Silver NP has been used for higher light accessibility (Hossain et al., 2019). A similar silver NP was used by Musa and Ikhajiagbe (2020) to improve the growth of rice under varying conditions through rhizoinoculation. During this research, increased microalgae population was observed which is likely due to the silver NP application. Magnetic NP is a potential candidate to consider for harvesting algal biofuel (Ayesha and Malik, 2019). This method has attracted interest in microalgae culturing due to its low cost and efficiency. Iron oxide nanoparticles have been used in this respect. Iron oxide NPs so made were used to cultivate two species of microalgae, *Scenedesmus ovalternus* and *Chlorella vulgaris*. This method showed an increase in harvesting efficiency by 95% for both species (Gracia et al., 2018).

In a work conducted by Ling et al. (2011), they studied the efficiency of magnetic NPs (Fe_3O_4) as separating agents in the disassembly of cells during biofuel production. The magnetite was prepared by precipitating FeO and dissolving hydrated 3-4-Fe_3Cl_4 and FeCl under nitrogen pressure in water. After heating the solution, NH OH was added, while magnetite was precipitated further. Certain species of algae were grown in a modified medium. When the desired biomass was reached, magnetite was added. Under different stirring speeds it was noted that harvested microalgae were greatly adsorbed over the magnetite. The recovery efficiency reached 99.9%. This indicates that nanoparticles increase the recovery of microalgae and hence can provide more biomass for biofuel production (Xu et al., 2011). In a similar work,

Ting et al. (2016) prepared magnetite with polymer coating with polyamidoamine (PAMAM) for efficient flocculation (Wang et al., 2016).

The cells of algae has to be disrupted in order to extract the lipid and carbohydrate to be used in biofuel production (Xu et al., 2011). With the use of nanoparticles, the algal cell wall can be restricted for easy disruption and extraction (Kumar et al., 2015). This has allowed the application of nanoparticles in biotechnology to carve out its own niche in the renewable energy sector, especially at it improves the processes involved in algal biofuel production (Wang et al., 2016).

Furthermore, nanoparticles are recognized to be effective catalytic agents in chemical and biological sciences (Andre et al., 2010). Due to nanoparticles' tiny structure, they penetrate and easily interact with the biomolecules acting upon them efficiently. Andre et al. (2010). prepared silver nanoparticles, and *Chlorella vulgaris* was harvested using the nanoparticles. Their work indicated that the concentration of carbohydrate and lipid increased in the presence of silver nanoparticles. This work clearly indicated that nanoparticles have the capacity to pierce through the cell and rupture them so as to extract the carbohydrate, lipids, and other metabolites needed for production. Since nanoparticles are developed from biological substrates, for example, as Ikhajiagbe and Musa (2020) used *Hibiscus sabderiffa* to develop silver nanoparticles (AgNPs), the process is ecofriendly, cost effective, and non-toxic in nature (Andre et al., 2010). Nanomaterials have been efficiently been used for enzyme immobilization for biofuel production, but are still in their infancy (Puri et al., 2013).

9.6 FUTURE PROSPECTS

The application of nanotechnology in the field of biotechnology to improve biofuel production is becoming an increasingly important economic asset, especially in the energy sector. With the growth in the demand for energy, especially using environmentally friendly processes, the world production of biofuel is expected to gain more consideration than ever (Kumar et al., 2015). The development of genetically engineered algae capable of reproducing more rapidly is imperative. However, an efficient regeneration protocol must be in place before genetic transformation studies can be initiated. The improvement and continuous supply of nutrients in growing algae may also be considered; enabling environments and situations that would improve algal multiplication would help in improving biofuel production. Since microalgae are very sensitive to environmental pollution (Wang et al., 2016), environments meant for growing algae can be monitored using biotechnological means prior to production. Some of the most important products are needed in biofuel production are the extracted carbohydrates and lipids; synthesized bionanoparticles should be used to improve the rapid extraction. The successful introduction of foreign genes into plant cells is primarily governed by two factors, optimization of the culture conditions of target plant cells/tissues and transformation procedure. In algal production, only a few reports are available regarding transformation and transgenic development since people are just getting into the field, but for nanoparticles, transgenic nanoparticles have been documented (Zhang et al., 2013). Further inputs are required for the establishment of efficient transgenic algae that would improve the supply of products needed in biofuel production. The judicious improvement of nanoparticles for the biotechnological production of biofuel from microalgae can bring more benefits to mankind throughout the world and reduce environmental pollution through the use of fossil fuels.

9.7 CONCLUSION

While global energy demand is increasing, people are shifting away from petroleum products, especially because of their significant risk to humans and the environment. The amount of fuel globally consumed as well as the demand are expected to grow rapidly, and the use of fossil energy causes significant problems and harmful impacts to the environment on earth. The current global energy crisis has brought significant attention to this problem. Renewable energy sources are critical in solving the global energy issue. Biofuels are an excellent example of renewable energy that can be produced using biological organisms and can reduce the dependence on fossil fuels. Several sources of biofuel have been employed previously; however, the use of algae as biomass in biofuel production is considered the most effective, natural, and environmentally friendly. Although photosynthesis is very important in increasing the biomass of algal for biofuel production, the production of biofuel from these algae comes with various challenges such as the disorganization of the algal cell wall to release carbohydrate and lipids needed for production. Application of nanoparticles have been used to improve the synthesis of biofuel from algae. Silver nanoparticles have shown potential for boosting the production of biofuel in the world. In the future, genetic improvement strategies can be used to improve the algal life cycle, to bring about greater algae reproduction for biofuel synthesis. The application of genetics in NP synthesis is considered a future giant for biofuel production.

REFERENCES

Abubakar, L. U., Mutie, A. M. and Muhoho, A. (2012). Characterization of algae oil (oilgae) and its potential as biofuel in Kenya. *J. Appl. Hydrotech. Environ. San.* 1(4), 147–153.

Acien, G., Fernandez-Sevilla,   and Molina, J. (2015). Direct supercritical methanolysis of wet and dry unwashed marine microalgae (*Nannochloropsis gaditana*) to biodiesel. *Appl. Energy* 148, 210–219. https://doi.org/10.1016/j.apenergy.2015.03.069.

Adesanya, V. O., Cadena, E., Scott, S. A. and Smith, A. G. (2014). Life cycle assessment on microalgal biodiesel production using a hybrid cultivation system. *Bioresour. Technol.* 163, 343–355. https://doi.org/10.1016/j.biortech.2014.04.051.

André, A., Diamantopoulou, P., Philippoussis, A., Sarris, D., Komaitis, M. and Papanikolaou, S. (2010). Biotechnological conversions of bio-diesel derived waste glycerol into added-value compounds by higher fungi: Production of biomass, single cell oil and oxalic acid. *Ind. Crop. Prod.* 31(240), 74–76.

Anoop, S., Stig, I., Olsena, O. and Poonam, S. (2011). A viable technology to generate third-generation biofuel. *J. Chem. Technol. Biotechnol.* 86(11), 1349–1353.

Awudu, V. and Zhang, B. (2012). Increase in Chlorella strains calorific values when grown in low nitrogen medium. *Enzyme Microb. Technol.* 27, 631–635.

Ayesha, M. and Malik, B. (2019). Magnetic nanoparticles: Eco-friendly application in bio-fuel production. In Abd-Elsalam, K., Mohammed, M., and Prasad, R. (Eds.), *Magnetic Nanostructures. Nanotechnology in the Life Sciences.* Springer, Cham, pp. 3–7.

Bacon, R. and Silvana, T. (2004). *Crude Oil Prices: Predicting Price Differentials Based on Quality.* Note No. 275, World Bank, Washington, DC. http://siteresources.worldbank.org/EXTFINANCIALSECTOR/Resources/282884-1303327122200/275-bacon-tordo.pdf.

Bitog, J. P., Lee, I.-B., Lee, C.-G., Kim, K.-S., Hwang, H.-S., Hong, S.-W., Seo, I. -H., Kwon, K. -S. and Mostafa, E. (2011). Application of computational fluid dynamics for modeling and designing photobioreactors for microalgae production: A review. *Comput. Electron. Agric.* 76(2), 131–147. https://doi.org/10.1016/j.compag.2011.01.015.

Bligh, E. G. and Dyer, W. J. (1959). A rapid method of total lipid extraction and purification. *Can. J. Biochem. Physiol.* 37(8), 911–917.

Boumesbah, R., Jason, W., Cherrington, T. and Danquah, M. (2015). Exploring alkaline pre-treatment of microalgal biomass for bioethanol production. *Appl Energy.* 10, 12–19.

Bradley, T., Maga, D. and Antón, S. (2015). Unified approach to life cycle assessment between three unique algae biofuel facilities. *Appl. Energy* 154, 1052–1061. https://doi.org/10.1016/j.apenergy.2014.12.087.

BrancoVieira, M., San Martin, S., Agurto, C., Freitas, M., Mata, T. M., Martins, A. A. and Caetano, N. (2018). Biochemical characterization of *Phaeodactylum tricornutum* for microalgae-based biorefinery. *Energy Procedia* 153, 466–470.

Brennan, L. and Owende, P. (2010). Biofuels from microalgae – A review of technologies for production, processing, and extractions of biofuels and co-products. *Renew. Sustain. Energy Rev.* 14(2), 557–577. https://doi.org/10.1016/j.rser.2009.10.009.

Bruhn, A., Dahl, J., Nielsen, H., Nikolaisen, L., Rasmussen, M., Markager, S., Olesen, B., Arias, C. and Jensen, P. D. (2011). Bioenergy potential of *Ulva lactuca*: Biomass yield, methane production and combustion. *Bioresour. Technol.* 102(3), 2595–2604.

Burgess, G. and Fernandez–Velasco, J. (2007). Materials, operational energy inputs, and net energy ratio for photobiological hydrogen production. *Int. J. Hydr. Energy* 32(9), 1225–1234.

Chanj, D., Yang, Z., Guo, R., Xu, X., Fan, X. and Luo, S. (2017). Fermentative hydrogen production from lipid-extracted microalgal biomass residues. *Appl Energy.* 2, 12–23.

Chew, K. W., Chia, S. R., Show, P. L., Yap, Y. J., Ling, T. C. and Chang, J.-S. (2018). Effects of water culture medium, cultivation systems and growth modes for microalgae cultivation: A review. *J. Taiwan Inst. Chem. Eng.* 91, 332–344.

Chiang, Y. D., Dutta, S., Chen, C. T., Huang, Y. T., Lin, K. S., Wu, J. C., Suzuki, N., Yamauchi, Y. and Wu, K. C. (2015). Functionalized Fe_3O_4 silica core–shell nanoparticles as micro-algae harvester and catalyst for biodiesel production. *ChemSusChem.* 8(5), 789–794.

Chisti, Y. (2007). Biodiesel from microalgae. *Biotechnol. Adv.* 25(3), 294–306.

Choi, S. P., Nguyen, M. T. and Sim, S. J. (2010). Enzymatic pretreatment of *Chlamydomonas reinhardtii* biomass for ethanol production. *Bioresour. Technol.* 101(14), 5330–5336.

Collet, P., Spinelli, D., Lardon, L., Hélias, A., Steyer, J. P. and Bernard, O. (2014). Life-cycle assessment of microalgal-based biofuels. In Pandey, A., Chisti, Y., Lee, D. J., and Soccol, C. R. (Eds.), *Biofuels from Algae.* Elsevier, London, pp. 287–312.

Converti, A., Casazza, A. A., Ortiz, E. Y., Perego, P. and Del Borghi, M. (2009). Effect of temperature and nitrogen concentration on the growth and lipid content of *Nannochloropsis oculata* and *Chlorella vulgaris* for biodiesel production. *Chem. Eng. Process. Process Intensif.* 48(6), 1146–1151.

Coyle, S. (2007). Enzymatic pretreatment of Chlamydomonas reinhardtii biomass for ethanol production. *Bioresource Technol.* 101, 5330–5336.

Cruz, Y. R., Aranda, D. G. and Seidl, P. R. (2018). Cultivation systems of microalgae for the production of biofuels. In *Biofuels - State of Development.* Chirmanda Edet, 23–39, New York State of Development.

Dautzenberg, G., Gerhardt, M. and Kamm, B. (2011). Bio-based fuels and fuel additives from lignocellulose feedstock via the production of levulinic acid and furfural. *Holzforschung* 65, 43–49.

David, S., Zelda, E., Penton, F. and Kitson, G. (2011). Hydrocarbons. In Sparkman, O. D., Penton, Z. E., and Kitson, F. G. (Eds.), *Gas Chromatography and Mass Spectrometry* (2nd ed.). Academic Press, New York, pp. 331–339.

Demirbas, A. (2008). *Biodiesel a Realistic Fuel Alternative for Diesel Engine*. Springer-London Limited, London.

Demirel, N., Bartlett, J., Kannangara, G., Milev, A., Volk, H. and Wilson, M. (2015). Catalytic upgrading of biorefinery oil from micro-algae. *Fuel*. 189, 26–74.

Dioha, I. J., Ikeme, C. H., Nafi'u, T., Soba, N. I. and Yusuf, M. B. S. (2013). Effect of carbon to nitrogen ratio on biogas production. *Int. Res. J. Natl. Sci.* 1(3), 1–10.

Dresselhaus, M. and Thomas, L. (2001). Alternative energy technologies. *Nature* 4(6861), 332–337.

Dyni, J. R. (2006). Geology and resources of some world oil-shale deposits. *Scientific Investigations Report*. 5294. US Geological Survey, Reston, VA.

Earle, S. (2015). *Physical Geology*. BCcampus, Victoria, BC. Retrieved from https://opentextbc.ca/geology/.

Enzler, A. (2019). Materials, operational energy inputs, and net energy ratio for photobiological hydrogen production. *Int. J. Hydrogen Energy*. 32, 1225–1234.

Fischer, C., Klein-Marcuschamer, R. and Stephanopoulos, G. (2008). Selection and optimization of microbial hosts for biofuels production. *Metab. Eng.* 10(295), 30–40.

Folch, J., Lees, M. and Stanley, G. (1957). A simple method for the isolation and purification of total lipids from animal tissues. *J. Biol. Chem.* 226(1), 497–509.

Gong, M., Hu, Y., Yedahalli, S. and Bassi, A. (2017). Oil extraction processes in microalgae. *Recent Advances in Renewable Energy*, 1, pp. 377–411.

Gouveia, L. and Oliveira, A. (2009). Microalgae as a raw material for biofuels production. *J. Ind. Microbiol. Biotechnol.* 36(2), 269–274.

Gracia, M., Lampis, S., Brignoli, P. and Vallini, G. (2018). Bioaugmentation and biostimulation as strategies for the bioremediation of a burned woodland soil contaminated by toxic hydrocarbons: A comparative study. *Journal of Environmental Management*. 153, 121–131.

He, S., Singh, R. N. and Sharma, S. (2016). Development of suitable photobioreactor for algae production—a review. *Renewable and Sustainable Energy Reviews*, 16, 2347–2353.

Harun, R., Jason, W., Cherrington, T. and Danquah, M. (2011). Exploring alkaline pre-treatment of microalgal biomass for bioethanol production. *Appl. Energy* 10, 12–19.

Hirano, A., Hon-Nami, K., Kunito, S., Hada, M. and Ogushi, Y. (1998). Temperature effect on continuous gasification of microalgal biomass. Theoretical yield of methanol production and its energy balance. *Catal. Today* 45(1–4), 399–404.

Hossain, S., Ahman, T. and Seraj, M. F. (2019). Available Approaches of Remediation and Stabilisation of Metal Contamination in Soil: A Review. *Ameri-can Journal of Plant Sciences*. 9, 2033–2052.

Huang, G., Chen, F., Wei, D., Zhang, X. and Chen, G. (2010). Biodiesel production by micro-algal biotechnology. *Appl. Energy* 87(1), 38–46.

Huerlimann, R., Nys, R. and Heimann, K. (2010). Growth, lipid content, productivity, and fatty acid composition of tropical microalgae for scale-up production. *Biotechnol. Bioeng.* 107(2), 245–257.

Huesemann, M. H., Hausmann, T. S., Bartha, R., Aksoy, M., Weissman, J. C. and Benemann, J. R. (2009). Biomass productivities in wild type and pigment mutant of *Cyclotella* sp. (diatom). *Appl. Biochem. Biotechnol.* 157(3), 507–526.

Ikhajiagbe, B. and Musa, S. I. (2020). Application of biosynthesized nanoparticles in the enhancement of growth and yield performances of rice (*Oryza sativa* var. Nerica) under salinity conditions in a ferruginous ultisol. *Fudma Sci.* 4(1), 120–132.

Illman, A. M., Scragg, A. H. and Shales, S. W. (2000). Increase in Chlorella strains calorific values when grown in low nitrogen medium. *Enzyme Microb. Technol.* 27(8), 631–635.

Isitua, C. C., Anadozie, S. O. and Ibeh, I. N. (2018). Bioethanol production from Cassava (*Manihot esculenta*) peels. *FACsalud UNEMI* 2(2), 40–45.

Jakob, M. and Hilaire, J. (2015). Climate science: Unburnable fossil-fuel reserves. *Nature* 517(7533), 150–152.

Janda, A., Hon-Nami, K., Kunito, S., Hada, M. and Ogushi, Y. (2012). Temperature effect on continuous gasification of microalgal biomass. Theoretical yield of methanol production and its energy balance. *Catal Today.* 45, 399–404.

Johnson, K.F., Tan, K.T., Abdullah, A.Z. and Lee, K.T. (2003). Life cycle assessment of palm biodiesel: revealing facts and benefits for sustainability. *Appl Energy.* 86, S189–96.

Jonvan, W., Hu, B., Li, Y., Min, M., Mohr, M., Du, Z., Chen, P. and Ruan, R. (2008). Mass cultivation of microalgae on animal wastewater: a sequential two-stage cultivation process for energy crop and omega-3-rich animal feed production. *Appl. Biochem. Biotechnol.* 168, 348–363.

Kamar, A., Dahl, J., Nielsen, H., Nikolaisen, L., Rasmussen, M. and Markager, S. (2018). Bioenergy potential of Ulva lactuca:Biomass yield, methane production and combustion. *Bioresource Technol.* 102, 2595–2604.

Kirby, J. and Keasling, J. D. (2009). Biosynthesis of plant isoprenoids: Perspectives for microbial engineering. *Annu. Rev. Plant Biol.* 60, 335–355.

Kothari, R., Pandey, A., Ahmad, S., Kumar, A., Pathak, V. V. and Tyagi, V. V. (2017). Microalgal cultivation for value-added products: A critical enviro-economical assessment. *3 Biotech* 7 (243), 1–15.

Kugler, F., Skarka, J., Rösch, C. and Davey, M. P. (2015). Environmental life cycle assessment (LCA) of microalgae production at. In *Crops*. Enterprise Hub, Wales, UK, pp. 23–29.

Kumar, P., Suseela, M. R. and Toppo, K. (2011). Physico-chemical characterization of algal oil: A potential biofuel. *Asian J. Exp. Biol. Sci.* 2(3), 493–497.

Kumar, R. R., Rao, P. H. and Arumugam, M. (2015). Lipid extraction methods from microalgae: A comprehensive review. *Front. Energy Res.* 2, 1–9.

Lee, S. J., Yoon, B. D. and Oh, H. (1998). Rapid method for the determination of lipid from the green alga *Botryococcus braunii*. *Biotechnol. Tech.* 12(7), 553–556.

Lewis, L. (2005). Seaweed to breathe new life into fight against global warming. *The Times.* London. Retrieved 11 February 2008.

Ling, C. C., Chikere, C. B. and Okpokwasili, G. C. (2011). Bioremediation techniques-classification based on site of application: principles, advantages, limitations and prospects. *World Journal Microbiology and Biotechnology.* 32(11), 180–190.

Mata, T. M., Caetano, N. S., Costa, C. V., Sikdar, S. K. and Martins, A. A. (2013). Sustainability analysis of biofuels through the supply chain using indicators. *Sustain. Energy Technol. Assess.* 3, 53–60.

Mata, T. M., Martins, A. A. and Caetano, N. S. (2010). Microalgae for biodiesel production and other applications: A review. *Renew. Sustain. Energy Rev.* 14(1), 217–232.

Mc Aloon, Y., Yumurtaci, M. and Kecebas, A. (2000). Renewable energy and its university level education in Turkey. *Energy Educ Sci Technol Part B.* 3, 143–52.

Mcglade, M. and Ekins, C. (2015). The geographical distribution of fossil fuels unused when limiting global warming. *Nature* 7(5), 90–111.

Mercer, P. and Armenta, R. (2011). Developments in oil extraction from microalgae. *Eur. J. Lipid Sci. Technol.* 113(5), 539–547.

Miao, X. and Wu, Q. (2004). High yield bio-oil production from fast pyrolysis by metabolic controlling of Chlorella protothecoides. *J. Biotechnol.* 110(1), 85–93.

Minowa, T. and Sawayama, S. (1999). A Novel microalgal system for energy production with nitrogen cycling. *Fuel* 78(10), 1213–1215.

Omar, E., Haitham, A. and Frede, B. (2014). Renewable energy resources: Current status, future prospects and their enabling technology. *Renew. Sustain. Energy Rev.* 39, 748–764.

Park, J., Craggs, R. J. and Shilton, A. N. (2011). Wastewater treatment high rate algal ponds for biofuel production. *Bioresou Tech.* 102(1), 35–42.

Poirreck, G., Minowa, T. and Sawayama, S. (1983). Novel microalgal system for energy production with nitrogen cycling. *Fuel.* 78, 1213–1215.

Poirreck, S. J., Sharma, N. and Yoon, B. D. (1984). Oh HM Rapid method for the determination of lipid from the green alga Botryococcusbraunii. *Biotechnology Techniques.* 12(7), 553–556.

Puri, J., Kazy, S. K., Gupta, A., Dutta, A., Mohapatra, B., Roy, A., Bera, P., Mitra, A. and Sar, P. (2013). Biostimulation of indigenous microbial community for bioremediation of petroleum refinery sludge. *Frontiers of Microbiology.* 7, 1407.

Radakovits, R., Jinkerson, R. E., Darzins, A. and Posewitz, M. C. (2010). Genetic engineering of algae for enhanced biofuel production. *Eukaryot. Cell* 9(4), 486–501.

Ramsurn, L. and, Gupta, V. (2013). Biodiesel production by microalgal biotechnology. *Applied Energy.* 87, 38–46.

Razaghifard, R. (2013). Algal biofuels. *Photosynth. Res.* 117(1–3), 20–29.

Rodionova, M., Poudyal, R., Tiwari, I., Voloshin, R., Zharmukhamedov, S., Nam, H., Zayadan, B., Bruce, B., Hou, H. and Allakhverdiev, S. (2016). Biofuel production: Challenges and opportunities. *Int. J. Hydr. Energy* 2, 1–12.

Rodolfi, L., Chini Zittelli, G., Bassi, N., Padovani, G., Biondi, N., Bonini, G. and Tredici, M. R. (2009). Microalgae for oil: Strain selection, induction of lipid synthesis and outdoor mass cultivation in a low-cost photobioreactor. *Biotechnol. Bioeng.* 102(1), 100–112.

Schindler, J., Zittel, W. and Wank, C. (2008). *Crude Oil.* The Supply Outlook. Energy Watch Group, Ottobrunn, Germany. 12, 222.

Sheehan, J., Aden, C., Riley, K., Paustian, K., Killian, J., Brenner, D., Lightle, R., Nelson, M., Walsh, N. and Cushman, J. (2002). *Is Ethanol from Corn Stover Sustainable? Adventures Incyber-Farming: A Life Cycle Assessment of the Production of Ethanol from Corn Stover Foruse in a Flexible Fuel Vehicle.* National Rewable Energy Laboratory, Golden. 23, 12–18.

Sheehan, J., Dunahay, T., Benemann, J. and Roessler, P. (2002). A Look Back at the U.S Department of Energy's Aquatic Species Program – Biodiesel from Algae. . Vol. Zaslavskaia LA, Lippmeier JC, Shih C, Ehrhardt D, Grossman AR, Apt KE. Trophic conversion of an obligate photoautotrophic organism through metabolic engineering. *Science.* 292 (5524).

Sialve, B., Bernet, N. and Bernard, O. (2009). Anaerobic digestion of microalgae as a necessary step to make microalgal biodiesel sustainable. *Biotechnol. Adv.* 27(4), 409–416.

Spolaore, P., Joannis-Cassan, C., Duran, E. and Isambert, A. (2006). Commercial application of microalgae. *J. Biosci. Bioeng.* 101(2), 87–96.

Teresa, M., Mata, A. A., Martins, B., Neto, M. L., Martins, R. R., Salcedo, C. A. and Costa, L. (2012). Tool for sustainability evaluations in the pharmaceutical industry. 23, 12–17.

Thirumurugaveera, K., Zhang, X.L., Yan, S., Tyagi, R.D. and Surampall, R. (2018). Biodiesel production from heterotrophic microalgae through transesterification and nanotechnology application in the production. *Ren. and Sust. Energy Rev.* 26, 216–223.

Ting, M., Pasadakis, N., Norf, H. and Kalogerakis, N. (2016). Enhanced *ex situ* bioremediation of crude oil contaminated beach sand by supplementation with nutrients and rhamnolipids. *Mar Pollut Bull.* 77, 37–44.

Tran, N., Bartlett, J., Kannangara, G., Milev, A., Volk, H. and Wilson, M. (2010). Catalytic upgrading of biorefinery oil from micro-algae. *Fuel* 189, 26–74.

Tyagi, V. (2007). Microalgal cultivation for value-added products: A critical enviro-economical assessment. *Biotech* 7 (243), 1–15.

Vonshak, A. (1986). Laboratory techniques for the cultivation of microalgae. In Richmond, A. (Ed.), *Handbook of Microalgal Mass Culture*, pp. 1–30.

Wang, B., Li, Y., Wu, N. and Lan, C. Q. (2008). CO2 bio-mitigation using microalgae. *Appl. Microbiol. Biotechnol.* 79(5), 707–718.

Wang, T., Yang, W., Hong, Y. and Hou, Y. (2016). Magnetic nanoparticles grafted with amino-richeddendrimer as magnetic flocculant for efficient harvesting of oleaginous microalgae. *Chem. Eng. J.* 297, 304–314.

Wilhelm, W. W., Johnson, J., Hatfield, J., Voorhees, W. and Linden, D. (2003). Cropand soil productivity response to corn residue removal: A review of the literature. *Agron. J. Rev.* 5, 56–59.

Wu, H., Miao, X. and Wu, Q. (2018). High quality biodiesel production from a microalga Chlorella protothecoides by heterotrophic growth in fermenters. *J Biotechnol.* 126, 499–507.

Xu, Y., Highina, B. K. and Zubairu, A. (2011). Bioremediation: A Solution to Environmental Pollution-A Review. *American Journal of Engineering Research* (AJER). 7(2), 101–109.

Yang, Z., Guo, R., Xu, X., Fan, X. and Luo, S. (2011). Fermentative hydrogen production from lipid-extracted microalgal biomass residues. *Appl. Energy* 2, 12–23.

Zhang, J. R. (2009). *Scientific Investigations Report. 5294.* US Geological Survey; VA, USA: 2006. Geology and resources of some world oil-shale deposits.

Zhang, X. L., Yan, S., Tyagi, R. D. and Surampall, R. (2013). Biodiesel production from heterotrophic microalgae through transesterification and nanotechnology application in the production. *Ren. Sust. Energy Rev.* 26, 216–223.

Zhang, Y. (2009). A sweet out-of-the-box solution to the hydrogen economy: Is the sugar-powered car science fiction? *Energy Environ. Sci.* 22(7), 22–32.

10 Bio-Mimetic and Bio-Templating of Natural Materials

An Approach to Sustainable Development Using Bionanotechnology

Vipin K. Sharma, Dinesh K. Sharma
and Sachin Sharma

CONTENTS

DOI: 10.1201/9781003316374-10

10.1 INTRODUCTION

The system of natural selection has brought about the progress of abundant substances and microstructures, models and processes which have been optimized for a vast variety of capabilities. Bio-mimicry or bio-inspiration entails studying nature's best ideas in constructing surprisingly complicated and state-of-the-art engineering models at diverse duration scales and utilizing expertise to solve crucial challenges faced by humankind (Lepora et al., 2013; Naik and Singamaneni, 2017). It is taken into consideration that natural materials and systems have advanced to carry out a vast variety of tasks like structural support, sensing and actuating devices, signal controlling elements, self-cleaning devices and light absorbing devices for obtaining various types of gains (Wegst et al., 2015; Huang et al., 2017). The various categories related to gains may be in the form of electrical gain, chemical gain, mechanical gain and multiplicity of gain, and the most important is the sustainability gain. Learning from this extensive solution provider manual or maybe immediately making use of the design principles can be a surprisingly effective method for solving various important international challenges related to food, water, safety, public health and clean energy.

Nature-inspired materials have played a very vital role in improving the lifestyle of human beings due to their properties that are suited to sustainable growth. As nature is a role model, materials inspired by nature have the capacity to produce sustainable products with better properties. These materials can be used to develop smart and innovative products with incredibly state-of-the-art techniques to solve problems. One such example is a fiber-based product that offers a tailored combination of properties including structural colors, self-healing properties and the ability to act as thermal insulation (Fratzl, 2007). The field of bio-mimetic materials is growing continuously due to its excellent capabilities in providing solutions. Also, it provides a platform to improve recent and sustainable fibers in a more effective way. The idea of producing fabric is based on the mimicry of functional values of cucumber found in the deep sea. These creatures have placid connective tissue with mutable mechanical houses; the animal can switch between low and excessive stiffness within seconds. With inspiration from the above characteristics of cucumber, a nano-composite has been prepared with collagen fibrils which provide a low modulus to its matrix. A narrow protein varies the degree of binding and strain switch among adjoining fibrils to govern the macroscopic homes of the gadget (Fratzl, 2007).

Life on earth in the form of flora and fauna has evolved over billions of years and provides the best solutions to problems related to man through nature-inspired processes in sustainable ways like super-hydrophobicity, self-cleaning and self-healing phenomena, the conservation of energy and the minimization of drag force (Eggermont, 2007). These nature-inspired processes help us to find solutions to daily problems faced by human beings. One such famous example is the creation of fishing nets inspired by the spider's web. The mechanical properties, in terms of strength and stiffness, of the hexagonal honeycomb might also be used as a low-density and light-weight structural component for aerospace applications. Similarly, Leonardo da Vinci designed ships and planes by looking at creatures like fish and birds. The design of airplanes was improved after figuring out that birds do not flap their wings constantly; as a substitute they float on air currents (Eggermont, 2007). The focus of the present endeavor and the primary aim of the current book chapter is related to future generations' engineering problems in achieving certain types of gain. These gains in the form of mechanical, electrical, chemical, sustainability and multiplicity of gains could be achieved via bio-mimetic or bio-inspired materials.

10.2 CASE STUDIES ON BIO-MIMETIC MATERIALS AND BIONANOTECHNOLOGY

Nature-inspired materials are being developed with the aim of harnessing a positive kind of capability, which permits us to tap into a specific type of benefit (Katiyar et al., 2021). Categorizations of nature-inspired materials by virtue of the gain they offer are shown in Figure 10.1 with different schemes mentioned in Table 10.1. The kind of benefit might be (i) electrical, (ii) biological, (iii) chemical, (iv) mechanical, (v) sustainable, (vi) or a multiplicity of gains. These types of gains generally provide solutions to problems related to the bio-medical field, aerospace applications, the defense field, electronics industries based on semiconductors, electrical industries based on E-vehicles, etc. These fields require low-cost materials with better properties that match nature-inspired processes. For example, the electric eel became a source of inspiration for the generation of artificial organs with self-powered characteristics. Another example is the discovery of self-healing material using concepts of bionanotechnology in aerospace applications. These smart materials with self-healing features are inspired by plants that exist in nature. So all these nature-inspired

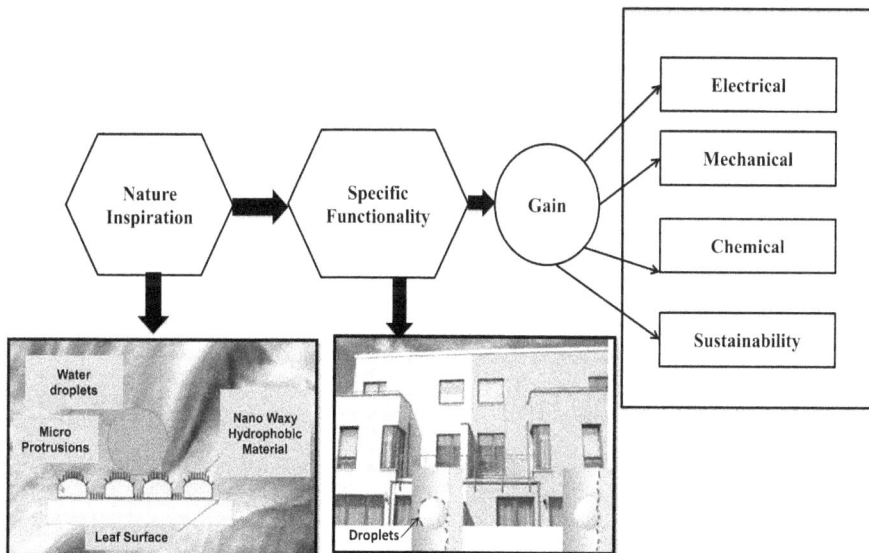

FIGURE 10.1   Gains associated with nature-inspired materials.

TABLE 10.1
Nature-Inspired Materials along with Their Functionalities

Nature-Inspired Features	Targeted Area of Application
Materials that became a source of inspiration in achieving "electrical gain"	
• The conversion of the chemical energy of food into electrical energy in electric eels.	• Body organs made of artificial material inspired by eels.
• Wind dispersed seeds.	• 3-D electronic micro fliers.
• Biological micro-motors based on bacteria.	• Nature-inspired nano-motor.
• Super-hydrophobicity in lotus leaves.	• TiO_2 coating used in self-cleaning buildings.
Materials that became the source of inspiration in achieving "biological gain"	
• Self-healing tunable polymers.	• Body tissues repair.
• Seeds with the capability of self-folding on swelling.	• Self-shaping ceramic.
• Coral-algal symbiosis.	• To grow microalgae with high spatial cell densities.
Materials that became the source of inspiration in achieving "chemical gain"	
• Stratified micro- and nano-scale quality of diatom in bionanotechnology system.	• Diatomite membrane with better wettability characteristics.
• Bio-mineralization.	• Organic composite-based bone and teeth.
• Self-healing phenomenon.	• Body parts of airplane.
Materials that became the source of inspiration in achieving "mechanical gain"	
• Mimicked from Damascus steel rings.	• High-strength steel with hard and soft layers.
• Response to peripheral stimuli.	• Self-actuated muscles.
• Spider dragline silk.	• Synthetic spider silks.
• Deep sea creature.	• Self-powered soft robot.
• Fish scales and osteoderms.	• Flexible bullet-proof material.
Materials that became the source of inspiration in achieving "sustainability gain"	
• Recyclability.	• Biodegradable materials with low density that can be reused and recycled for aerospace applications.
• Biodegradable plastic.	• Used in diapers, plastic bags and packaging.
Materials that became the source of inspiration in achieving "multiplicity of gains"	
• Spider silk.	• Humanoid robotics.
• Macroscopically ordered rod-like nanoapatites.	• Aqueous liquid crystal, aqueous $Mg (OH)_2$ and $Mg_3(PO_4)_2$ LCs.
• Bio-mimicked antireflective properties (insect compound eyes).	• Solar energy harvesting.
• Artificial urushi (wetting).	• Coatings.

Source: Katiyar et al., 2021.

processes and materials provide a certain type of gain that improves human life in the most effective and efficient way.

To provide a clear overview of the current applications of bio-mimetic materials, a few case studies are discussed below. The first case study describes how the butterfly has become a source of inspiration for developing solar power. The second case study relates to the hydrophobicity phenomenon in lotus leaves. The third and fourth case studies are related to the electric eel and bio-inspired robots. Detailed analysis are given below.

10.2.1 CASE STUDY RELATED TO THE WHITE BUTTERFLY AND SOLAR PANELS: A BEST WAY TO USE BIONANOTECHNOLOGY

10.2.1.1 Introduction

Day by day, the harvesting of photovoltaic energy deploys extensive arrays of solar panels (Soudi et al., 2020). A study headed by a scientist of Indian origin found a way to overcome the problem. The study showed that, by mimicking the v-formed posture of the white butterfly belonging to the family Pieridae to heat up its flight muscular tissues before takeoff, the quantity of electricity produced by way of solar panels could be increased by almost 50%. This type of feature helps in reducing the cost of solar panels. The study was led by Radwanul Siddique, a bioengineer at California Institute of Technology; along with his team, he studied the wings underneath an electron microscope and created a 3-D model of his observations. The tiny holes in the butterfly wings are random in length, distribution and shape, says Siddique. The research led to the creation of a model which concluded that shape did not matter in the absorption of light but the order and position played an important role. Next, they created a comparable structure using extraordinarily thin sheets of hydrogenated amorphous silicon, which was a tremendous achievement.

10.2.1.2 Problems Faced

The solar cells consumed too much power in the harvesting of solar energy, making it too expensive for people to afford. Also, the arrays of the solar panels were deploying day by day due to extensive power consumption. Cabbage white butterflies take flight earlier than all of the other species of butterflies on cloudy days, which are restricted by how fast the insects can use the energy from the sun to warm their flight muscle groups. It was thought that this ability was due to the v-shaped posture they use to maximize the attention of solar rays on their thorax, which allows for flight. Also, the unique sub-systems of their wings allowed the sun's rays to reflect more efficiently, ensuring the warming of the flight muscle tissue to the greatest temperature. The crew then researched a way to create a reproduction of the butterflies' wings to develop a brand new, light-weight reflective material for solar panels. They found that there was an optimum angle at which the butterfly held its wings to increase the temperature of its body, about 17°, which increased the temperature by 7°C. Three degree centigrade compared to that once the wings had been held flat. The scientists additionally found that by replicating the mono-layered cells of the wings in solar panels there was a significant improvement in the strength-to-weight ratios of future solar concentrators, which were relatively lighter and more efficient.

10.2.1.3 Applications

This finding allowed the creation of more efficient, light-weight, bio-mimetic photovoltaic cells, which were of feasible cost (Zhao et al., 2011; Rodriguez et al., 2018).

10.2.1.4   Value Addition

Enhanced power conversion energy; the solar cell developed was two times better at harvesting solar energy as shown in Figure 10.2.

10.2.2   HYDROPHOBICITY IN LOTUS LEAVES: A BEST WAY TO USE BIO NANOTECHNOLOGY SYSTEMS IN SELF-CLEANING BUILDINGS

This is another interesting case study related to the phenomenon of hydrophobicity in lotus leaves.

10.2.2.1   Inspiration

Lotus flower (*Nelumbo nucifera*).

10.2.2.2   Introduction

Detter and Johnson first studied the phenomenon of hydrophobicity in 1964 by using rough hydrophobic surfaces. After some time using this phenomenon, Barthlott and Elher described the self-cleaning and hydrophobic properties of the lotus in 1977. The term hydrophobic comes from the ancient Greek word "Hydrophobos" meaning "having a fear of water". Hydrophobic actually means "the fear of water". Hydrophobic molecules and surfaces repel water. Hydrophobic fluids together with

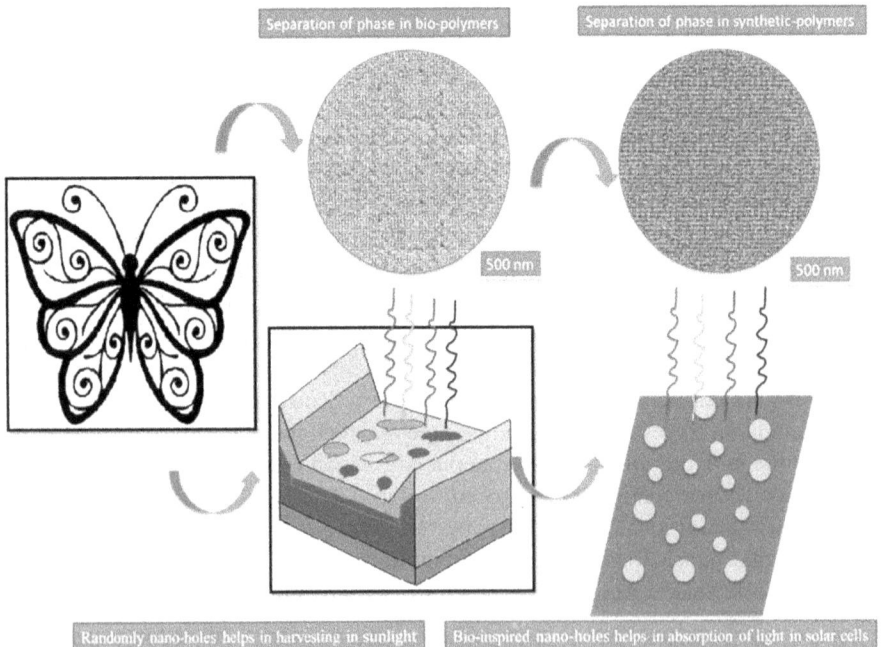

FIGURE 10.2   The butterfly is a source of inspiration for the generation of solar power.

oil will become independent from water. Hydrophobic molecules are typically non-polar, which means the atoms that make up the molecules do not produce a static electric field. Hydrophobic materials are often used to remove oil from water and manage oil splits and chemical separation processes that require the removal of a non-polar substance from a polar compound.

The idea of hydrophobicity among the strong-liquid interfaces has been described by diverse researchers with the usage of their very own mathematical fashions in phrases of different important characteristics like surface energy and contact angle (Laurent et al., 1987; Sharma et al., 2021). After an exhaustive literature review of the assessment of interaction energy, Thomas Young model was found to be an accurate mathematical model for it. In his model, Thomas Young makes use of the Laplace equation to discover the relation between the contact angle and surface energy in order to define the hydrophobicity. Following are the equations used to decide the shape of the droplet and the contact angle role in various boundary conditions as indicated in Figure 10.3. Before defining the expression of interaction energy within a liquid and solid, the assumption is made that the solid surface is smooth, inflexible and homogenous (Hashim et al., 2001).

At equilibrium condition, the contact angle (θ) is defined by the Young–Dupre equation as indicated in Equation 10.1:

$$\gamma_{sv} = \gamma_{sl} + \gamma_{lv} \cos\theta \qquad (10.1)$$

In Equation 10.1, the terms γ_{sv}, γ_{sl} and γ_{lv} are the specific energies of the solid-vapor, solid-liquid and liquid-vapor interfaces, respectively. Here, each specific energy or surface tension force can be taken as the tension force per unit area. As demonstrated in Figure 10.3(a), the fall of a fine liquid drop on the sample surface replaces a bit of the interface formed between solid-vapor by a liquid-vapor and liquid-solid interface. As per the condition stated above, under equilibrium conditions a decrease in the free energy of the system will lead to the phenomenon of liquid spreading (Kwok and Neumann, 1999; Kalantarian et al., 2009). The bonding force between the liquid and solid, i.e. the adhesion force (F_a), is defined as

$$F_a = \gamma_{lv} + \gamma_{sv} - \gamma_{sl} \qquad (10.2)$$

Combining Equations 10.1 and 10.2, we get

$$F_a = \gamma_l (1 + \cos\theta) \qquad (10.3)$$

It is clearly evident from the above equations that the surface tension of the liquid (γ_{lv}) and the contact angle can be used to express the bonding force between the liquid and solid phase. This can also be seen in Figure 10.3(b–e). It is clearly seen from Figure 10.3(b) that for absolute wetting, the angle of contact $\theta=0$ and $\gamma_{sv} = \gamma_{lv} + \gamma_{sl}$. Likewise, for good wetting, the contact angle θ should be between 0° and 90°. As indicated by Figure 10.3(c), with an increase in trend of contact angle, the surface

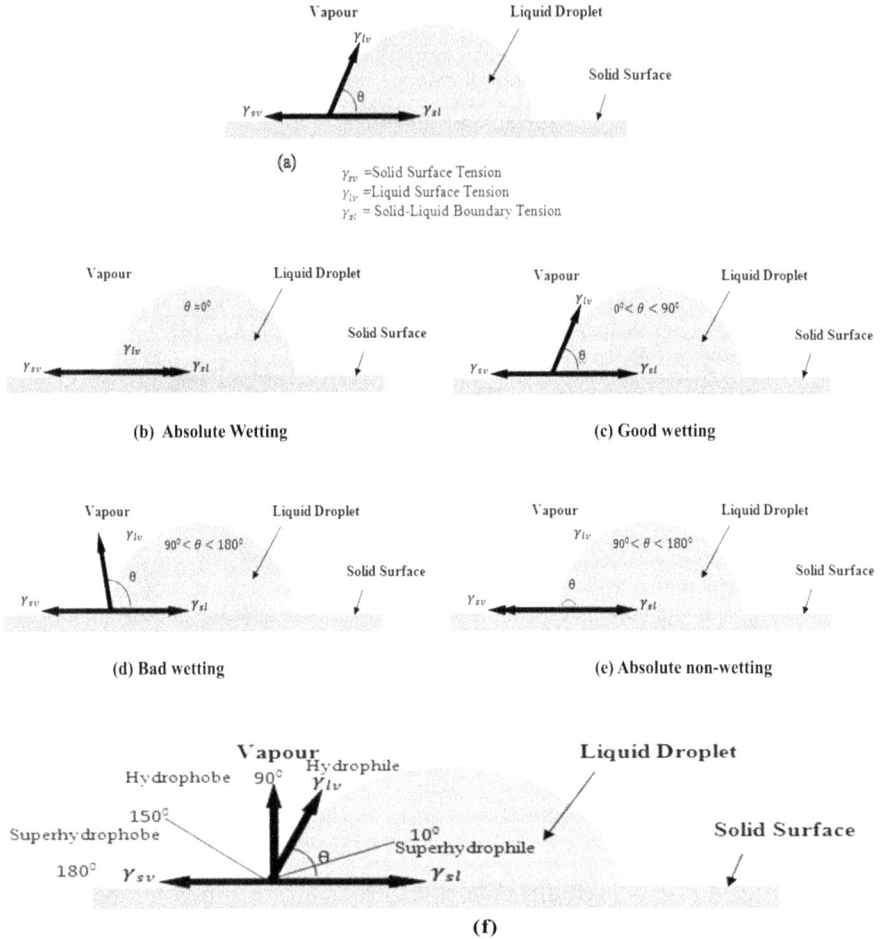

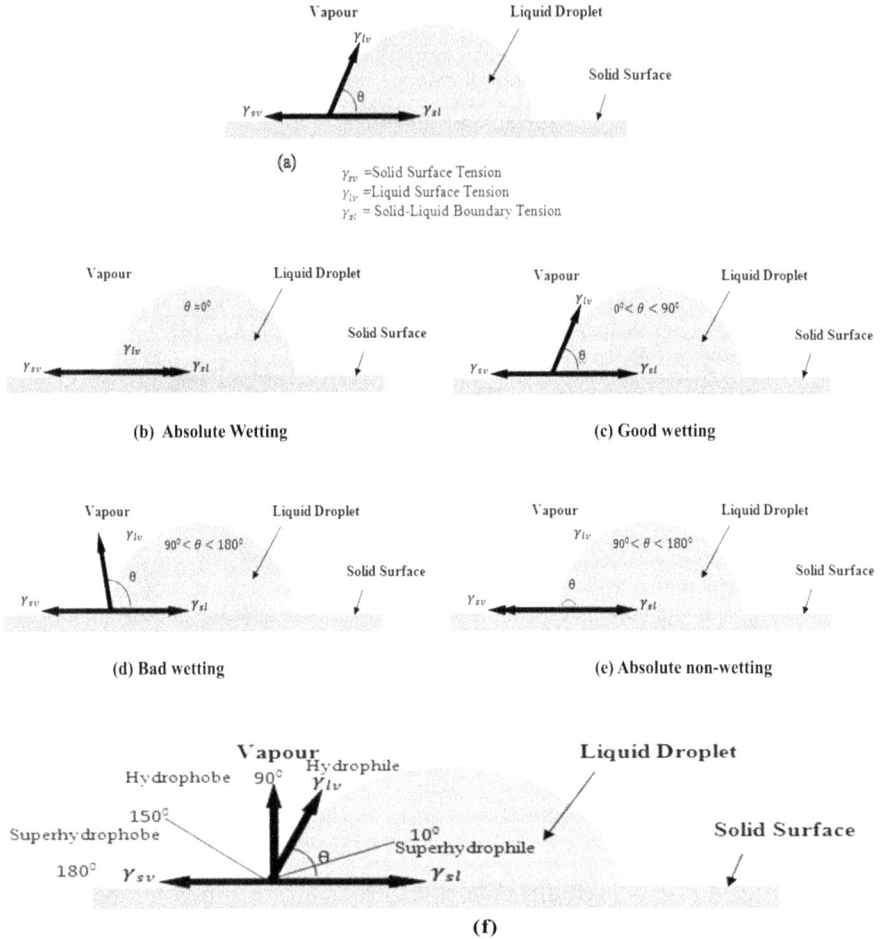

FIGURE 10.3 (a–e) Hydrophobicity based on surface energy and contact angle, (f) wetting behavior in terms of contact angle.

tension relation becomes $\gamma_{sv} > \gamma_{sl}$. From Figure 10.3(d) it is observed that the contact angle lies between 90° and 180° and as of a very large contact angle $(\theta), \gamma_{sv} < \gamma_{sl}$ An increase in the value of surface energy at the solid-vapor interface as compared to the solid-liquid interface is an indication of bad wetting (Kwok and Neumann, 1999; Kalantarian et al., 2009). For non-wetting conditions, the contact angle should be 180° and $\gamma_{sv} = \gamma_{lv} - \gamma_{sl}$ as seen from Figure 10.3(e). Hence, surface energy and contact angle measurement are the two factors used as a basis for defining the degree of hydrophobicity of the surface.

On the basis of interaction between water and solids in particular the surfaces can be classified as hydrophilic or hydrophobic as shown in Figure 10.3(f) (Bird et al., 2013).

10.3 CASE STUDY RELATED TO THE LOCOMOTION PATTERN OF BIOLOGICAL SPECIES AND ROBOTS

Another interesting case study is related to the different locomotion patterns in species that have become a source of inspiration for the design and development of bio-inspired robots that are widely used in today's era of Industry 4.0. Bio-stimulated robotic locomotion is a fairly new subcategory of bio-inspired locomotive pattern. It studies principles from nature and makes use of them in the design of real-international engineered structures. Bio-mimicry and bio-stimulated layout both are very confusing in terms of definition. Bio-mimicry is copying from nature, while bio-inspired is analyzing nature and developing a mechanism that is simpler and more effective than the system found in nature. Bio-mimicry also includes the improvement of a unique area of robotics known as gentle robotics. Organic structures have been optimized for specific responsibilities in line with their habitat. However, they are multifunctional and are not designed for only one unique functionality. Bio-stimulated robotics studies organic systems, and searches for mechanisms that can remedy problems in engineering. The design engineers have continuously worked on them in order to simplify and analyze the mechanism for identifying the solution of particular problem. Bio-inspired roboticists are typically interested in bio-sensors (like eyes), bio-actuators (like muscles) or bio-substances (like spider silk). Most of the robots have some sort of locomotion system. Thus, in this case, special modes of animal locomotion are examined and some examples of the corresponding bio-inspired robots are given. The framework's shape matches different bio-inspiration fashions described in, however additionally information the unique goals related to the improvement of bio-inspired morphologies for on foot robots as shown in Figure 10.4.

In the first actual step of biological models, designers begin with a couple of biological solutions of interest as proposed assets with the purpose of using them to resolve a particular problem. This step involves the selection of biological species which carry out a desired function in an effort to be analyzed best by its morphological observe. The biological species is chosen on the basis of shared characteristics that can be only defined by the taxonomy. Those species with different locomotion modes like crawling, swimming, flying and jumping become a model for the development of robots. The different types of biological species with planar and vertical locomotion are shown in Table 10.2.

In the second phase related to design, the biological model becomes a source of inspiration in developing a model of a robot. Initially, a conceptual design of the robot is made by considering different focal points of the desired functions. In this phase, the engineer is getting to know enough about the biology and the biologist is gaining knowledge related to engineering laws and restrictions to recognize the useful information for the project, duties with a regularly underestimated time duration.

During the third phase related to the implementation, a prototype of the model is prepared. Initially an optimization technique is implemented to optimize

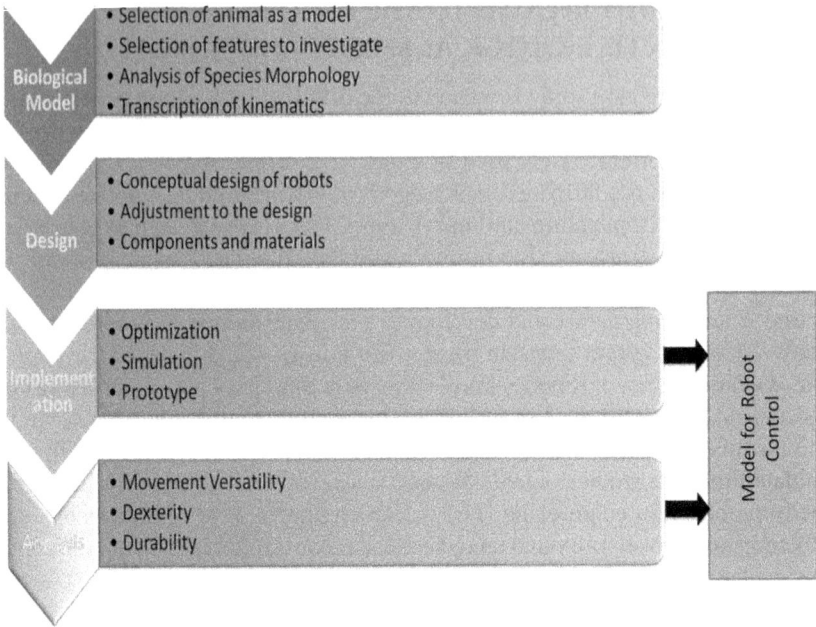

FIGURE 10.4 Phases involved in the development of robots (Billeschou et al., 2020).

parameters like the weight and load of the actuator that controls the robot's movement. After that, software and physical models may be used to check the leg design and scaling aspect of the animal's range of motion and also to check the overall performance of the movement. Simulations of the robot are used to perform preliminary assessments of the robot's overall performance. Additional optimizations might be done to improve mobility. Subsequently, the prototype is constructed as a physical robot. Finally, advanced software may be used to initiate the improvement of the robot controller in analogous amid the subsequent steps.

In the final phase related to analysis, the morphology's static and dynamic behavior is analyzed within the context of the morphological and mechanical layout. The evaluation can guide biomechanical hypotheses and design engineers on how to improve gait styles and reduce the fatigue-inducing impulse from foot strikes. A template model may be used as a reference to suggest the differences that the anchor morphology induces. Billeschou et al. (2020) proposed a development framework for transferring animal morphologies to robots and substantiated it with a replication of the ability of the dung beetle species *Scarabaeus galenus*, using the same morphology for both locomotion and object manipulation. Detailed information on the development of a robot from the initial phase to the final phase is shown in Figure 10.5.

TABLE 10.2
Species with Planar and Vertical Locomotion Modes

Biological Species as Model	Planar Locomotion	Vertical Locomotion
Spider	 Spider with walking locomotion	 Spider with climbing locomotion
Snake	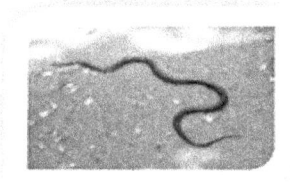 Snake with walking locomotion	 Snake with climbing locomotion
Gecko	 Gecko with walking locomotion	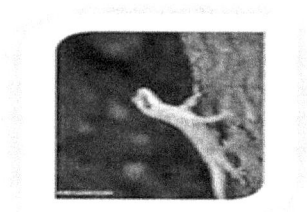 Gecko with climbing locomotion
Apes	 Apes with walking locomotion	 Apes with climbing locomotion

Source: Tan et al., 2019.

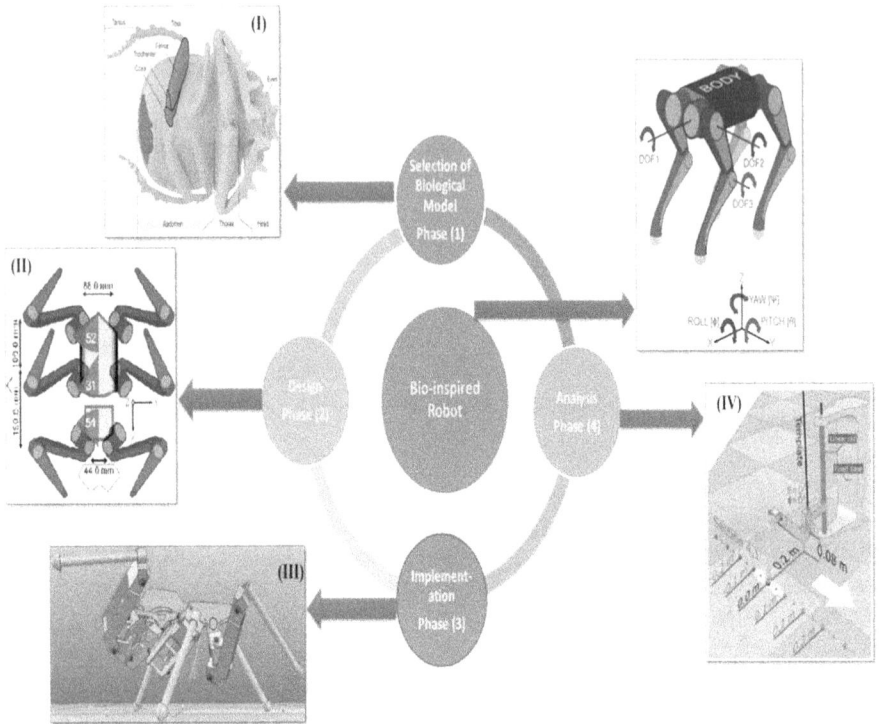

FIGURE 10.5 Development of new robot inspired by beetle species *Scarabaeus galenus* (Billeschou et al., 2020).

10.4 CASE STUDY RELATED TO THE GENERATION OF ELECTRICITY IN ELECTRIC EELS THAT BECAME A SOURCE OF INSPIRATION FOR ARTIFICIAL ELECTRIC ORGANS

The capacity of electric eels to stun victims with a high-voltage zap is renowned. Scientists have adopted the eel's astonishing secret to create a squishy, flexible new way to generate electricity, inspired by the species (Gotter et al., 1998; Schroeder et al., 2017). Their novel artificial electric "organ" could provide power in conditions where traditional batteries would fail. The new artificial organ, which is made primarily of water, can function in moist environments. As a result, a device like this may be used to power soft-bodied robots that can swim or move like genuine animals. It may potentially be used to power a heart pacemaker within the body. It also creates power with a simple motion: a squeeze. Electrolytes, or specialized cells, are used by electric eels to generate

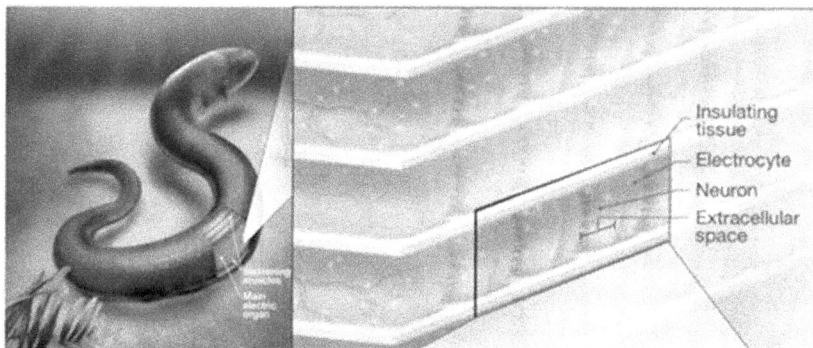

FIGURE 10.6   Electric eels along with their functionality (Schroeder et al., 2017).

their electric charge. Those cells occupy the majority of an eel's 2-meter body. Thousands of these cells are lined up in rows, resembling rows and rows of stacked hot-dog buns as shown in Figure 10.6.

They resemble muscles; however, they don't assist the animal in swimming. To generate electricity, they direct the flow of charged particles known as ions. Like pipes, tiny tubes link the cells. These channels allow positively charged ions to flow outward from both the front and back of a cell most of the time. When the eel seeks to deliver an electric shock, however, its body opens some channels while closing others. This now allows positively charged ions to flow in one direction and out the other, much like an electric switch. These ions build a positive electric charge in some spots as they migrate. This generates a negative charge in other parts of the body. Each electrocyte generates a small amount of electricity as a result of the charge differential. With so many electrolytes, even small amounts pile up. They can combine to create a jolt powerful enough to stun fish or knock a horse down.

Electrolytes are used in a unique way by the new artificial organ. It has no resemblance to an eel or a battery. Instead, two sheets of translucent plastic are covered in colorful dots. The whole thing looks like a bunch of bright, fluid-filled bubble wrap sheets. Each colored dot represents a distinct gel. There are red and blue dots on one sheet. The main element in the red dots is saltwater. Freshwater is used to make the blue spots. The dots on the second sheet are green and yellow. Positively charged particles are present in the green gel. As seen in the Figure 10.7, the yellow gel contains negatively charged ions.

Place one sheet on top of the other and press to create electricity. On one sheet, the red and blue dots will nestle between the green and yellow dots on the other sheet. The red and blue spots represent the electrocytes channels. Charged particles will be allowed to travel between the green and yellow spots. This charge movement creates a tiny trickle of electricity, just like in an eel. A lot of dots together, just as in an eel, might give you a big jolt. The scientists were able to generate 100 volts in the lab. A 3-D printer can be used to create artificial organ-charged gels. This technique is also inexpensive because the major ingredient is water. It's also a little rough. The gels are still functional after being pushed, compressed and stretched. An eel isn't

FIGURE 10.7   Large complementary arrays of printed hydrogel lenses.

very productive. To produce a little jolt, the eel requires a lot of energy in the form of food. As a result, eel-based cells are unlikely to replace other renewable energy sources like solar and wind power. But that doesn't rule out the possibility of them being useful. They are appealing in situations where a modest amount of power is required without a lot of metal waste. Soft robots, for example, may be able to operate on very little energy. These devices are built to withstand extreme conditions. They might go under the sea or climb volcanoes. They might look for survivors in disaster zones. It's critical in cases like these that the power source doesn't die if it gets wet or crushed. The squishy gel grid concept could generate electricity from a variety of unexpected sources, including contact lenses. Gels are simple and long-lasting; however, they only create modest currents that aren't useful. Researchers have come up with a solution by making a vast grid of gel dots. The gels can imitate the eel's channels and ions by dividing the dots between two sheets. The researchers are now looking for ways to improve the organ's performance.

10.5   BIO-MIMETICS AND BIO-TEMPLATING OF NATURAL MATERIALS FOR A SUSTAINABLE MANUFACTURING APPROACH TO BIONANOTECHNOLOGY

Sustainable development is a globally recognized mandate, and it consists of green or eco-friendly production practices as shown in Figure 10.8. Such practices configured the self-recovery and self-replenishing functionality of herbal ecosystems. Green production encompasses synthesis, processing, fabrication and system optimization, and also testing, overall performance evaluation and reliability (Paris et al., 2010; Sarkar et al., 2008). The opportunity of adopting sustainable materials as alternatives to

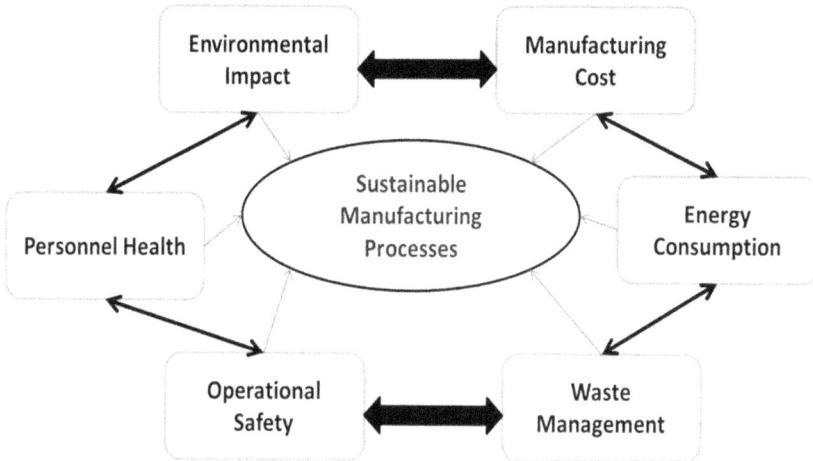

FIGURE 10.8 Sustainable manufacturing processes.

contemporary technological materials whose fabrication is primarily based on the use of non-renewable resources including oil, coal and natural gas is a global challenge. One of the major advantages associated with green manufacturing is related to the bio-template of the living organism. This is the major reason that material scientists are increasingly interested in bio-mimetic materials. Bio-mimetic materials studies, often called material bionics or bio-inspired substances studies, an old subject, has now started to broaden very dynamically. Indeed, bio-inspired ideas no longer result from the general study of structures alone; they instead require an investigational approach related to its characteristic in relation with biological substances.

Green and sustainable materials processing and production are seen as a key enabler of sustainable development. The various natural materials that are found in nature and can be treated as sustainable materials are hemp, recycled cotton, organic linen, cork, bamboo and coconut; bio-plastics and straw are shown in Figure 10.9. Since many nations have started lifting their ban on growing hemp, it has improved its reputation. It is fast turning into an ordinary sustainable cloth, utilized in food merchandize, cloth merchandize and even as an eco-friendly building material. Due to the above facts, hemp is very much used nowadays as an eco-friendly material as compared to cotton. Cotton is a really perfect and sustainable material. However, in terms of sustainability, cotton farming at the scale and depth that is required to keep up with the demand from the fashion industry. Another sustainable material is recycled cotton which is widely used in textile industries due to its plentiful availability. Cotton farming is water-intensive, and the crop requires the use of insecticides and a massive land footprint. It isn't a totally green crop at all. That's where recycled cotton comes in. There are already great volumes of cotton fabric in the world. The majority is in landfills, as off-cuts and waste from factories or complete articles dumped by the fashion industry or discarded by consumers. Recycling reduces waste and reduces the need for intensive cotton farming.

FIGURE 10.9   Various types of green materials.

Now days, bamboo is very much widely used by all manufacturers and companies due to its variety of properties, and in recent times you'll find all kinds of products made from bamboo. Therefore, new growing materials are meeting our needs and accordingly helping to achieve sustainability in our way of life. The various nature-inspired materials for sustainable development are mentioned in Table 10.3.

10.6   CHALLENGES AND FUTURE PROSPECTS

In today's era of Industry 4.0, the scope of advanced materials that have characteristics and features similar to those of living and non-living organisms is rising continuously. As nature-inspired materials have better characteristics, they are widely used in engineering applications. These types of materials also have a wide scope in sustainable development. Nature-inspired materials have excellent features that have been widely used in the development of novel techniques and processes related with the fabrication of bio-template materials for sustainable development. Therefore, nature-inspired materials have been consistently used in every field of engineering like bio-mechanics, aviation industries, automobile industries and defense applications.

TABLE 10.3
Various Nature-Inspired Materials for Sustainable Development

Nature-Inspired Features	Targeted Area of Application
Materials that became a source of inspiration in achieving "sustainability gain"	
Cob	• Environmentally friendly.
	• Smart texture.
	• Act as insulating and efficient materials.
Recycled steel	• Maintains properties after recycling.
	• Excellent durability.
	• Saves large portion of energy costs.
Sheep's wool	• Saves large portion of energy costs.
Reclaimed, recycled or sustainable wood	• Biodegradable.
	• Good aesthetic features.
Bamboo	• Good mechanical strength.
Recycled plastic	• Excellent durability.
	• Robust material.
	• Excellent sound retaining capability.
Clay brick	• Acts as composite wall to maintain the room temperature during summer and winter.
Newspaper wood	• Biodegradable.
	• Good aesthetic features.

It is now very important to adopt a loose definition of bio-mimetic material with material chemistry; this is due to the fact that there is lots of conceptual overlap between the techniques which are stimulated with the aid of biology and those which are based on molecular engineering in material chemistry. The major task in this area is how to translate the harsher regimes of modern fabrication generation into temperature and chemical reaction-based techniques like bio-mineralization. Unfortunately, we do not now know exactly which problem is to be solved. It may be to provide a study material and additionally to meet some exclusive biological constraints. This means that we won't prevail if we observe without modifications the answers found by means of nature as the ultimate solution for a certain unknown requirement. So, we must cautiously observe the biological machine and understand the shape-function properties of the biological material in the context of its physical and biological constraints. Careful investigation of a biological system with same features is essential for bio-mimetic fabrication studies.

Another important area that researchers have to keep in mind is bio-mineralization. In bio-mineralization, large amounts of poisonous chemicals are released that become a concern in achieving sustainable growth. So, careful study in this regard is required and therefore careful study of bio-mineralization while following the nature-inspired processes and materials. Lastly, this new perception is helpful while developing a new kind of biocompatible substances and interfaces which will

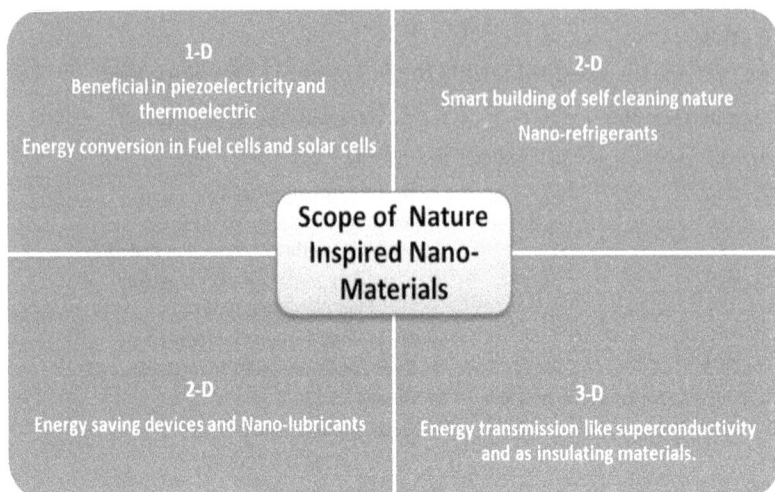

FIGURE 10.10   Scope of nature-inspired materials.

accelerate the creation of probes, sensors and implants to be used within the bionic machine; this may depend on advances in our knowledge of local biological nano-materials, which includes bio-minerals as shown in Figure 10.10.

10.7   CONCLUSION

Natural materials not only effectively provide a renewable supply of substances and energy, but in addition they may function as direct templates or as a suggestion for the expansion of novel materials of technical relevance. Scientific study of novel categories of materials also tells about the development of nature-inspired processes and their classifications, nature-inspired methods as well as nature-inspired materials. It also suggests various types of gains like electric gain, biological gain, mechanical gain and sustainability gain for sustainable manufacturing approaches in bionanotechnology. This chapter discussed case studies related to nature that open vistas for new developments like bionic cars, wind turbines, self-cleaning building materials, the hydrophobicity phenomenon, the Shinkansen train, power sources and solar power. The chapter also suggests how different locomotive patterns of species can be used in the development of new vehicles with ergonomic design. Also, study related to the sensory behavior of plants opens vistas for the development of new sensors and transducers used in engineering applications. The role of bionanotechnology in the development of sensors is also beneficial for future smart machines with special features. At last, study related to bio-mimicry and bio-template materials opens a channel for material scientists as well as bionanotechnologists to work for sustainable growth and societal benefits.

REFERENCES

Billeschou, P., Bijma, N.N., Larsen, L.B., Gorb, S.N., Larsen, J.C., and Manoonpong, P. (2020). Framework for developing bio-inspired morphologies for walking robots. *Appl. Sci.* 10(19), 6986.

Bird, J.C., Dhiman, R., Kwon, H.M., and Varanasi, K.K. (2013). Reducing the contact time of a bouncing drop. *Nature.* 503(7476), 385–388.

Eggermont, M. (2007). Biomimetics as problem-solving, creativity and innovation tool. *Proceedings of the Canadian Engineering Education Association (CEEA)*, July 22 – 24, Manitoba.

Fratzl, P. (2007). Biomimetic materials research: What can we really learn from nature's structural materials? *J. R. Soc. Interface.* 4(15), 637–642.

Gotter, A.L., Kaetzel, M.A., and Dedman, J.R. (1998). Electrophorus electricus as a model system for the study of membrane excitability. *Comp. Biochem. Physiol. A Mol. Integr. Physiol.* 119(1), 225–241.

Hashim, J., Looney, L., and Hashmi, M.S. (2001). The wettability of SiC particles by molten aluminium alloy. *J. Mater. Process. Technol.* 119(1–3), 324–328.

Huang, G., Li, F., Zhao, X., Ma, Y., Li, Y., Lin, M., Jin, G., Lu, T.J., Genin, G., and Xu, F. (2017). Functional and biomimetic materials for engineering of the three-dimensional cell microenvironment. *Chem. Rev.* 117(20), 12764–12850. https://doi.org/10.1021/acs.chemrev.7b00094.

Kalantarian, A., David, R., and Neumann, A.W. (2009). Methodology for high accuracy contact angle measurement. *Langmuir.* 25(24), 14146–14154.

Katiyar, N.K., Goel, G., Hawi, S., and Goel, S. (2021). Nature-inspired materials: Emerging trends and prospects. *NPG Asia Mater.* 13(1), 1–6.

Kwok, D.Y., and Neumann, A.W. (1999). Contact angle measurement and contact angle interpretation. *Adv. Coll. Interface Sci.* 81(3), 167–249.

Laurent, V., Chatain, D., and Eustathopoulos, N. (1987). Wettability of SiC by aluminium and Al-Si alloys. *J. Mater. Sci.* 22(1), 244–250.

Lepora, N.F., Verschure, P.F., and Prescott, T.J. (2013). The state-of-the-art in biomimetics. *Bioinspir Biomim.* 8(1), 1–11. https://doi.org/10.1088/1748-3182/8/1/013001.

Naik, R.R., and Singamaneni, S. (2017). Introduction: Bioinspired and biomimetic materials. *Chem. Rev.* 117(20), 12581–12583.

Paris, O., Burgert, I., and Fratzl, P. (2010). Biomimetics and biotemplating of natural materials. *MRS Bulletin.* 35(3), 219–225.

Rodríguez, R.E., Agarwal, S.P., Kazyak, E., Das, D., Shang, W., Skye, R., Deng, T., and Dasgupta, N.P. (2018). Biotemplated Morpho butterfly wings for tunable structurally colored photocatalysts. *ACS Appl. Mater. Interf.* 10(5), 4614–4621.

Sarkar, P., Phaneendra, S., and Chakrabarti, A. (2008). Developing engineering products using inspiration from nature. *J. Comput. Infor Sc. Eng.* 8(3), 1–9.

Schroeder, T.B., Guha, A., Lamoureux, A., VanRenterghem, G., Sept, D., Shtein, M., Yang, J., and Mayer, M. (2017). An electric-eel-inspired soft power source from stacked hydrogels. *Nature.* 552(7684), 214–218.

Sharma, V.K., Kumar, V., and Joshi, R.S. (2021). Manufacturing of stable hydrophobic surface on rare-earth oxides aluminium hybrid composite. *Proc. Inst. Mech. Eng. E.* 235(4), 899–912.

Soudi, N., Nanayakkara, S., Jahed, N.M., and Naahidi, S. (2020). Rise of nature-inspired solar photovoltaic energy convertors. *Sol. Energy.* 208, 31–45.

Tan, N., Sun, Z., Mohan, R.E., Brahmananthan, N., Venkataraman, S., Sosa, R., and Wood, K. (2019). A system-of-systems bio-inspired design process: Conceptual design and physical prototype of a reconfigurable robot capable of multi-modal locomotion. *Front. Neurorobot.* 13, 78.

Wegst, U.G., Bai, H., Saiz, E., Tomsia, A.P., and Ritchie, R.O. (2015). Bioinspired structural materials. *Nat. Mater.* 14(1), 23–36.

Zhao, Q., Fan, T., Ding, J., Zhang, D., Guo, Q., and Kamada, M. (2011). Super black and ultra-thin amorphous carbon film inspired by anti-reflection architecture in butterfly wing. *Carbon.* 49(3), 877–883.

11 Bionanotechnology and Microbial Synergism in Biohydrogen Generation and Wastewater Treatment

Priya Singh, Rachita Sharma, Naveen Dwivedi, and Sanjeev Maheshwari

CONTENTS

DOI: 10.1201/9781003316374-11

11.1 INTRODUCTION

Recent work on nanotechnology will give us a broad range of advances in different sectors. The application of nanotechnology in biohydrogen synthesis will enhance the biohydrogen produced by microorganisms. Different inorganic and organic nanoparticles play an important role in increasing the efficiency of product formation by using bio-waste matter such as agricultural waste or food industries waste as feed. The overall hydrogen production will depend upon the nature of the catalytic function of the enzymes, the metabolic pathway used by the microorganism, and the electron transport system used. Nanotechnology provides us with many pathways which are much more efficient than the conventional techniques. The nanoparticles used in nanotechnology have a large surface area to volume ratio which makes them superior to other techniques (Baruah et al., 2019). This high surface area to volume ratio gives rise to properties like high absorbing capacity and high interacting and reaction capabilities. As energy demand is increasing in recent years there is a fear of the exhaustion of non-renewable sources like fossil fuels from nature. Fossil fuels have a negative impact on nature as well as on humans due to toxic gaseous emissions. To reduce the dependency on fossil fuels, we need a more convenient, renewable, and eco-friendly source of energy like biofuels. Biofuels include bio-ethanol, bio-methanol, biodiesel, and hydrogen (H_2) and are produced by using biological waste material. Hydrogen is said to be a clean fuel as no carbon dioxide is released into the atmosphere while burning. Hydrogen is produced using different sources including fossil fuels, coal and natural gas. Out of all methods, biohydrogen synthesis is the most desired and eco-friendly.

Biohydrogen is an environmentally friendly, third-generation biofuel synthesized by microorganisms using waste organic materials. There are different methods for biological hydrogen production like dark fermentation, photo-fermentation, direct bio-photolysis, indirect bio-photolysis, and bio-electrolysis utilizing different prokaryote (cyanobacteria and bacteria) and eukaryote (algae) organisms. The reaction for biohydrogen production is mainly catalyzed by the enzymes hydrogenase and nitrogenase. To increase the dihydrogen production rate and yield, various approaches are taken, i.e., improving the microbial strain or using mixed strains, optimizing media, pH and temperature, and other external conditions. Some of the improved microbial strains that are used for biohydrogen production are *Escherichia coli*, *Clostridium butyricum*, *Clostridium beijerinckii*, *Enterobacter aerogenes*, and *Enterobacter asburiae*. Sometimes mixed strains are also used comprising two or more bacterial species for improving microbial dihydrogen synthesis.

Water is a renewable or conventional source of energy and is essential for the continuation of life. But still, it is necessary to conserve water because everything comes at a price. This water waste can be in the form of sewage water, water effluents from industries and homes, which makes it unsafe for human use. This wastewater is usually released into nature without any treatment which not only will cause severe illness in humans but also harms other animals. Due to these reasons, water can be used again after some treatment with wastewater treatment technologies to make it more potable and usable. Wastewater treatment technology filters and treats the water by removing contamination, harmful toxic ions, and disease-causing microbes. Effluents from industries and from agriculture (in the form of fertilizers, pesticides) include many toxic ions. To remove these toxic ions, various electro-chemicals, oxidation processes, and valorization techniques are used to make waste-water more sustainable (Gupta and Shukla, 2020). But these techniques are very costly, and therefore we have to find other ways to treat the wastewater which are less costly as compared to the above given techniques.

11.2   BIOHYDROGEN SYNTHESIS

Dihydrogen (H_2) is synthesized using various approaches and different fuel sources like fossil fuel, coal, renewable energy, etc. Out of all approaches, microbial synthesis is considered eco-friendly and the most desired method for dihydrogen production. The biohydrogen synthesis process is mainly classified into three parts as shown in Figure 11.1, i.e., photo-fermentation, dark fermentation, direct bio-photolysis, indirect bio-photolysis, and electrolysis.

11.2.1   Photo-Fermentation

In the photo-fermentation process, the H_2 and CO_2 are produced by photosynthesis bacteria consuming the organic substrate in the presence of solar energy. The bacteria are able to capture light as the source energy for light-dependent hydrogen production. The microorganism completely consumes organic acid (butyric acid, acetic acid) as a substrate for the product formation, i.e., hydrogen and CO_2. H_2 generation by photo-fermentation is catalyzed by hydrogenase and nitrogenase enzymes. The oxygen that is formed during the photolysis will show an inhibitory effect on both of the enzymes. It has a comparatively lower hydrogen production rate than dark fermentation due to the nitrogenase activity as well as dependency on the light.

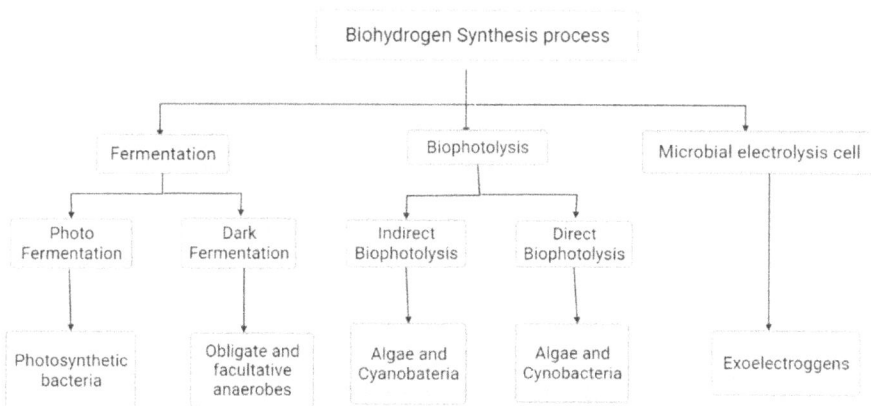

FIGURE 11.1   Classification of biohydrogen synthesis process.

Some bacterial species involved in photo-fermentation are *Chlorobium,* *Halobacterium, Rhodospirillum, Rhodopseudomonas,* etc.
    Reaction:

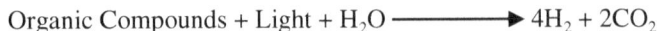

$$\text{Organic Compounds} + \text{Light} + H_2O \longrightarrow 4H_2 + 2CO_2$$

11.2.2 DARK FERMENTATION

In the dark fermentation process, the bacteria can consume a variety of organic substrates to produce biohydrogen in the absence of light energy. Either obligate or facultative anaerobes are used in the dark fermentation process which can use various types of organic material in the absence of oxygen. This method is more preferred than photo-fermentation due to easy maintenance and handling, simple reactor technology, beneficial by-product formation, and no light dependency. The only limitation is that it uses only one-third of the substrate for hydrogen formation while two-thirds are consumed for other fermentation products (by-products). In recent years, a large number of researchers have been working to overcome this limitation through the application of nanotechnology.
    Bacteria involved are *Escherichia coli, Enterobacter aerogenes, Clostridium* *paraputrificum, Clostridium beijerinckii,* etc.
    Reaction:

$$C_6H_{12}O_6 + 2\ H_2O \longrightarrow 4H_2 + CO_2 + 2CH_3COOH$$

11.2.3 BIO-PHOTOLYSIS

In bio-photolysis, algae or cyanobacteria break the water into hydrogen and oxygen by utilizing light energy. Water is used as the substrate for the reaction which is abundant and inexhaustible. It is classified into the direct bio-photolysis process and the indirect bio-photolysis process. Microbial species used in photolysis are cyanobacteria and microalgae.

11.2.3.1 Indirect Bio-Photolysis

In this process, the cyanobacteria decompose the water into hydrogen and oxygen in the presence of sunlight by photosynthesis. In this process, the electrons released by H_2O splitting are accumulated into carbohydrates which on dark fermentation produce H_2. Indirect bio-photolysis occurs mainly in vegetative cells and heterocysts, and both hydrogenase and nitrogenase enzymes play an important catalytic role in biohydrogen synthesis.
    Reaction:

$$6CO_2 + 6H_2O \longrightarrow C_6H_{12}O_6 + 6O_2$$

$$C_6\ H_{12}O_6 + 6H_2O \longrightarrow 12\ H_2 + 6CO_2$$

11.2.3.2 Direct Bio-Photolysis

In direct bio-photolysis, the water directly decomposes into hydrogen and oxygen in the presence of light with the help of algae species. There is no intermediate step occurs in direct bio-photolysis in the biohydrogen synthesis process. Both photosystems, i.e., PSI and PSII, are directly involved in splitting water into H_2 and O_2 with the help of the hydrogenase enzyme. The H_2 generation is highly affected by the catalytic function of hydrogenase which is highly sensitive to O_2.

Reaction:

$$2H_2O \longrightarrow 2H_2 + O_2$$

11.2.4 MICROBIAL ELECTROLYSIS CELL

Microbial electrolysis is a newly developed technology that uses the bio-electrochemical system for the synthesis of biohydrogen by exoelectrogens bacteria using organic substrate from food industries and agricultural farms. In this process, the electric current is applied for the conversion of a wide range of organic substrates (glucose, starch, acetate, lactate, etc.) into hydrogen (H_2). Microbial electrolysis cells are mainly derived by using the technology of the microbial fuel cells.

The basic structure of microbial electrolysis is shown in Figure 11.2. It consists of an anode, cathode, membrane, exoelectrogen bacterial species, and external power source. In the presence of a small potential difference, the exoelectrogen bacteria present at the anode consume the organic matter, forming and releasing electrons,

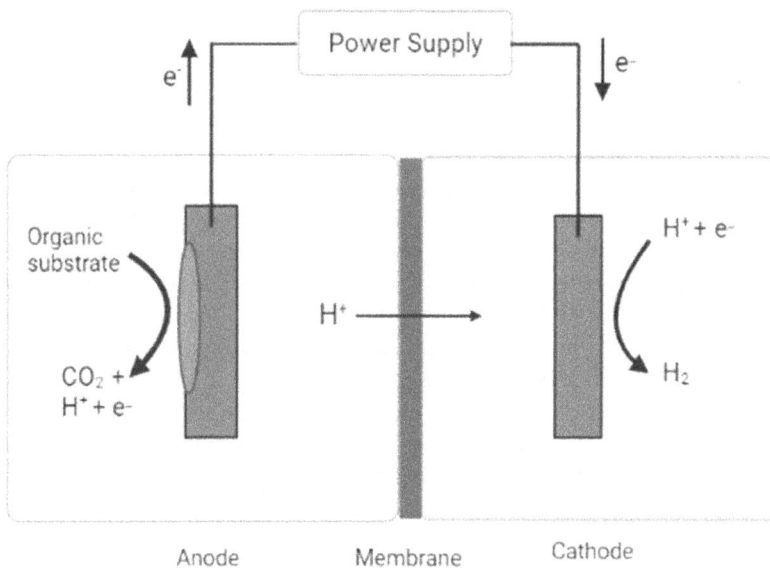

FIGURE 11.2   Microbial electrolysis cell.

protons, and CO_2. On the cathode side, the electron combines with the proton forming hydrogen.

There are various advantages of using microbial electrolysis cells for biohydrogen generation; these include:

(i) Using organic waste material from agricultural farms, food industries, and water treatment plants.
(ii) High efficiency of hydrogen production as compared to other methods.

11.3 NANOTECHNOLOGICAL INTERVENTION IN MICROBIAL HYDROGEN SYNTHESIS

The application of nanotechnology for the production of environmentally friendly biofuels has been increasing dramatically in recent years. Due to the smaller size and high surface area of nanoparticles, they help to enhance the production of biohydrogen. For the production of nanoparticles, a green approach for synthesizing is preferred over the chemical and physical approach due to less chemical toxicity and being environmentally friendly. The high surface area and quantum effect will help in greater absorption of electrons and their transfer.

Nowadays, a large amount of research occurs in the field where the implementation of various nanoparticles is used for increasing the production of biohydrogen. Dark fermentation is the most favored method for the synthesis of dihydrogen because of its light-independent behavior and the variety of substrates used as energy sources (Patel et al., 2018). The formation of by-products from the substrate will decrease the biohydrogen production efficiency. The mixed condition (light fermentation and dark fermentation) also shows a high yield for dihydrogen production (Zhao et al., 2011).

Microbial electrolysis cells are the most recent and sustainable method for the preparation of biohydrogen. Nanoparticles when used as anode or cathode material in microbial electrolysis cells will increase the synthesis of hydrogen by microbes. At the anode, the breakdown of H_2O takes place due to exoelectrogenic bacteria. If the nanoparticles are used as the material for the anode, it will enhance the electrocatalytic activity of the anode present in microbial electrolysis cells (Qiao et al., 2007). Qiao et al. (2007) used carbon nanotube/polyaniline composite as the material for the anode in the microbial fuel cell. They reported increased performance in the catalytic activity of *Escherichia coli* due to the high electrochemical activity and power density of the carbon nanotubes/polyaniline (CNT/PANI) anode (Qiao et al., 2007). Certain transitional metal phosphide nanoparticles when used as cathode catalysts show promising performance in hydrogen synthesis. Kim et al. (2019) reported that phase-pure nickel phosphide nanoparticles are used as cathode catalysts for hydrogen production in microbial electrolysis cells (MEC). The hydrogen produced over a 24-hour cycle is in the range of 0.29–0.04 L in a 1 L reactor. This amount is found to be greater than normal electrolysis cells (Kim et al., 2019).

11.4 GREEN SYNTHESIZED NANOPARTICLES AFFECTING BIOHYDROGEN SYNTHESIS

Various green synthesized metal nanoparticles (silver, gold, iron, palladium, nickel), metal oxides (iron oxide, nickel oxide), and carbon nanotubes have shown the increasing effect of biohydrogen synthesis. The intervention of bio-nanoparticles will increase the enzyme catalytic activity and microbial bioactivity which will enhance biohydrogen production (Sivagurunathan et al., 2018). The increase in hydrogen yield with the application of bio-nanoparticles is shown in Table 11.1.

11.4.1 SILVER BIO-NANOPARTICLES

According to the work done by Yildirim et al. (2021), hydrogen production is enhanced by an appropriate concentration of silver nanoparticles. Green synthesized nanoparticles are formed by using *Chlorella* sp. microalgae of size 85 nm is used in the biohydrogen synthesis process by dark fermentation. In their work, they used *Clostridium* sp. in a dark fermentation reaction using glucose feed. Different concentrations of silver nanoparticles (100–600 µg/L) were used to study their effect on biohydrogen synthesis. The maximum yield of biohydrogen found was 2.44 mole/mole glucose at Ag concentration (400 µg/L). The hydrogen production efficiency was increased by 17% with the application of bio-nanoparticles (Yildirim et al., 2021). It was also found that the presence of Ag-NPs also decreases the lag phase of hydrogen production responsible for higher H_2 yield (Zhao et al., 2013).

11.4.2 GOLD BIO-NANOPARTICLES

Gold nanoparticles increase the catalytic function of biohydrogen synthesis enzyme, i.e., hydrogenase enzyme, and improve the biohydrogen production efficiency of microbes (Zhang and Shen, 2007). Khan et al. (2013) reported enhancement in the production of biohydrogen from decomposing sodium acetate by electrochemically active biofilm (EAB) using biologically synthesized Ag nanoparticles in their study. The gold nanoparticles will enhance the catalytic function of decomposing sodium acetate by EAB releasing photons and electrons. They concluded in their experiment that the maximum rate of biohydrogen produced 105 ± 2 mL/L is observed in single chamber reactor (Khan et al., 2013). The silver nanoparticles with a diameter in the range of 6 nm to 12 nm show enhancement in hydrogen production by 17% and 9% respectively due to the increased bioactivity of microbes (Guangzhen et al., 2006).

11.4.3 PALLADIUM BIO-NANOPARTICLES

Mohanraj et al. (2014) investigated the effect of palladium ion and phytogenic palladium nanoparticles in the fermentation hydrogen synthesis process. In this experiment, *Coriandrum sativum* leaves are used for the green synthesis of palladium nanoparticles. They found that $PdCl_2$ and Pd nanoparticles increased the production of biohydrogen from glucose sources using the mixed culture and *Enterobacter*

TABLE 11.1

Biohydrogen Generation Using Nanotechnology Approach Associated with Microorganisms

S.No.	Nanoparticles	Biological Material for Nanomaterial	Feed Used	Organism for H$_2$ Production	Maximum Yield/ Maximum Rate	Conc. of Nanoparticles	Reference
1.	Silver nanoparticles	*Chlorella sp.* microalgae	Glucose	*Clostridium* sp.	2.44 mole/mole glucose	400 µg/L	Yildirim et al., 2021
2.	Gold nanoparticles	Biological material	Glucose	EAB	105 ± 2 mL/L day	-	Khan et al., 2013
3.	Palladium nanoparticles	*Coriandrum sativum* leaf extract	Glucose	*Enterobacter cloacae*	1.48 ± 0.04 mol/mole glucose	5 mg/L	Mohanraj et al., 2014
		Coriandrum sativum leaf extract	Glucose	*Mixed culture*	2.48 ± 0.09 mol/mole glucose	5 mg/L	Mohanraj et al., 2014
4.	Iron nanoparticles	*Syzygium cumini* leaf extract	Glucose	*Enterobacter cloacae*	1.9 mol/mol glucose	100 mg/L	Nath et al. 2015
5.	Iron oxide nanoparticles	*Murraya koenigii* leaf extract	Glucose	*Clostridium acetobutylicum*	2.33 ± 0.09 mol/mole glucose	175 mg/L	Mohanraj et al., 2014
6.	Nickel nanoparticles	*Eichhornia crassipes* leaf extract	Glucose		101.45 ± 3.32 mL/g glucose	20 mg/L	Zhang et al., 2021

cloacae. The observed hydrogen production yields using *Enterobacter cloacae* and mixed cultures were 1.48 ± 0.04 moleH$_2$/glucose and 2.48 ± 0.09 moleH$_2$/glucose. This means that the hydrogen yield produced using mixed culture is higher than *E. cloacae*. Using only PdCl$_2$ causes a metabolite shift which results in decreasing the hydrogen production (Mohanraj et al., 2014).

11.4.4   IRON AND IRON OXIDE BIO-NANOPARTICLES

The specific concentration of green synthesized iron nanoparticles can enhance the yield of hydrogen production by different synthesis approaches. Dhrubajyoti et al. (2015) demonstrated the effect on iron nanoparticles of size 20–25 nm produced by a biological approach using an aqueous leaf extract of *Syzygium cumini* as a capping and stabilizing agent in biohydrogen synthesis. The dark fermentation reaction uses glucose as feed by mesophilic bacteria *Enterobacter cloacae* for the emission of biohydrogen. They obtained the maximum hydrogen yield, 1.9 mol/mol glucose at 100 mg/L concentration of nanoparticles. They concluded that iron nanoparticles will increase the hydrogen production by enhancing microbial bioactivity and biocatalytic function (Nath et al., 2015).

The optimum temperature, concentration of nanoparticles, and pH play important roles in increasing the biohydrogenation process by enhancing its biocatalytic functions (Engliman et al., 2017). Mohanraj et al. (2014) reported the influence of green synthesized iron oxide nanoparticles on biohydrogen production using *Clostridium acetobutylicum* bacteria. *Murraya koenigii* leaf extract is used for the green synthesis approach for the production of iron nanoparticles of particle size of about 59 nm. The iron nanoparticles show excellent results in hydrogen production compared to FeSO$_4$ because of the higher activity of ferredoxin. The maximum yield of biohydrogen observed was 2.33 ± 0.09 mol-H$_2$/mol glucose in a closed reactor using glucose as feed. The optimum concentration of iron(II) nanoparticles at which the maximum yield was observed was 175 mg/L. The increased hydrogen production rate and hydrogen concentration observed were 34 ± 0.8–$52 \pm 0.8\%$ and 23–25 mL/hour (Mohanraj et al., 2014).

11.4.5   CARBON NANOTUBE

Using carbon nanotube in biohydrogen synthesis will increase the yield and rate of biohydrogen production by immobilizing the bacterial species. Boshagh et al. (2019) demonstrated the influence of various concentrations of COOH-functionalized multi-walled carbon nanotube (COOH-MWCNT) on hydrogen production by immobilizing *Enterobacter aerogenes* bacterial species. The various concentrations of COOH-MWCNT were 0.3 mg/ml, 0.6 mg/ml, and 1.2 mg/ml. At a 1.2 mg/ml concentration of COOH-MWCNT, a maximum yield of 2.2 mol/mol glucose and a maximum hydrogen production rate of 2.72 L/L hour were observed. The efficiency of utilization of glucose is increased by 96.20% using a carbon nanotube. The high yield, the high production rate of hydrogen, and increased efficiency are due to the low log phase of bacteria by COOH-MWCNT (Boshagh et al., 2019).

11.4.6 OTHER NANOPARTICLES

Some other nanoparticles affect biohydrogen biosynthesis by increasing the catalytic function of enzymes (hydrogenase enzyme). The effect of green synthesized nickel oxide nanoparticles on biohydrogen production is explained by Zhang et al. (2021) using the fermentation process. A green synthesis of nickel oxide nanoparticles is developed using the leaf extract of *Eichhornia crassipes* of diameter (9.1 ± 2.6 nm). The observed maximum yield and maximum hydrogen concentration using glucose feed is 101.45 ± 3.32 mL/g glucose and 4842.19 ± 23.43 mL/L respectively with 20 mg/L concentration of nickel oxide nanoparticles. Due to the intervention of nickel oxide nanoparticles the efficiency of biohydrogen synthesis is increased by 63% due to increasing hydrogenase activity enhancing hydrogen production. It also improves the consumption efficiency of organic substrate increasing hydrogen generation (Zhang et al., 2021). Nano-titanium oxide particles can also be used to increase the efficiency of biohydrogen synthesis due to the increased efficiency in protein and carbohydrate breakdown by TiO_2 nanoparticles (Zhao and Chang, 2011).

Mixed nanoparticles also have applications in enhancing biohydrogen production by using microbes. The addition of more than one nano bio-particle will show more hydrogen production efficiency than independent nanoparticles as it will promote the activity of ferredoxin oxidoreductase and hydrogenase enzyme (Gadhe et al., 2015).

11.5 WASTEWATER TREATMENT THROUGH BIONANOTECHNOLOGY

Generally, wastewater treatment is divided into three categories as shown in Figure 11.3. These are:

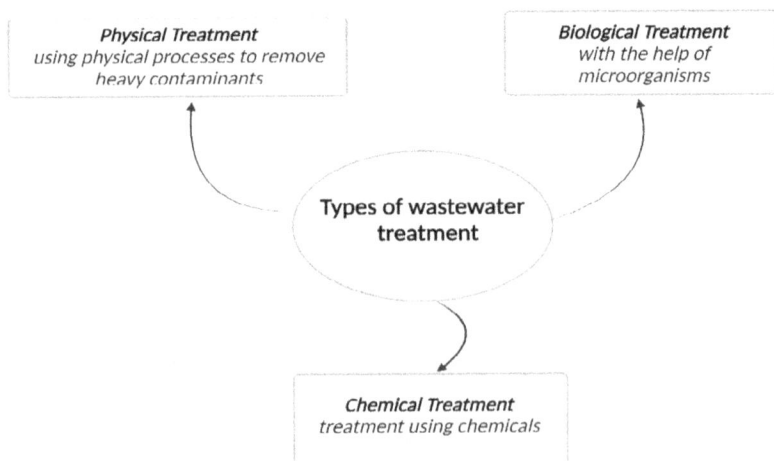

Physical Treatment
using physical processes to remove heavy contaminants

Biological Treatment
with the help of microorganisms

Types of wastewater treatment

Chemical Treatment
treatment using chemicals

FIGURE 11.3 Types of wastewater treatment.

- Physical treatment
- Chemical treatment
- Biological treatment

Physical methods are used for the cleaning of wastewater. These methods consist of processes like screening and sedimentation used to remove the solids. No use of chemicals is involved in these treatment methods. Chemical water treatment involves the use of various chemicals for the treatment of wastewater. In the biological treatment process the degradation of organic contaminants is done using microorganisms. This treatment relies on bacteria, nematodes, algae, fungi, protozoa, and rotifers to break down unstable organic waste, using normal cellular processes, into stable inorganic forms. Based on the process, the biological treatment of wastewater is largely classified into two types as follows:

- Biological aerobic treatment (in the presence of oxygen)
- Biological anaerobic treatment (in the absence of oxygen)

11.5.1 GREEN NANOTECHNOLOGY IN WASTEWATER TREATMENT

Nanoparticles synthesized from microorganisms, green nanoparticles, aid in the eco-friendly removal of pollutants, contaminants, toxic ions, and organic matter from wastewater. These bio-nanoparticles provide a cost-effective, efficient, and easy method for the removal of contaminants. There have been many recent advances in different nanoparticles for the removal of contaminants from wastewater. These include (Figure 11.4):

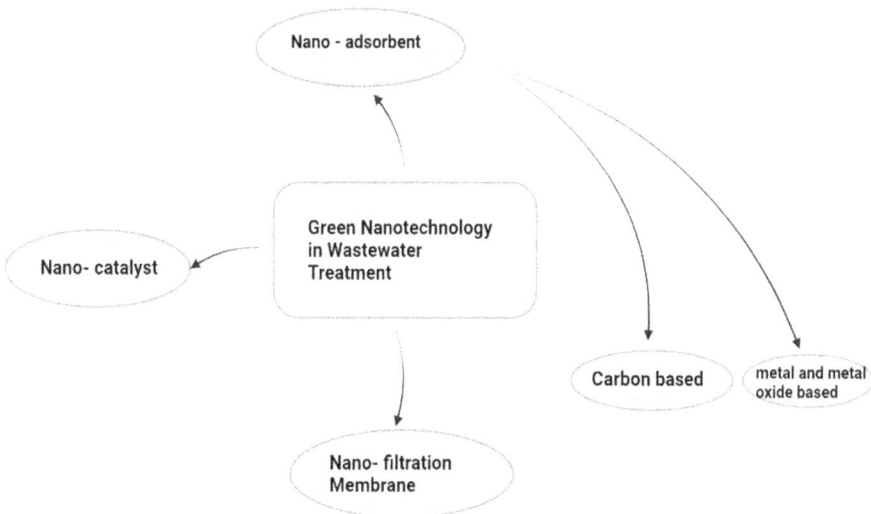

FIGURE 11.4 Different methods of wastewater treatment using a green/bionanotechnological approach.

- Nano-adsorbents
- Nanofiltration membranes
- Nano-catalysts

11.5.1.1 Nano-Adsorbents

Nano-adsorbents are nanomaterials which adsorb harmful contaminants on their surface. These are widely used nanoparticles for the removal of organic and inorganic pollutants from wastewater (Kumari et al., 2019). These nano-adsorbents have very high adsorption capacity with wide application in water purification, remediation, and treatment processes. The most commonly used nano-adsorbent in wastewater remediation is chitosan nanoparticles. It is a polysaccharide synthesized from chitin deacetylation. It is mostly used because of its special properties like dissolvability in different solvents and capacity to frame metal chelates and films, as well as biodegradability, biocompatibility, and non-toxicity, which lead to its increasing use in biomedicine and bioremediation (Kaur and Dhillon, 2014). Its good adsorbent property helps in the removal of harmful toxic contaminants from wastewater. The chitosan nanoparticle has a large surface area and high adsorption capacity which leads to effective decomposition and removal of dyes from wastewater. Tanhaei et al. (2015) reported chitosan/magnetic iron oxide nanoparticles which were then later used in the removal of methyl orange dye (Tanhaei et al., 2015). Magnetic chitosan has also been prepared which acts as a good adsorbent, used for the removal of methyl orange dye. The capacity can further be enhanced by modifying the surface with the help of some functional groups. Pesticides and fertilizers, used for crop production and the killing of pests on crops, also cause harm to the water, which due to being toxic will harm the aquatic animals/ecosystem also. The accumulation of these toxic compounds will harm the food chain cycle and lead to soil pollution. For the removal of these toxins from the ecosystem, adsorbents are used. In a recent study chitosan nanoparticles mixed with zinc oxide were used to remove pests in place of toxic pesticides. These nanoparticles have great capacity in pest removal and are better than the use of pesticides (Dehaghi et al., 2014).

These nano-adsorbents are categorized as:

- Carbon-based nanoparticles
- Metal and metal oxide nanoparticles

11.5.1.1.1 Carbon-Based Nanoparticles

Carbon-based nanoparticles include carbon nanotubes (CNTs), activated carbon, graphene, and fullerenes. Carbon nanotubes like MWCNTs act as adsorbents for toxic chemicals in manufacturing and the pharmaceutical industry. They can also adsorb metals from wastewater. These nanoparticles have high surface area, excellent chemical resistance, mechanical strength, and good adsorption capacity (Lee et al., 2012).

The use of CNTs has been a great method for the removal of heavy metals from wastewater due to their hollow and layered structure along with large surface area, although the use of carbon nanotubes does not give optimum performance due to the presence of some impurities like carbonaceous substances or residues from metal catalysts produced during the synthesis of the carbon nanotubes. These surface impurities can be removed by treating these CNTs with acids and alkaline solutions. The use of acids and alkalis will further strengthen the CNTs by attaching an additional functional group to the surface of CNTs. This functional group will increase the solubility of CNTs in any solvent (Georgakilas et al., 2002). CNTs coated with carbon are used in the effective removal of contaminants from wastewater leading to a clean environment. Mubarak et al. (2012) compared the use of functionalized CNTs with non-functionalized CNTs for better adsorption of Cu^{2+} from aqueous solution. According to the results, CNTs functionalized with H_2SO_4 and HNO_3 show a high removal percentage of about 94.5% of Cu^{2+} from the aqueous solution (Mubarak et al., 2012).

Sometimes biomaterials produced from agricultural waste will be used as carbon-based nano-adsorbents for wastewater treatment. These biomaterials are effective at low metal concentration and are reusable. Biomaterials such as saw dust, crab shells, and bagasse can be used for the synthesis of carbon-based nanoparticles. The use of these adsorbents is not very efficient; however their efficiency can be increased by increasing their surface area or reducing their particle size, or they can be modified into nano-porous activated carbon by functionalization. Mangun et al. (2001) used synthesized nano-porous activated carbon fibers with an average pore size of 1.16 nm and surface area within a range of 171 m^2/g to 483 m^2/g to determine the adsorption capacity by adsorbing benzene, toluene, p-xylene, and enthylbenzene. In their results, the obtained adsorption data fit the Freundlich adsorption isotherm well and it was also found that the nano-porous synthesized carbon fibers have a higher adsorption capacity than any granular activated carbon (Mangun et al., 2001).

Activated carbon is added to nanomagnets which are used for the removal of ions from wastewater. Here the principles of magnetic separation and sorption are utilized together for the removal of heavy metal toxic ions (Nogueira et al., 2019). Takmil et al. (2020) prepared nanocomposites which remove up to 97.4% of fluoride ions from synthetic wastewater by a sorption method (Takmil et al., 2020). The nanomagnets produced from activated carbon are prepared by coating the magnetic nanoparticles with specific ligands over the surface which shows affinity to specific contaminants (Apblett et al., 2001). Also the coating of magnetic nanoparticles with activated carbon helps to increase their adsorption efficiency. Graphene-based nanoparticles are used to remove nickel ions by an ion floatation process. Hoseinian et al. (2020) used the graphene-based nanoparticles to remove up to 100% of nickel ions from wastewater (Hoseinian et al., 2020). Zhao et al. (2011) used graphene oxide nano-sheets, synthesized from graphene, for the removal of Cd^{2+} and Co^{2+} from an aqueous solution using the adsorption method. They studied the effects of pH on the adsorption of metal ions on the graphene oxide nano-sheets. Based on their study, the maximum sorption capacity of Cd^{2+} and Co^{2+} on graphene oxide nano-sheets is seen to be at pH 6.0 and is about 106.3 mg/g and 68.2 mg/g respectively at a temperature of about 303 K (Zhao et al., 2011).

Apart from the removal of heavy metal ions, graphene can also be used in the removal of textile dyes from wastewater. Sometimes graphite is oxidized through the

Hummers Offeman method to form graphite oxides. Although not that much change can be seen in respect to surface area in graphite oxide, several functional groups can be found to be attached on the surface which increase its adsorption capacity. Bradder et al. (2011) used graphite oxide for the removal of methylene blue dye and malachite green dye. The results show that the adsorption of methylene blue dye and malachite green dye on graphite oxide was much higher than on graphite. The adsorption capacity of dyes on graphite oxide is 351 mg/g whereas the adsorption capacity of dyes on graphite in 248 mg/g which is much less than graphite oxide (Bradder et al., 2011).

11.5.1.1.2 Metal and Metal Oxide Nanoparticles

Metal and metal oxide nanoparticle-based adsorbents also play a big role in the removal of contaminants from wastewater. The coating of magnetic nanoparticles with other supports led to an increase in their adsorption efficiency. Najafpoor et al. (2020) used coated magnetic nanoparticles for the treatment of chemical oxygen demand in wastewater. They showed that magnetic nanoparticles coated with silver remove the chemical oxygen demand (COD) from wastewater up to 36.56% which was 6.16% higher than the use of uncoated magnetic nanoparticles for the removal of chemical oxygen demand (Najafpoor et al., 2020). Almomani et al. (2020) used iron oxide magnetic nanoparticles for the removal of heavy metals from wastewater. They showed that iron oxide magnetic nanoparticles were able to remove nickel, copper, and aluminum from wastewater in 35 seconds (Almomani et al., 2020).

11.5.1.2 Nanofiltration Membrane

The removal of contaminants from wastewater by the use of membranes is a kind of physical process which operates by particle movement on the basis of concentration difference or difference in particle size on either side of the membrane. A membrane with high permeability and selectivity is always preferred due to lower energy consumption.

The nanofiltration membranes are used in the removal of nutrients from the industrial effluents. A nanofiltration membrane NF90 removes phosphorus up to a limit of 70% produced by the pulp and paper industry. Shalaby et al. (2020) prepared a nanofiltration membrane woven with gold nanoparticles. This nanofiltration membrane achieves high phosphorus recovery from wastewater (Shalaby et al., 2020).

Tiraferri et al. (2011) used thin-walled membranes incorporated with SWCNTs by covalent bonding. This incorporation leads to purification due to sidewall functionalities, as well as enhancement in cytotoxic properties. These SWCNT-coated thin-walled membranes were used to reach up to 60% inactivation of *E. coli* bacteria which led to a greater percentage of purified water than the use of only CNTs (Tiraferri et al., 2011). Fullerenes were also incorporated into membranes for increasing the removal of contaminants from wastewater. Jin et al. (2007) used a polymer membrane made of poly (2,6-dimethyl-1, 4-phenylene oxide) incorporated with fullerene C_{60} to study the adsorption of estrogenic compounds which are present as contaminants in wastewater. The results show that 10% weight C_{60} and polymer membrane were able to remove up to 95% of estrogenic compounds (Jin et al.,

2007); as the pore size increases, the diffusion of estrogenic compounds across the membrane increases as well.

11.5.1.3 Nano-Catalysts

Nano-catalysts are also widely used nanoparticles used for the treatment of wastewater. Due to their high surface area to volume ratio, they have high catalytic activity, due to which the reactivity as well as the degradation of contaminants capability are enhanced. Some examples of nano-catalysts are semiconductor materials, zero-valence metals, and bimetallic nanoparticles which are used for the removal of contaminants such as polychlorinated biphenyls (PCBs), azo dyes, halogenated aliphatic, organochlorine pesticides, halogenated herbicides, and nitro-aromatics (Zhao et al., 2011). Catalyst nanoparticles can also be used for the degradation of disease-causing microbes present in wastewater. These catalyst nanoparticles are Ag-nano-catalyst, N-doped TiO_2, and ZrO_2 nanoparticles (Chaturvedi et al., 2012). Sometimes nano-structure catalytic membranes are also used for wastewater treatment. They have more advantages in industrial scale up because of properties like the high uniformity of catalytic sites, capability of optimization and also allows provides ease in industrial scale-up. With great advances in nanotechnology, some additional properties like increased permeability, selectivity, and resistance to fouling are added to nano-structured catalytic membranes (Volodymyr, 2009).

The use of silver nano-catalysts is a popular strategy nowadays in wastewater treatment are they are reusable. These Ag nano-catalysts are also very efficient in controlling microbes present in wastewater. This removal of harmful microbes from wastewater will result in a decrease in chemical oxygen demand. Scientists have recently used Ag nanoparticles in combination with Al_2O_3 and carbon for wastewater treatment. This combination leads to properties like high mechanical strength; low acidity and presence of mesopores in carbon are highly beneficial in wastewater treatment (Chaturvedi et al., 2012). Halogenated organic compounds (HOCs) are important toxic contaminants found in industrial wastewater effluent which may cause serious health problems such as cancer and can cause mutagenic damage. These contaminants can be selectively degraded with the use of nano-catalysts. The most used nano-catalysts for the degradation of these HOCs are nanosized palladium (Pd) catalysts. The nano-catalysts used can be further recycled back from treated water due to their ferromagnetism property which further will help in treating the polluted water (Chaturvedi et al., 2012).

But there are some problems related to the chemical nature of nanomaterials which further can cause contamination in water. Therefore, scientists are now using microorganisms for the generation of nanoparticles. This microorganism-assisted nanotechnology greatly reduces the use of chemicals.

11.5.2 Microorganism-Assisted Nanotechnology

The fabrication of bio-nanoparticles with the help of microorganisms makes the process of wastewater treatment more sustainable and eco-friendly. Nanoparticles produced chemically have a great disadvantage due to the use of harmful chemicals in their formation. They also show self-agglomeration in the solution. And therefore, the green synthesis of nanoparticles from plant extracts and microorganisms (such as fungi, algae) is preferred as it does not involve any use of harmful chemicals. These bio-nanoparticles act as a reducing agent for the metal complex salts present in wastewater. These bio-nanoparticles also show solidity in the aqueous environment by adding proteinaceous and other bioactive elements to the outer surface of bio-nanoparticles, which proves to be an advantage in wastewater treatment. Mahanty et al. (2020) bio-fabricated iron oxide nanoparticles from *Aspergillus officinalis* which are able to remove up to 90% of heavy metals, i.e., Pb (II), Ni (II), Cu (II), and Zn (II), from wastewater. These also show a regeneration ability of up to five cycles meaning iron oxide nanoparticles can be re-used/recycled for up to five cycles (Mahanty et al., 2020). These iron oxide nanoparticles remove the toxic metal ions by adsorbing them on their outer surface. Govarthanan et al. (2020) observed the removal of about 91% of PO_4^{3-} and 85% of NH_4^+ with the use of bio-nanocomposites (Govarthanan et al., 2020). The fabrication of bio-nanoparticles has been a cost effective and eco-friendly process of wastewater treatment (Noman et al., 2020). Microorganisms sometimes also help in the production of useful products from the industrial effluents shown in Table 11.2.

Nanoparticles produced as a result of the endogenous method release some harmful gases and metal complexes, but still the use of bio-nanoparticles in synergism with microorganisms is a superior method in wastewater treatment. Microorganisms could also provide some catalytic enzymes along with bio-nanoparticles which further boost the treatment process and the remediation of industrial effluent. The use of enzymes together with nanoparticles makes them less harmful as it minimizes their cell interaction through steric hindrances and also decreases their surface energy (Dwivedi et al., 2018). It also makes them more efficient in the remediation of effluents in green energy production.

11.6   VALORIZATION OF WASTE USING MICROBIAL SYNERGISM

Nanotechnology has helped in enhancing the removal of toxic ions from wastewater as well as in the efficient conversion of waste into resources using microorganisms in synergism. This technology is used for the production of adsorbents, clinker, biogas, biohydrogen, biomolecules, and many more products from the industrial effluent. Kumar et al. (2019) studied bio-nanoparticles in the enhancement of dark fermentation reactions for increased biohydrogen generation from wastewater (Kumar et al., 2019). Different researchers work independently for the increased generation of biohydrogen from wastewater. Elreedy et al. (2019) used mixed culture media along with nanoparticles to generate biohydrogen. They determined that the use of multiple nanoparticles increases biohydrogen production up to 14% compared to the use of a single nanoparticle. Many nanoparticles increase the hydrogenase and dehydrogenase activity leading to the further increase in biohydrogen production (Elreedy et al., 2019). Gadhe et al. (2015) used additional nickel oxide and hematite along with nanoparticles which further increases biohydrogen production up to 1.2–4.5-fold compared to the use of a single nanoparticle (Gadhe et al., 2015).

TABLE 11.2

Remediation of Industrial Effluents Using Advanced Nanotechnology Associated with Microorganisms

Sr. No.	Types of Nanoparticles Used	Modification	Associated Microorganisms	Working Mechanism	Special Features	Reference
1.	NiO and MgO NPs	Embedded with silica	-	Physisorption of Cu^{2+} and Cr^{3+} and Zn^+ chemicals	Regeneration and sustainability	Abuhatab et al., 2020
2.	Electrospun nano Fibrous web	Bacterial encapsulation	Pseudomonas aeruginosa	Biological removal of dye	Genetic engineering	Sarioglu et al., 2017
3.	Mesoporous organosilica NPs	Ferrocene incorporation	-	Non-covalent interaction due to ferrocene provides more surface area	Hybrid nanomaterial	Yang et al., 2019
4.	Cobalt and cobalt oxide NPs	Microwave and reductive chemical heating	-	Irradiation and large surface area	Cost effective, photocatalytic degradation	Adekunle et al., 2020
5.	Electrospun cyclodextrin fibers	Bacterial encapsulation	Lysinibacillus sp.	Bioremediation of bacteria	Extra carbon source for bacterial growth	San Keskin et al., 2018
6.	Zirconia NPs	Synthesis from microbial cell free culture supernatant	Pseudomonas aeruginosa	Chemisorption	Green synthesis, bioremediation	Debnath et al., 2020
7.	Enzyme immobilized NPs	Laccase immobilization	Pseudomonas ostreatus	Oxidation facilitated by immobilized laccase	Enzyme reuse, cost effective	Ji et al., 2017
8.	Graphene oxide and carbon nanotubes	Ni NPs organ framework	-	High surface area and hydrophobic interactions	Nanocomposite interaction	Ahsan et al., 2020
9.	Silica NPs	Synthesis from actinomycetes	Actinomycetes	Photocatalytic degradation	Cost effective, sustainable	Mohanraj et al., 2020

Microorganisms can be a significant part of sewage treatment. These microorganisms can range from harmful biological pollutants such as antibiotic-resistant bacteria, viruses, biotoxins, and protozoans to environmentally friendly microbes like algae and pseudomonas which help in wastewater treatment. These algae species can grow in wastewater due to the presence of micronutrients, macronutrients, and vitamins which are essential for their growth (Abou-Shanab et al., 2013). This will result in nutrient consumption from wastewater and further algae biomass can be used for energy production such as in biofuels including biodiesel, bioelectricity, biohydrogen, etc. (Grima et al., 2003). Hu et al. (2015) incorporated polyvinylidene fluoride (PVDF) hollow fiber membranes with TiO_2 to form $PVDF/TiO_2$ nanocomposite membrane. These membranes were then tested on an algae membrane bioreactor for wastewater treatment. The results demonstrated the maximum nutrient removal and about 75% of phosphorus and nitrogen (Hu et al., 2015).

Biodiesel are fatty acid methyl esters produced from the trans-esterification of oils with alcohol in the presence of a catalyst (Bhatia et al., 2020). The steps required for biodiesel generation from algae biomass are algae biomass cultivation, drying of biomass, oil extraction from algae biomass, and trans-esterification to fatty acid methyl esters (Bindra et al., 2017). Kong et al. (2010) reported that the microalgae *Chlamydomonas reinhardtii* produced about 505 mg/L of biofuel from municipal waste (Kong et al., 2010). Liu et al. (2020) reported that *Chlamydomonas vulgaris* used in the treatment of wastewater through continuous supply of carbon dioxide, resulting the removal of a very high concentration of phosphorus (Liu et al., 2020). Nanotechnology incorporation with microalgae promised a very high yield of biodiesel. Silver nanoparticles incorporated with *Chlamydomonas reinhardtii* have increased microalgae harvesting up to 30%, which is very high biomass productivity. Also, calcium oxide nanoparticles increased the conversion yield from biomass to biodiesel up to 91% through catalytic trans-esterification (Torkamani et al., 2010; Safarik et al., 2016).

Biohydrogen generation from microalgae is an eco-friendly, sustainable process. Different methods like the direct and indirect photolysis of water and dark fermentation can be used for biohydrogen production from wastewater in addition to some volatile fatty acid production. Ruiz-Marin et al. (2020) reported the immobilized culture of *Scenedesmus obliquus* in manipulated light condition at about 140 μE/m^2s with sulfur deprivation at 30°C temperature and at pH of about 7.5. The result showed a very high production of biohydrogen of about 204 ml H_2 $L^{-1}day^{-1}$ (Ruiz-Marin et al., 2020).

11.7   CONCLUSION

Biohydrogen is a sustainable, renewable, and clean biofuel. The intervention of nanoparticles is found to enhance the synthesis of biological hydrogen using mixed bacterial culture or mixed culture. This is due to an increase in the catalytic properties of enzymes or a decrease in the log phase of bacteria. The high hydrogen yield or high hydrogen production rate will explain the enhancement of biohydrogen production with the application of nanoparticles. Nanotechnology has also been used in wastewater treatment because of its high potential. The removal of harmful contaminants like toxic metal ions, inorganic and organic matter from water has been done more efficiently with the nanotechnology approach due to nanoparticles' large surface area to volume ratio. This high surface area to volume ratio gives rise to properties like high absorbing capacity and high interacting and reaction capabilities. Nanotechnology and microorganisms integrated together gives more boost to the wastewater treatment being more economic, eco-friendly, and sustainable. Microbes such as algae and pseudomonas are eco-friendly and found in wastewater and can be used for microalgal biomass production which with nanotechnology intervention can be used for the generation of biofuels such as biodiesel and biohydrogen. They can also be used in the production of other useful resources like bio-plastics from wastewater. In addition to water purification, microalgae also help in nutrient utilization and hence promoting a bio-refinery approach, due to which other microbes would not be able to grow.

REFERENCES

Abou-Shanab, R. A., Ji, M. K., Kim, H. C., Paeng, K. J., & Jeon, B. H. (2013). Microalgal species growing on piggery wastewater as a valuable candidate for nutrient removal and biodiesel production. *J. Environ. Manag.* 115, 257–264.

Abuhatab, S., El-Qanni, A., Al-Qalaq, H., Hmoudah, M., & Al-Zerei, W. (2020). Effective adsorptive removal of Zn2+, Cu2+, and Cr3+ heavy metals from aqueous solutions using silica-based embedded with NiO and MgO nanoparticles. *J. Environ. Manag.* 268, 110713.

Adekunle, A. S., Oyekunle, J. A., Durosinmi, L. M., Oluwafemi, O. S., Olayanju, D. S., Akinola, A. S., ... & Ajayeoba, T. A. (2020). Potential of cobalt and cobalt oxide nanoparticles as nanocatalyst towards dyes degradation in wastewater. *Nano-Struc. Nano-Obj.* 21, 100405.

Ahsan, M. A., Jabbari, V., Imam, M. A., Castro, E., Kim, H., Curry, M. L., Noveron, J. C. (2020). Nanoscale nickel metal organic framework decorated over graphene oxide and carbon nanotubes for water remediation. *Sci. Total Environ.* 698, 134214.

Almomani, F., Bhosale, R., Khraisheh, M., & Almomani, T. (2020). Heavy metal ions removal from industrial wastewater using magnetic nanoparticles (MNP). *Appl. Surf. Sci.* 506, 144924.

Apblett, A. W., Al-Fadul, S. M., Chehbouni, M., & Trad, T. (2001). Removal of petrochemicals from water using magnetic filtration. In *Proceedings of the 8th Int. Environ. Petrol. Consort.* November 6, Oklahoma.

Baruah, A., Chaudhary, V., Malik, R., & Tomer, V. K. (2019). Nanotechnology based solutions for wastewater treatment. In *Nanotechnology in Water and Wastewater Treatment* Ed. Amimul Ahsan and Ahmad Fauzi Ismail (pp. 337–368). Amsterdam, Netherlands: Elsevier.

Bhatia, S. K., Gurav, R., Choi, T. R., Kim, H. J., Yang, S. Y., Song, H. S., ... & Yang, Y. H. (2020). Conversion of waste cooking oil into biodiesel using heterogenous catalyst derived from cork biochar. *Bioresour. Technol.* 302, 122872.

Bindra, S., Sharma, R., Khan, A., & Kulshrestha, S. (2017). Renewable energy sources in different generations of bio-fuels with special emphasis on microalgae derived biodiesel as sustainable industrial fuel model. *Biosci. Biotechnol. Res. Asia* 14(1), 259–274.

Boshagh, F., Rostami, K., & Moazami, N. (2019). Biohydrogen production by immobilized *Enterobacter aerogenes* on functionalized multi-walled carbon nanotube. *Int. J. Hydrog. Energy* 44(28), 14395–14405.

Bradder, P., Ling, S. K., Wang, S., & Liu, S. (2011). Dye adsorption on layered graphite oxide. *J. Chem. Eng. Data* 56(1), 138–141.

Chaturvedi, S., Dave, P. N., & Shah, N. K. (2012). Applications of nano-catalyst in new era. *J. Saudi Chem. Soc.* 16(3), 307–325.

Debnath, B., Majumdar, M., Bhowmik, M., Bhowmik, K. L., Debnath, A., & Roy, D. N. (2020). The effective adsorption of tetracycline onto zirconia nanoparticles synthesized by novel microbial green technology. *J. Environ. Manag.* 261, 110235.

Dehaghi, S. M., Rahmanifar, B., Moradi, A. M., & Azar, P. A. (2014). Removal of permethrin pesticide from water by chitosan–zinc oxide nanoparticles composite as an adsorbent. *J. Saudi Chem. Soc.* 18(4), 348–355.

Dhrubajyoti, N., Manhar, A. K., Gupta, K., Saikia, D. B., Das, S. K., & Mandal, M. (2015). Photosynthesized iron nanoparticles: Effects on fermentative hydrogen production by *Enterobacter cloacae* DH-89. *Bull. Mater. Sci.* 38(6), 1533–1538.

Dwevedi, A. (2018). *Solutions to Environmental Problems Involving Nanotechnology and Enzyme Technology*. United States: Academic Press.

Elreedy, A., Fujii, M., Koyama, M., Nakasaki, K., & Tawfik, A. (2019). Enhanced fermentative hydrogen production from industrial wastewater using mixed culture bacteria incorporated with iron, nickel, and zinc-based nanoparticles. *Water Res.* 151, 349–361.

Engliman, N. S., Abdul, P. M., Wu, S. Y., & Jahim, J. M. (2017). Influence of iron (II) oxide nanoparticle on biohydrogen production in thermophilic mixed fermentation. *Int. J. Hydrog. Energy* 42(45), 27482–27493.

Gadhe, A., Sonawane, S. S., & Varma, M. N. (2015). Influence of nickel and hematite nanoparticle powder on the production of biohydrogen from complex distillery wastewater in batch fermentation. *Int. J. Hydrog. Energy* 40(34), 10734–10743.

Georgakilas, V., Kordatos, K., Prato, M., Guldi, D. M., Holzinger, M., & Hirsch, A. (2002). Organic functionalization of carbon nanotubes. *J. Am. Chem. Soc.* 124(5), 760–761.

Govarthanan, M., Jeon, C. H., Jeon, Y. H., Kwon, J. H., Bae, H., & Kim, W. (2020). Non-toxic nano approach for wastewater treatment using *Chlorella vulgaris* exopolysaccharides immobilized in iron-magnetic nanoparticles. *Int. J. Biol. Macromol.* 162, 1241–1249.

Grima, E. M., Belarbi, E. H., Fernández, F. A., Medina, A. R., & Chisti, Y. (2003). Recovery of microalgal biomass and metabolites: Process options and economics. *Biotechnol. Adv.* 20(7–8), 491–515.

Guangzhen, L., Zhanfang, M. A., & Jianquan, S. (2006). Enhancement effect of gold nanoparticles on the bioactivity of Hydrogen-producing microbe. *J. Wuhan Uni. Tech. (Mater. Sci. Ed.)* 21(4), 16–18.

Gupta, G. K., & Shukla, P. (2020). Insights into the resources generation from pulp and paper industry wastes: Challenges, perspectives and innovations. *Bioresour. Technol.* 297, 122496.

Hoseinian, F. S., Rezai, B., Kowsari, E., Chinnappan, A., & Ramakrishna, S. (2020). Synthesis and characterization of a novel nanocollector for the removal of nickel ions from synthetic wastewater using ion flotation. *Sep. Purif. Technol.* 240, 116639.

Hu, W., Yin, J., Deng, B., & Hu, Z. (2015). Application of nano TiO2 modified hollow fiber membranes in algal membrane bioreactors for high-density algae cultivation and wastewater polishing. *Bioresour. Technol.* 193, 135–141.

Ji, C., Nguyen, L. N., Hou, J., Hai, F. I., & Chen, V. (2017). Direct immobilization of laccase on titania nanoparticles from crude enzyme extracts of *P. ostreatus* culture for micropollutant degradation. *Sep. Purif. Technol.* 178, 215–223.

Jin, X., Hu, J. Y., Tint, M. L., Ong, S. L., Biryulin, Y., & Polotskaya, G. (2007). Estrogenic compounds removal by fullerene-containing membranes. *Desalination* 214(1–3), 83–90.

Kaur, S., & Dhillon, G. S. (2014). The versatile biopolymer chitosan: Potential sources, evaluation of extraction methods and applications. *Crit. Rev. Microbiol.* 40(2), 155–175.

Khan, M. M., Lee, J., & Cho, M. H. (2013). Electrochemically active biofilm mediated biohydrogen production catalysed by positively charged gold nanoparticles. *Int. J. Hydrog. Energy* 38(13), 5243–5250.

Kim, K. Y., Habas, S. E., Schaidle, J. A., & Logan, B. E. (2019). Application of phase-pure nickel phosphide nanoparticles as cathode catalysts for hydrogen production in microbial electrolysis cells. *Bioresour. Technol.* 293, 122067.

Kong, Q. X., Li, L., Martinez, B., Chen, P., & Ruan, R. (2010). Culture of microalgae *Chlamydomonas reinhardtii* in wastewater for biomass feedstock production. *Appl. Biochem. Biotechnol.* 160(1), 9–18.

Kumar, G., Mathimani, T., Rene, E. R., & Pugazhendhi, A. (2019). Application of nanotechnology in dark fermentation for enhanced biohydrogen production using inorganic nanoparticles. *Int. J. Hydrog. Energy* 44(26), 13106–13113.

Kumari, P., Alam, M., & Siddiqi, W. A. (2019). Usage of nanoparticles as adsorbents for waste water treatment: An emerging trend. *Sustain. Mater. Technol.* 22, e00128.

Lee, X. J., Lee, L. Y., Foo, L. P. Y., Tan, K. W., & Hassell, D. G. (2012). Evaluation of carbon-based nanosorbents ynthesized by ethylene decomposition on stainless steel substrates as potential sequestrating materials for nickel ions in aqueous solution. *J. Environ. Sci.* 24(9), 1559–1568.

Liu, X., Chen, G., Tao, Y., & Wang, J. (2020). Application of effluent from WWTP in cultivation of four microalgae for nutrients removal and lipid production under the supply of CO2. *Renew. Energy* 149, 708–715.

Mahanty, S., Chatterjee, S., Ghosh, S., Tudu, P., Gaine, T., Bakshi, M., & Chaudhuri, P. (2020). Synergistic approach towards the sustainable management of heavy metals in wastewater using mycosynthesized iron oxide nanoparticles: Biofabrication, adsorptive dynamics and chemometric modeling study. *J. Water Process Eng.* 37, 101426.

Mangun, C. L., Yue, Z., Economy, J., Maloney, S., Kemme, P., & Cropek, D. (2001). Adsorption of organic contaminants from water using tailored ACFs. *Chem. Mater.* 13(7), 2356–2360.

Mohanraj, R., Gnanamangai, B. M., Poornima, S., Oviyaa, V., Ramesh, K., Vijayalakshmi, G., & Robinson, J. P. (2020). Decolourisation efficiency of immobilized silica nanoparticles synthesized by actinomycetes. *Mater. Today: Proceedings*, 48.

Mohanraj, S., Anbalagan, K., Kodhaiyolii, S., & Pugalenthi, V. (2014). Comparative evaluation of fermentative hydrogen production using *Enterobacter cloacae* and mixed culture: Effect of Pd (II) ion and phytogenic palladium nanoparticles. *J. Biotechnol.* 192(A), 87–95.

Mohanraj, S., Kodhaiyolii, S., Rengasamy, M., & Pugalenthi, V. (2014). Green synthesized iron oxide nanoparticles effect on fermentative hydrogen production by *Clostridium acetobutylicum*. *Appl. Biochem. Biotechnol.* 173(1), 318–331.

Mubaraka, N., Daniela, S., Khalid, M., & Tana, J. (2012). Comparative study of functionalize and non-functionalized carbon nanotube for removal of copper from polluted water. *Int. J. Chem. Environ. Eng.* 3(5).

Najafpoor, A., Norouzian-Ostad, R., Alidadi, H., Rohani-Bastami, T., Davoudi, M., Barjasteh-Askari, F., & Zanganeh, J. (2020). Effect of magnetic nanoparticles and silver-loaded magnetic nanoparticles on advanced wastewater treatment and disinfection. *J. Mol. Liq.* 303, 112640.

Nogueira, H. P., Toma, S. H., Silveira, A. T., Carvalho, A. A., Fioroto, A. M., & Araki, K. (2019). Efficient Cr (VI) removal from wastewater by activated carbon superparamagnetic composites. *Microchem. J.* 149, 104025.

Noman, M., Shahid, M., Ahmed, T., Niazi, M. B. K., Hussain, S., Song, F., & Manzoor, I. (2020). Use of biogenic copper nanoparticles synthesized from a native *Escherichia* sp. As photocatalysts for azo dye degradation and treatment of textile effluents. *Environ. Poll.* 257, 113514.

Patel, S. K., Lee, J. K., & Kalia, V. C. (2018). Nanoparticles in biological hydrogen production: An overview. *Ind. J. Microbiol.* 58(1), 8–18.

Qiao, Y., Li, C. M., Bao, S. J., & Bao, Q. L. (2007). Carbon nanotube/polyaniline composite as anode material for microbial fuel cells. *J. Power Sources* 170(1), 79–84.

Ruiz-Marin, A., Canedo-López, Y., & Chávez-Fuentes, P. (2020). Biohydrogen production by *Chlorella vulgaris* and *Scenedesmus obliquus* immobilized cultivated in artificial wastewater under different light quality. *AMB Expr.* 10(1), 1–7.

Safarik, I., Prochazkova, G., Pospiskova, K., & Branyik, T. (2016). Magnetically modified microalgae and their applications. *Crit. Rev. Biotechnol.* 36(5), 931–941.

San Keskin, N. O., Celebioglu, A., Sarioglu, O. F., Uyar, T., & Tekinay, T. (2018). Encapsulation of living bacteria in electrospun cyclodextrin ultrathin fibres for bioremediation of heavy metals and reactive dye from wastewater. *Colloids Surf. B Biointerfaces* 161, 169–176.

Sarioglu, O. F., San Keskin, N. O., Celebioglu, A., Tekinay, T., & Uyar, T. (2017). Bacteria encapsulated electrospun nanofibrous webs for remediation of methylene blue dye in water. *Colloids Surf. B Biointerfaces* 152, 245–251.

Shalaby, M. S., Abdallah, H., Cenian, A., Sołowski, G., Sawczak, M., Shaban, A. M., & Ramadan, R. (2020). Laser synthesized gold-nanoparticles, blend NF membrane for phosphate separation from wastewater. *Sep. Purif. Technol.* 247, 116994.

Sivagurunathan, P., Kadier, A., Mudhoo, A., Kumar, G., Chandrasekhar, K., Kobayashi, T., & Xu, K. (2018). Nanomaterials for biohydrogen production. *Nanomater. Biomed. Environ. Engg. App.* 217–237.

Takmil, F., Esmaeili, H., Mousavi, S. M., & Hashemi, S. A. (2020). Nano-magnetically modified activated carbon prepared by oak shell for treatment of wastewater containing fluoride ion. *Adv. Powder Technol.* 31(8), 3236–3245.

Tanhaei, B., Ayati, A., Lahtinen, M., & Sillanpaa, M. (2015). Preparation and characterization of a novel chitosan/Al2O3/magnetite nanoparticles composite adsorbent for kinetic, thermodynamic and isotherm studies of methyl orange adsorption. *Chem. Eng. J.* 259, 1–10.

Tiraferri, A., Vecitis, C. D., & Elimelech, M. (2011). Covalent binding of single-walled carbon nanotubes to polyamide membranes for antimicrobial surface properties. *ACS Appl. Mater. Interf.* 3(8), 2869–2877.

Torkamani, S., Wani, S. N., Tang, Y. J., & Sureshkumar, R. (2010). Plasmon-enhanced microalgal growth in miniphotobioreactors. *Appl. Phys. Lett.* 97(4), 043703.

Volodymyr, T. V. (2009). Multifunctional nanomaterial-enabled membranes for water treatment. In *Nanotechnology Applications for Clean Water* Ed. Anita Street et al. (pp. 59–75). William Andrew Publishing, Boston.

Yang, S., Chen, S., Fan, J., Shang, T., Huang, D., & Li, G. (2019). Novel mesoporous organosilica nanoparticles with ferrocene group for efficient removal of contaminants from wastewater. *J. Colloid Interface Sci.* 554, 565–571.

Yildirim, O., Tunay, D., Ozkaya, B., & Demir, A. (2021). Effect of green synthesized silver oxide nanoparticle on biological hydrogen production. *Int. J. Hydr. Energy.* https://doi.org/10.1016/j.ijhydene.2021.11.176.

Zhang, Q., Xu, S., Li, Y., Ding, P., Zhang, Y., & Zhao, P. (2021). Green-synthesized nickel oxide nanoparticles enhances biohydrogen production of *Klebsiella* sp. WL1316 using lignocellulosic hydrolysate and its regulatory mechanism. *Fuel* 305, 121585.

Zhang, Y., & Shen, J. (2007). Enhancement effect of gold nanoparticles on biohydrogen production from artificial wastewater. *Int'l J. Hydr. Energy* 32(1), 17–23.

Zhao, G., Li, J., Ren, X., Chen, C., & Wang, X. (2011). Few-layered graphene oxide nanosheets as superior sorbents for heavy metal ion pollution management. *Environ. Sci. Technol.* 45(24), 10454–10462.

Zhao, W., Zhang, Y., Du, B., Wei, D., Wei, Q., & Zhao, Y. (2013). Enhancement effect of silver nanoparticles on fermentative biohydrogen production using mixed bacteria. *Bioresour. Technol.* 142, 240–245.

Zhao, X., Lv, L., Pan, B., Zhang, W., Zhang, S., & Zhang, Q. (2011). Polymer-supported nanocomposites for environmental application: A review. *Chem. Eng. J.* 170(2–3), 381–394.

Zhao, Y., & Chen, Y. (2011). Nano-TiO2 enhanced photofermentative hydrogen produced from the dark fermentation liquid of waste activated sludge. *Environ. Sci. Technol.* 45(19), 8589–8595.

12 Bionanotechnology in a Biogas-Based Power Generation System Using Lignocellulosic Biomass

Deepa Sharma

CONTENTS

12.1 INTRODUCTION

Agro-based lignocellulosic biomass (LB) has arisen as a key solution to environmental and energy challenges since it is rich in feedstock that can be converted to biofuels. The bioconversion of lignocellulosic biomass to sugar is a quite complex system. There are various techniques that have been utilized in the bioconversion process, viz., physical, chemical, and biological approaches. Every method has its own merits and demerits when used on a large scale, including the high cost of processing, the

DOI: 10.1201/9781003316374-12

development of harmful inhibitors, and the detoxification of the inhibitors that have been produced. Altogether, these restraints hamper the efficiency of current solutions and demand the invention of a new, creative, cost-effective, and eco-sustainable technique for LB processing.

The main composition of lignocellulosic biomass, which itself is a generic term used to label the cell wall of plants, consists of (Sankaran et al., 2021):

- Cellulose (38–50%) which includes pentose and hexose sugar monomers
- Hemicellulose (17–32%), pentose, and hexose sugar monomers
- Lignin (15–30%) polyphenol aromatics

The structure-based functions of lignocellulose can be explained by the way that cellulose forms the structure of the cell walls, while hemicellulose assists in the cross-linking between the non-cellulosic and cellulosic polymers via covalent bonding as shown in Figure 12.1. In cellulose, glucose units are linked together by β-(1 → 4) glycosidic bonds and form a homopolysaccharide, while hexose and pentose units constitute heteropolymeric, branched hemicellulose. D-Glucose, d-mannose, and d-galactose are the main components of the hexosans (mannans), while d-xylose and l-arabinose are the elements of pentosans, which arexylans. Lignin is the second most abundantly present polymer which constitutes the plant cell wall complex called lignocellulose (Sun and Chang, 2002).

In recent years, research on lignocellulosic material and its transformation into bioenergy has been increasing at a persistent pace owing to the declining trend in

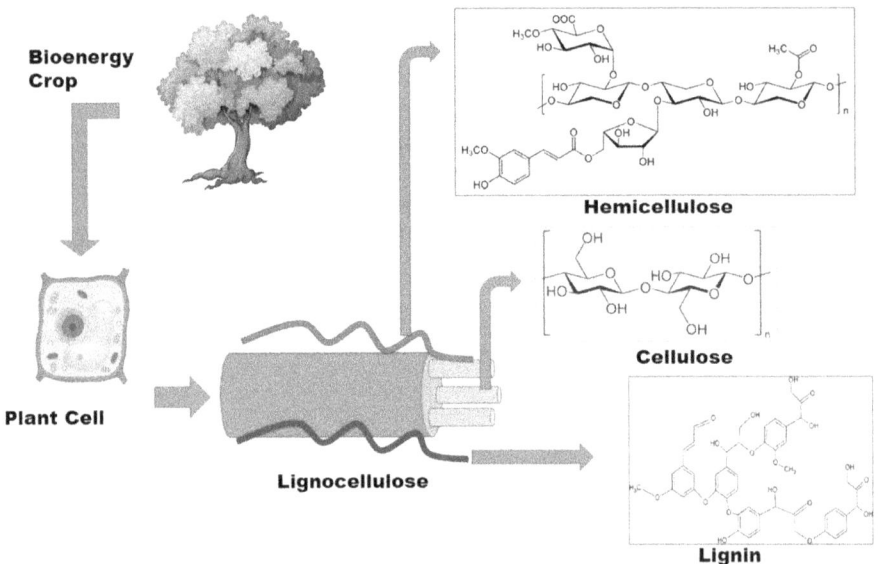

FIGURE 12.1   Schematic designs showing the main molecules in bioenergy crop (lignocellulosic biomass).

the use of fossil fuel reserves, and the consequence is the declining trend of environmental pollution associated with the exploitation of these resources. In an attempt to increasingly transit toward a bio-based economy, several renewable energy resources have been discovered to date, such as solar, wind, hydrothermal, and biomass. However, biomass-derived energy has persisted as the only renewable energy source that can potentially replace petroleum-based fuel for transportation or to produce chemicals. The high chemical energy content of lignocellulosic biomass owing to the 75% carbohydrates from cellulose and hemicellulose is ideal for bioenergy use (Chen, 2014; Sanusi et al., 2021).

The entire conversion process of lignocellulosic biomass to value-added products involves the following steps:

- Pretreatment of LB to disrupt biomass structure and make it suitable for the enzymatic process
- Enzymatic hydrolysis of components to release sugars for fermentation
- Fermentation to convert monomer sugars into biofuel or other biochemicals depending on the application
- Valorization process

The different steps involved in the production of bioenergy *via* lignocellulose biomasses are shown in Figure 12.2. The lignocellulosic biomass can be converted into bioethanol and/or biochemical products using a variety of pretreatment processes, including physical, chemical, and biological ones. However, these conventional

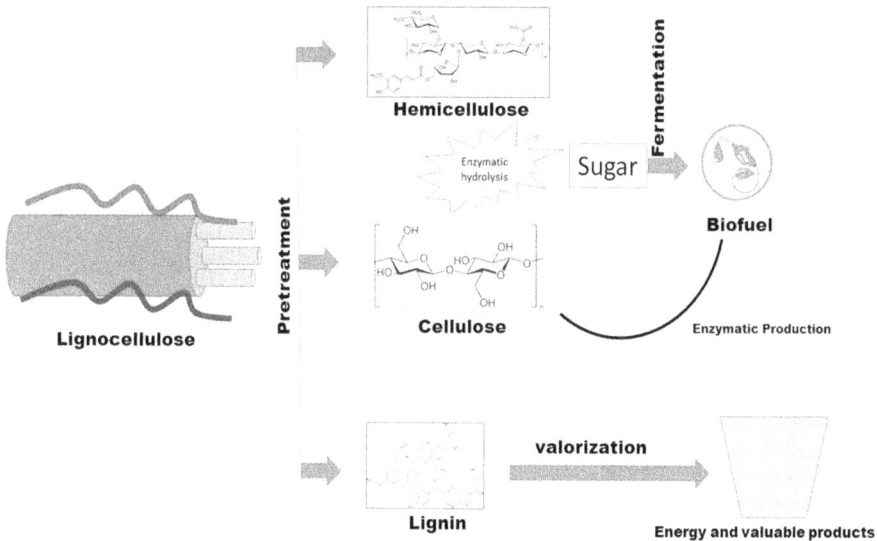

FIGURE 12.2 Schematic designer steps involved in the production of bioenergy using lignocellulosic biomasses.

methods are costly, inconvenient, and have some limitations. These limitations have been sidestepped by introducing a unique, cost-effective, efficient, and environmentally friendly method, namely, bionanotechnology. The application of bionanotechnology presents a promising alternative to existing pretreatment methods in which bionanoparticles are prepared from agro-based feedstock and these bionanoparticles are also used for the pretreatment of lignocellulosic biomass for energy production (Zanuso et al., 2021).

The technological demerits associated with the production of bioenergy from lignocellulosic biomass such as high cost and insufficiency in the existing infrastructure have posed major challenges in attaining high quality and yield of bioenergy from lignocellulosic material. As an interpretation of these limitations, various studies have reported many optimization strategies in pretreatment stages and also at the enzymes and fermentation stages to augment bioenergy production in an energy-efficient and cost-effective manner. In recent times, with the increasing interest in bionanomaterials, their excellent properties have been immensely exploited to enhance bioenergy generation. Producing materials from lignocellulosic biomass is a value-added proposition compared with the fuel-centric approach. For example, the production of high-value nanoscale polymeric materials (NPMs), such as cellulose nanocrystals (CNCs), cellulosic nanofibrils (CNFs), and lignin nanoparticles (LNPs), from lignocellulosic biomass also has potential to attain saleable success.

12.2 BIOGAS PRODUCTION TECHNOLOGY

Biogas has long been recognized as a renewable energy source which generates energy using animal waste by the process of anaerobic digestion; agricultural and organic waste are used for this purpose. Anaerobic digestion is the process in which microorganisms break down the biomass in the absence of oxygen. The composition of biogas is shown in Table 12.1.

Biogas, composed of biogenic components, is seen as a renewable source of energy worldwide. Biogas production is an appealing unconventional energy source with regards to energy yield as shown in Figure 12.3. The cumulative biomass supply in 2014 was calculated to have increased by 2.6% from the previous year to 59.2

TABLE 12.1
Different Gaseous Components Present in Biogas

Components Present in Bio-Gas	%
Methane	50–75
Carbon dioxide	25–50
Nitrogen	0–10
Hydrogen	0–1
Hydrogen sulfide	0–3
Oxygen	0–1

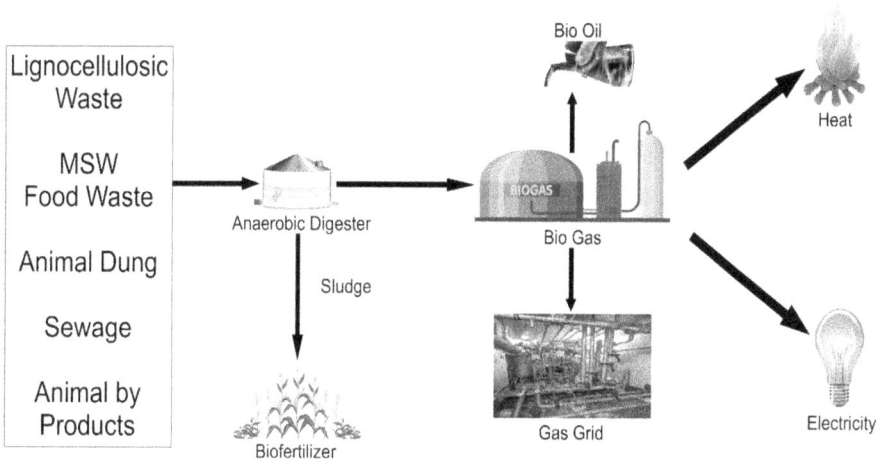

FIGURE 12.3 Classical production diagram of biogas generation and its utilization in various forms.

EJ which represents about 10.3% of the total energy supplied globally (Weiland, 2006). The major sectors of solid waste contribution are agriculture, forestry, and the organic fraction of municipal solid waste contributed 10, 87, and 3%, respectively, to the biomass supplied (WBA, 2021).

Various types of anaerobic digesters have been constructed, and used for biogas production, but regardless of the digester type engaged, there is a need to measure the performance of the design configuration in use thoroughly to prevent any unexpected change that may occur during the process. Different production parameters as mentioned below can alter the process adversely and affect the effectiveness of the process if not well managed.

- pH
- Temperature
- Volatile fatty acids
- Shear stress
- Organic loading rate
- Role of inhibitors (NH_3, hydroxymethyl furfural, furfural, etc.)
- Total solids
- Initial concentration
- Mixing
- Hydraulic retention time

There are two categories of biogas production, i.e. dry and wet processes. The differences between both the processes are based on their solid concentrations; wet digestion process has solid concentration less than 10% in the fermenter, while dry process has up to 35% of the same. The wet digestion process is more stable. The process of the preparation of biogas involves four steps, hydrolysis, acidogenesis, acetogenesis, and methanogenesis, as shown in Figure 12.4. The whole process is elaborated in Figure 12.5 (Deublein et al., 2008).

12.2.1 BIOSENSORS AND BIOGAS PRODUCTION

Biogas production by anaerobic digestion can be linked successfully with industrial waste management also, but the process needs regular checks for the accumulation of intermediates leading to unsteady progression (Komemoto et al., 2009). This accumulation of organic acids (including formate, lactate, and alcohols, and fatty acids like butyrate, propionate, and acetate) results in acidification of the reactor which may lead to an imbalance in the process (Montag et al., 2016). The composition and biodegradability of the waste play an important role in deciding the yield of biogas as some components of waste like sugar have much faster rates of degradation than

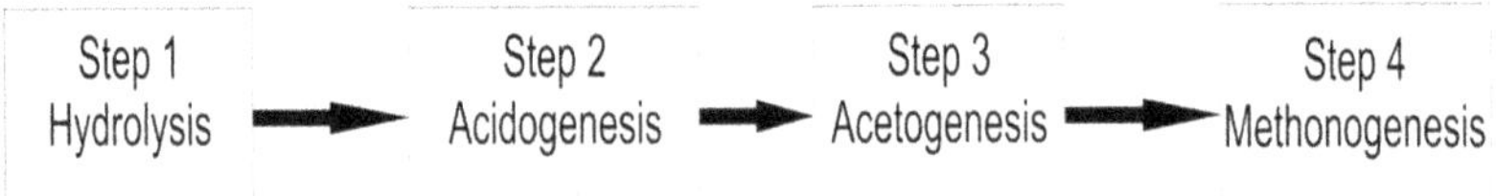

FIGURE 12.4   Flow chart of biogas generation.

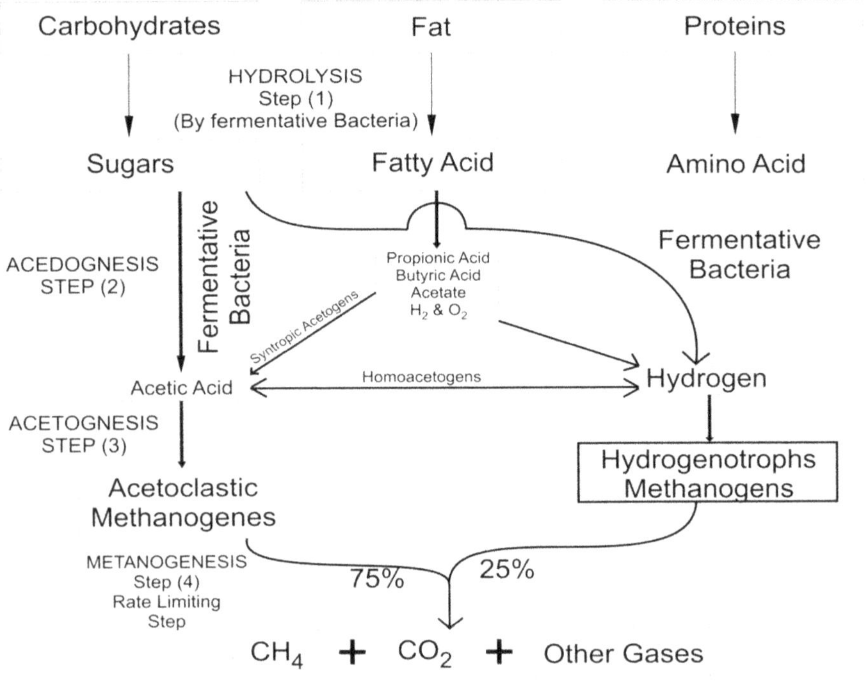

FIGURE 12.5   Flow chart of biogas production by anaerobic digestion.

lignocelluloses. The estimation of acid is an essential requirement for this purpose, for which various conventional techniques like spectroscopy, chromatography, and HPLC are used, but all these procedures are expensive. As an alternative, biosensors are used for the purpose which analyze the compounds quickly and accurately, apart from these, enzyme-based sensor analyzed acetate and propionate (Goriushkina et al., 2009, Rathee et al., 2016, Sode et al., 2008).

Still, there are certain limitations to the proper use of biosensors, such as unstable fermentation, poor biodegradation, and low methane production (Yang et al., 2015); to combat these limitations, certain advancements are introduced, such as,

1. Pretreatment of sludge (Figure 12.6)
2. Addition of additives to sludge
3. Addition of additives to digester (Zhang et al., 2015)
4. Modification in digester to increase biogas yield as shown in Figure 12.7

The effect of pretreatment on anaerobic digestion was studied by Achinas et al. (2017) and depicted as a graph.

The designs of biogas plants play an important role in increasing the rate of biogas production; recently, innovations are going on in biogas plants by altering the design parameters such as gas storage volume and cost effectiveness.

12.2.2 MODIFICATIONS IN CURRENT BIOGAS TECHNOLOGY

The novel anaerobic digesters include some modifications in the technologies with new reactor designs, as follows.

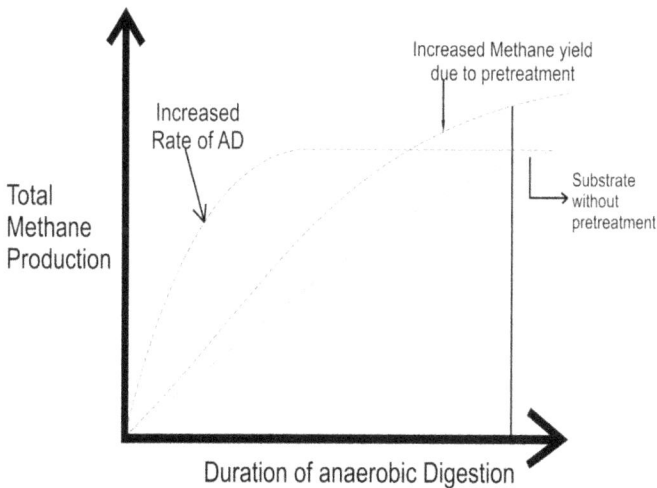

FIGURE 12.6   Graphic representation of the effect of pretreatment on anaerobic digestion and CH$_4$ yield.

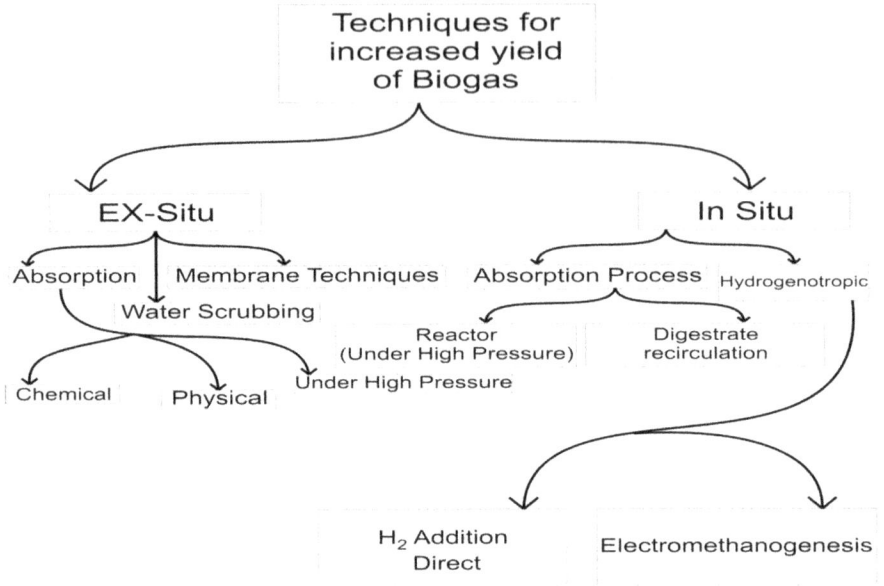

FIGURE 12.7 Biogas improvement techniques.

12.2.2.1 High-Rate Anaerobic Reactors/Up Flow Anaerobic Sludge Blanket (UASB)

In this type of anaerobic digestion (AD), anaerobic microorganisms self-immobilize their cells and form granules; granulation is the main characteristic feature of UASB, which makes it more efficient with its compact size, high loading rates, low sludge production, and cheap and high methane production.

Furthermore, a modification is made by separating hydrolysis and acidolysis from methanogenesis in the form of two stage processes, firstly the treatment of municipal solid waste in a continuous stirrer tank reactor and then an up flow anaerobic sludge blanket reactor (Aslanzadeh et al., 2014).

12.2.2.2 Anaerobic Membrane Bioreactors

Such bioreactors involve a semi-permeable membrane which forms a barrier to permit certain components to pass and to retain biological components.

12.2.2.3 Membrane Integrated High-Rate System

Such reactors involve the integration of anaerobic membrane technology with the production of methane and are suitable for high-strength industrial waste and municipal wastewater.

The biomass specific surface can be increased by reducing the biomass particle size; this is carried out to reduce cellulose fiber organization during biomass fibrillation by a milling-based pretreatment, and this is measured by a decline in crystallinity (Kumar et al., 2009).

Bionanotechnology plays an important role in the enhancement of economically leveraged renewable energy production and energy efficiency. Active sludge and some algal biomass showed the nanomaterial interactions. An adverse or increased yield was visible (Victor et al., 2011) with regards to the inhibition of bioenergy production. The assessment is done by calculating the particle surface area to volume ratio, leading to variation in the severity of the effects. NPs have an effect on energized sludge systems. The outcomes of the NPs impact energized sludge can be seen in a few related characteristics of NPs and energized sludge including aggregation, size of nanoparticles, and reciprocation of microorganisms.

12.3 METHOD OF PREPARATION OF NANOPARTICLES FROM CELLULOSE AND LIGNIN

From the viewpoint of biogas production, nanomaterials can be projected in many aspects to meet the rapidly growing global energy requirements due to their exceptional properties as follows (Zhang et al., 2011; Hamawand et al., 2020).

- Crystallinity
- Durability
- Large surface area
- Adsorption capability
- Stability
- High catalytic activity
- Efficient storage

All the above-mentioned properties of nanoparticles can enhance the productivity, hydrolysis, and stability of cellulase enzymes for the production of various types of bioenergy. Generally, NPs derived from cellulosic structures for biogas production can be beneficial in the pretreatment of biomass, waste management, or sugar and biohydrogen production. A short review is presented in Table 12.2 which focuses on the utilization of nanoparticles in the generation of biogas.

12.3.1 CELLULOSE NANOMATERIALS

Today, cellulose nanomaterials (CNMs) denotes mainly two types of materials:

CNC: produced primarily using concentrated sulfuric acid hydrolysis
CNF: produced by mechanical fibrillation

The steps of the production procedure of CNM and CNF from cellulosic biomass are illustrated in Figure 12.8.

12.3.2 LIGNIN NANOPARTICLES (LNP)

Nanomaterials of diverse shapes and size can be prepared from lignin as shown in Figure 12.9.

TABLE 12.2
Use of Nanoparticles for the Generation of Biogas

S. No.	Nanomaterial/ Nanoparticles	Biomass	% Yield Increase	Citation
1	Zn	Natural cellulosic matter	-	Abraham et al., 2014
2	Si coated Fe/Fe$_2$O$_3$ core	Natural cellulosic matter	-	Wang et al., 2015
3	Si nanocatalysts β-cyclodextrin- Fe$_2$O$_3$	Cellulose	-	Huang et al., 2015
4	ZnO	Cellulose	-	Srivastava et al., 2016 Lima et al., 2017
5	Multiwalled carbon nanotubes	Cellulose	-	Ahmed et al., 2018
6	Ni, Co	*Chenopodium album*	23.75	Ali et al., 2020
7	Ni, Co	*Parthenium hysterophorus*	17.66	Tahir et al., 2020
8	Co	Spent tea	28	

Very extensive research is going on to prepare LNPs. One normally used process for the preparation of LNPs is based on the dissolution of lignin in an organic solvent or water-organic solvent mixture followed by precipitation resulting from the increased concentration of water (anti solvent). In various literatures, these methods are referred to as "solvent shifting", "self-assembly", "nanoprecipitation", or "solvent exchange" (Zhu et al., 2021).

- The solvent-shifting methods
- Three-solvent system (THF:ethanol:water)
- Acid hydrotropes fractionation
- Milling
- Ultrasonication
- High shear homogenization
- Mechanical disintegration of lignin macroparticles

"Nanomaterials" have an external dimension or internal or surface structure on the nano scale ranging from 1 to 100 nm in size (ISO/TS, 2008).

Nanoparticles are categorized into four groups: inorganic, organic, composite, and carbon NPs, on the basis of their special chemical, physical, and optical characteristics. At the nano scale, the properties of the particles change unpredictably, making them behave in different ways to the same substance at the macro scale. Nanoparticles are ideal in areas such as energy, electronic, and medical commercial products due to their high reactivity and other features. Nanoparticles possess different chemical and physical properties from their macro counterparts, which makes them more useful. Nanoparticles possess large surface area resulting in higher chemical reactivity; large surface area provides a greater number of reaction sites. This can be explained by the example of gold

Cellulosic Fibers

↓

Hydrolysis

↓

Centrifugation

↓

Dialysis

↓

Centrifugation

Preparation of Fibrous Cellulosic
Solid Residue (FCSR)

Carboxylated
Nano-CrystalsCNC

Carboxylated Nano-Fibrils (CNF)

Followed by Different Drying Techniques
→ Air Oven Drying
→ Freeze Drying
→ Supercritical CO$_2$ - Drying
→ Spray Drying

↓

Dewatering

Used for pretreatment of Lignocellulosic
Biomass for Biogas Production

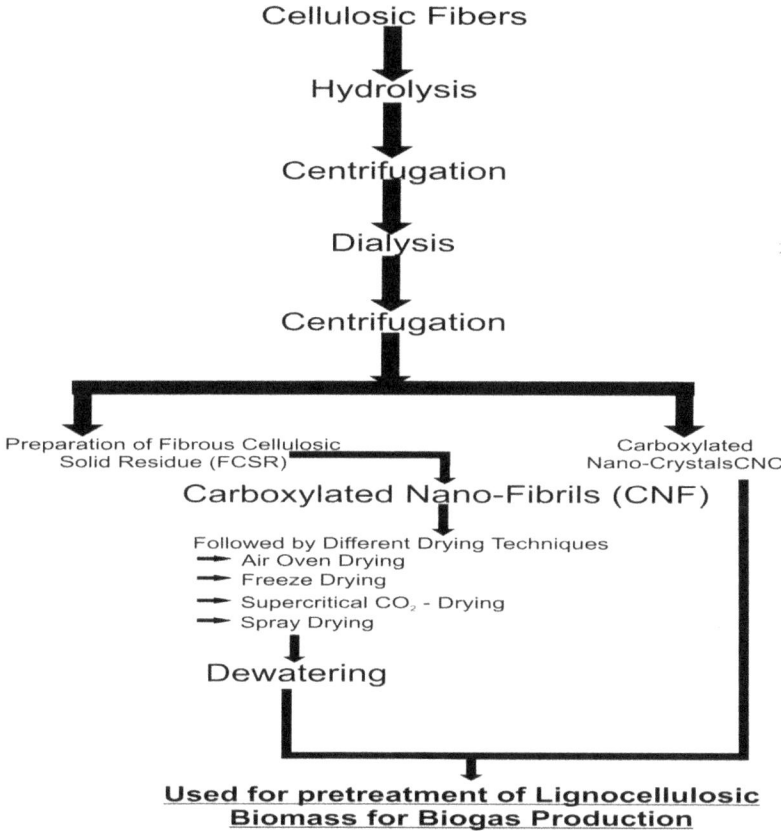

FIGURE 12.8 Production steps of carboxylated nano-fibrils (CNF) and carbonylated nano-crystals (CNC).

(Au) nanoparticles. Gold does not react with many chemicals at the macro scale and behaves as an inert element, but at the nano scale, gold becomes very reactive and behaves as a catalyst to accelerate reactions. The ratio between the mass and open area of nanoparticles contributes to this highly reactive property of nanoparticles. The surface area to volume ratio in the human digestive system is a biological example of AD processes where microorganism activity aids AD digestion. Nanoparticles have been acquired from both anthropogenic and natural resources. In waste sludge, a very high concentration of NPs could accumulate. The anaerobic digestion system is not stimulated by nanoparticles always; rather some nanoparticles inhibit the production rate considerably when compared with a control sample.

The production rate of the anaerobic digestion system is greatly affected by the types and concentration of nanoparticles. For illustration, the exposure concentration of ZnO at 1000 mg/L resulted in inhibition to 65.3% biogas volume and 47.7% methane composition in comparison with a control sample.

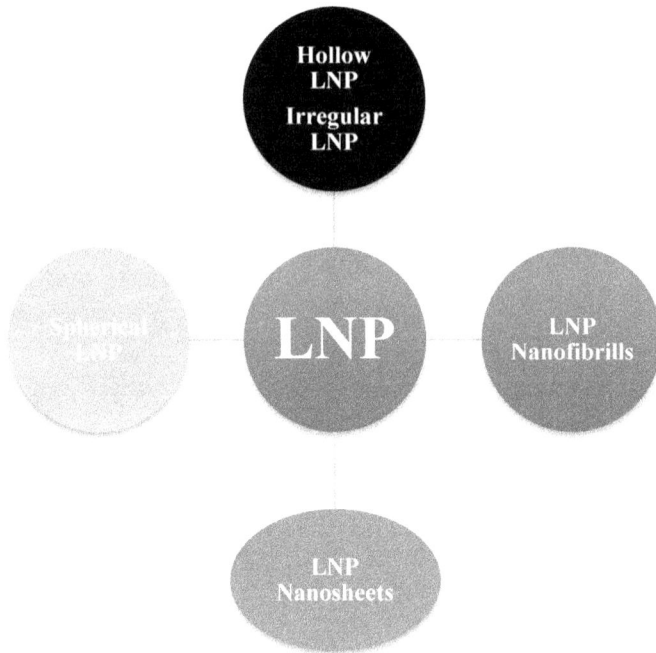

FIGURE 12.9   Different types of nanoparticles generated from lignin biomass.

12.3.3   NANOPARTICLES AND BIOMASS

Nanoparticles have significant effects on microorganisms. According to an overview of their antimicrobial properties, NPs strain has detrimental effects on wastewater microorganisms. Statistical data on the NPs effect on wastewater microorganisms during aerobic digestion is significant (Batley et al., 2012).

Minimized efficiency of anaerobic digestion processes, the absolute collapse of treatment and environmental pollution from contaminated effluents, and the utilization of bio-solids for changes in soil texture may result due to NPs and microbial community contact (Hoffman et al., 2001). A broad range of microorganisms are affected by the silver ion. Bacterial growth in a variety of medical treatments, including dental work, catheters, and healing burn wounds, has been controlled in recent years by silver ions (Kalsen et al., 2000). Concentration and contact time influencing the mechanism of release of ions from Ag was exhibited by *Escherichia coli*. The leaking of reducing sugar and protein, enzyme inhibition, cell obstruction, and dispersed vesicles, which slowly disintegrate, are the detrimental effects inhibiting cellular respiration and cell growth.

12.3.4   NANOPARTICLES AND ALGAL BIOMASS

In order to fulfill the future energy requirements of the world, algal biomass is potentially the next energy resource for bioenergy production. It also has the potential to serve as a resource of high value for basic chemicals and extracts. Being a member of aquatic systems, algae serve as food sources and hold a key role in photosynthesis. In fresh water, the impact of nanoparticles on microalgae can be found in the form of silver ions, silver chloride (AgCl) and silver sulfide (Ag_2S). Silver ion is the most toxic form of silver NPs (Ribeiro et al., 2014). With an increase in concentration in an aquatic environment, these nanoparticles can damage and affect the biota (Angel et al., 2013). In drinking water, ground water, and surface water, the concentration of silver nanoparticles has already been found to be above 5 g/L (WHO, 2003). Among many of the possible reasons for the toxicity of nanoparticles, three are very obvious: (i) a high dissolution rate due to a high surface area to volume ratio (Angel et al., 2013), (ii) aggregation behavior due to bioavailability, and (iii) coating with the stabilized organic material. Dramatic alterations in the algal cell wall characterized by pithophoral degradation and ruptured cell walls were revealed in SEM micrographs when treated with nanoparticles.

12.3.5 NANOPARTICLES AND AQUATIC PLANTS BIOMASS

Zada et al. (2013) demonstrated with water hyacinth that the fermentative production of hydrogen and ethanol is a feasible and sustainable process. The production of ethanol and hydrogen was significantly affected by iron nanoparticles. Fermentative hydrogen production and ethanol production were enhanced by iron nanoparticles. In addition to concentrations already present in dry plant biomass, the optimum iron nanoparticle concentration in fermentative hydrogen production and ethanol production was 250 mg/L and 150 mg/L, respectively. Of the plant biomass, maximum hydrogen yield was 57 mL/g, which comprises 85.50% of the theoretical maximum hydrogen yield. The maximum ethanol yield was 0.0232 mL/g of the plant biomass, which is 90.98% of the maximum theoretical yield. This study concluded that different types of nanoparticles accumulate in water hyacinth containing metal and carbonaceous nanomaterials. Nanofibers and carbon nanotubes (CNTs) also show an affirmative result. Due to a high surface area to volume ratio, better results have been revealed with nanomaterials than other water purification techniques. Monitoring, adsorption, microbial control, photocatalysis, membrane processes, disinfection, monitoring, and sensing are potential and current applications of nanotechnology for wastewater and water treatment (Qu et al., 2013). But the facts and data related to the toxicity of nanomaterials are still not sufficient. There are two main factors responsible for the antibacterial activity of nanoparticles: (i) the physiochemical properties of nanoparticles, and (ii) the nature of bacteria. It has been observed that the antibacterial effects of silver nanoparticles are enhanced by treating coliform bacteria with ultrasonic radiation for a short period prior to nanoparticle treatment. Ag NPs also displayed major activity against biofilms of bacteria. The antibacterial effect and Ag NPs' concentrations depend upon the class of bacteria to be treated.

Previous research has demonstrated that *Vibrio cholera* and *Pseudomonas aeruginosa* showed more resistance than *Salmonella typhi* and *Escherichia coli*, but bacterial growth was completely abolished at concentrations above 75 mg/L

(Zang et al., 2014). The antimicrobial activity of silver NPs against *Staph aureas* and *Escherichia coli* has also been studied, and at low concentrations, *Escherichia coli* was inhibited but *Staphylococcus aureus* was less inhibited (Wu et al., 2014). Ag NPs also showed significant adverse effects on filamentous green algae (Dash et al., 2012). The biomass to bioenergy conversion by the interaction of nanoparticles has shown the importance of nanomaterials in research. This conversion could be either a biological, chemical, or thermal process. The conversion process is affected by inorganic contaminants obtained from organic biomass and molecular size. Functionalized nanoparticles are acquired from both natural and synthetic sources.

12.3.6 EFFECT OF NANOMATERIALS ON BIOMASS

According to Theivasanthi et al. (2011), nanoparticles synthesized through electrolysis exhibit antibacterial activities against both *Gram-positive bacteria* and *Gram-negative bacteria*. The antibacterial activity of copper is enhanced by an alteration in the surface area to volume ratio. The antibacterial activity of copper NPs synthesized by electrolysis is higher than copper NPs synthesized using a chemical reduction method against *Escherichia coli*. When using electric power for the synthesis of copper NPs, antimicrobial activity is increased. The chemicals required for the synthesis of NPs are easily accessible, cheap, and non-toxic. Minimum infrastructure is required for the technology implementation. It is experimentally proven that this material could be used in antibacterial packaging, water purification, air filtration, air quality management, etc. For the biochemical conversion of biomass, microorganisms play a vital role.

Various nanomaterials affect the performance of AD. The mechanism of interaction of these particles is very complex, and the way they interact with the biomass, the process of conversion, overcoming the adverse effects, and optimizing the positive effect, all need to be understood better. The rate of AD can be influenced by the particle size because it affects the surface area for biomass biodegradation. Regardless of the chemical constituents, all nanoparticles possess an extremely high surface area to volume ratio. Therefore, the atoms of the surface and capping agents dominate the physical properties of nanoparticles. For applications such as catalysis, a high surface area to volume ratio is vital. The greater the surface of the same material, the greater is the reactivity. Different nanoparticles stimulate different responses and interactions among microorganisms. Although a few studies concluded that copper NPs have significant potential as bactericidal agents, NPs such as iron oxide, silica and its oxides, gold, and platinum have not shown a bactericidal effect in *Escherichia coli* studies. By stimulating the bacterial growth, magnetite NPs (Fe_3O_4 NPs) can enhance methane production. Another application of NPs is as a fuel catalyst for the reduction of harmful engine combustion emissions. Recent studies found that the AD process, biodegradation, and nitrification are inhibited by NPs (Liu et al., 2011). The particle size, time, and concentration determine whether there is enhancement, inhibition, or adverse effects of energy conversion.

12.4   PRETREATMENT OF BIOMASS AND BIOGAS PRODUCTION

The process of the pretreatment of biomass could be generally classified into physical, chemical, and biological methods as shown in Table 12.3 (Kamperidou and Terzopoulou, 2021).

Biomasses are refractory in nature owing to their cellulose crystallinity and non-reactive lignin (Mosier et al., 2005). As such, pretreatment techniques are required to make the cellulose amenable to the enzymatic hydrolysis process that subsequently assists in the extraction of fermentable sugars prior to the biofuel generation processes. Besides, technological limitations such as high cost and inadequacy in the existing infrastructure have posed major challenges in attaining high quality and yield of bioenergy from lignocellulosic material. In view of these limitations, studies have reported various optimization strategies in pretreatment stages as well as the enzymes and fermentation stages to enhance bioenergy production in an energy-efficient and cost-effective manner. Recently, the growing interest in nanomaterials and their exceptional properties has been vastly exploited to enhance bioenergy generation.

12.5   FUTURE RECOMMENDATIONS

In the future,

- There is a need to study the efficiency of different nano materials in the process of pretreatment of lignocellulosic biomass and hence the production yield of biogas.
- Apart from the application of nanotechnology in the pretreatment of lignocellulosic biomass leading to the generation of bioenergy, certain areas such as immobilization via covalent binding, which prevents the leaching of enzymes, need to be studied along with the study of the interaction of lipase with carrier for the full utilization of biomass in bioenergy generation.
- Microwave reactions for chemical reactions must be explored to minimize the duration of chemical reaction to reduce the energy use.
- There is a need for the development of a nano-based catalytic system to process a wide range of biomasses for bioenergy production.

12.6   CONCLUSION

The depleting sources and increasing demand of fossil fuels have diverted the focus of environmentalists towards renewable energy sources like biogas. The option of the production of cheap and abundant bioenergy from lignocellulosic biomass is being explored nowadays. The pretreatment methods of lignocellulosic biomass were expensive and had certain disadvantages, but by applying nanotechnology the effectiveness of the pretreatment method is increased, with the advantages of easy recovery, cost effectiveness, and renewability of immobilized enzymes. It has been observed and evidenced by a number of studies that nanobiotechnology plays

TABLE 12.3

Key Steps with Merits and Demerits of the Most Generally Used Pre-Treatment Methods Available for Ligno-Cellulosic Biomass

Name of Pretreatment		Key Process	Advantages	Disadvantages
Physical pretreatment	Milling, grinding or chipping methods	Particle size reduction methods, increase the surface area of substrate and reduce the polymerization level and cellulose crystallinity	1. Reduces the cellulose crystallinity and particle size 2. Increases surface area, and substrate management is easy 3. Disruption of hydrogen bonds and cellulose crystallinity 4. Increases surface area 5. Fast heat transfer 6. Short reaction time	1. High energy demand 2. Scalability issues
	High pressure homogenizing	Cell wall breakage	1. Increases the surface area of substrate by decreasing the particle dimensions 2. Improves the accessibility of the substrate 3. Increases its susceptibility to microbial and enzyme attacks	
	Extrusion	Exposed to heat, compression, friction and shearing forces		High energy consumption
	Ultrasonic treatment	Destructs, through the ultrasound waves		
	High hydrostatic pressure	Pressure level can be altered		High energy demand Recalcitrant compounds formation
	Gamma ray irradiation	Irradiation is obtained from radioisotopes such as cesium-137 and cobalt-60		
	Pulse electric field	Pretreatment, biomass materials are subjected to a brief (nanoseconds–milliseconds) and sudden burst of high voltage of 5.0–20.0 kV/cm		

(Continued)

TABLE 12.3 (CONTINUED)

Key Steps with Merits and Demerits of the Most Generally Used Pre-Treatment Methods Available for Ligno-Cellulosic Biomass

Name of Pretreatment		Key Process	Advantages	Disadvantages
Thermal treatment	-	The biomass materials are exposed to elevated temperatures	Thermal treatment could be preferable in regions where the surplus heat coming from a neighboring industry is available to be utilized	The release of dangerous compounds like furfural, hydroxymethyl furfural, and phenolic acids because of the high temperatures applied could inhibit the process
	Liquid hot water treatment	Involves hot water (170–230°C) and pressure (till 5 MPa)	The pulp and paper industries applied liquid hot water treatment process for pretreatment	At temperature ranges of 200 to 210°C, biogas production declines
	Steam explosion	Biomass is subjected to steam for 30 s to 30 min at a specific temperature (120–260°C) and pressure (5–20 bars)	Requires low energy, limited chemicals, no cost of recycling, and is environmentally friendly	Inhibitors may be produced during the process
	Hydrothermal treatment	At 200°C and hydrogen catalyst is low-cost	Eco-friendly method of high improvement efficiency	The generation of inhibitory substances such as 5-hydroxymethylfural and furfural
	Microwave	Electromagnetic radiation that uses wavelengths of 1 mm–1 m, recorded between 300 and 300,000 MHz on the electromagnetic spectrum	Gradually scaled up to pilot scale and seems to be very promising concerning its future development	Low level of thermodegradation or production of by-products/intermediate products
Chemical treatment	Alkali activation	(NaOH, Ca(OH)$_2$) is the most effective chemical pretreatment	Efficient in lignin solubility	Downstream processing is not economical
	Acidic pretreatment	Pre-treatments using either diluted or pure acid (H$_2$SO$_4$, HNO$_3$, HCL)	Efficient technology	The generation of inhibitory substances
	Oxidizing agents	Ozone, FeCl$_3$, hydrogen peroxide, and oxygen/air, in order to degrade the polymers	Improves cellulose hydrolysis	Great loss of hemicelluloses is reported

(Continued)

TABLE 12.3 (CONTINUED)
Key Steps with Merits and Demerits of the Most Generally Used Pre-Treatment Methods Available for Ligno-Cellulosic Biomass

Name of Pretreatment		Key Process	Advantages	Disadvantages
	Ozonolysis	Ozone is considered effective for pretreatment	Does not generate toxic or inhibitory substances and, therefore, it is environmentally friendly	—
	Sulfite pretreatment	The biomass is exposed to magnesium or calcium sulfite to decrease all the lignin and hemicelluloses contents	Highly effective, easy and innovative pretreatment	High cost
	Organosolv	Use of organic solvents such as methanol, ethanol, ethylene glycol, and acetone	Many factors determine	Whole process cost
	Carbon dioxide explosion	Supercritical CO_2 with the gas in the role of hydrolysis agent	Appropriate for raw materials of high moisture content	High cost of reactor
	Ammonia fiber explosion	Ammonia in liquid form is utilized	Cost-effective and highly efficient	High cost
	Ionic liquids	Green solvents used for pretreatment	Easily copes with biomass	Increases the production cost
Biological pretreatment	Microbial	Strong hydrolytic action has been employed	1. Hydrolysis of lignin and hemicelluloses 2. Alteration of cellulose structure 3. No any inhibitory compound formation 4. Low energy consumption	1. The process is slow 2. There is carbon loss 3. Necessity of a large sterile area
	Enzymes	Using enzymes that have endogluconase, exogluconase, and glucosidase properties	1. Alteration of cellulose structure 2. Delignification 3. Partial hydrolysis of hemicellulose 4. The process is fast 5. The energy demand is low	1. The cost of enzymes is high 2. Continuous addition may be required

a significant role in the pretreatment process of lignocellulosic biomass thereby significantly increasing the yield of biogas.

REFERENCES

Abraham, R.E., Verma, M.L., Barrow, C.J., Puri, M. (2014). Suitability of magnetic nanoparticle immobilized cellulases in enhancing enzymatic saccharification of pretreated hemp biomass. *Biotechnol. Biofuels* 7(1), 90. https://doi.org/10.1186/1754-6834-7-90.

Achinas, S., Achinas, V., Euverink, G.J.W. (2017). A technological overview of biogas production from biowaste. *Engineering* 3(3), 299–307.

Ahmad, R., Khare, S.K. (2018). Immobilization of *Aspergillus niger* cellulase on multiwall carbon nanotubes for cellulose hydrolysis. *Bioresour. Technol.* 252(72–75). https://doi.org/10.1016/j.biortech.2017.12.082.

Ali, S., Shafique, O., Mahmood, S., Mahmood, T., Khan, B.A., Ahmad, I. (2020). Biofuels production from weed biomass using nanocatalyst technology. *Biomass Bioenergy* 139(May), 105595. https://doi.org/10.1016/j.biombioe.2020.105595.

Angel, B.M., Batley, G.E., Jarolimek, C.V., Rogers, N.J. (2013). The impact of size on the fate and toxicity of nanoparticulate silver in aquatic systems. *Chemosphere* 93(2), 359–365.

Aslanzadeh, S., Rajendran, K., Taherzadeh, M.J. (2014). A comparative study between single-and two-stage anaerobic digestion processes: Effects of organic loading rate and hydraulic retention time. *Int. Biodeterior. Biodegrad.* 95, 181–188.

Batley, G.E., Kirby, J.K., McLaughlin, M.J. (2012). Fate and risks of nanomaterials in aquatic and terrestrial environments. *Acc. Chem. Res.* 46(3), 854–864.

Chen, H. (2014). *Biotechnology of Lignocellulose*, Springer, Dordrecht.

Dash, A., Singh, A.P., Chaudhary, B.R., Singh, S.K., Dash, D. (2012). Effect of silver nanoparticles on growth of eukaryotic green algae. *Nano Microlett.* 4(3), 158–165.

Deublein, D., Steinhauser, A. (2008). *Biogas from Waste and Renewable Resources*, Wiley Online Library, Weinheim, Germany.

Goriushkina, T.B., Soldatkin, A.P., Dzyadevych, S.V. (2009). Application of amperometric biosensors for analysis of ethanol, glucose, and lactate in wine. *J. Agric. Food Chem.* 57(15), 6528–6535.

Hamawand, I., Seneweera, S., Kumarasinghe, P., Bundschuh, J. (2020). Nanoparticle technology for separation of cellulose, hemicellulose and lignin nanoparticles from lignocellulose biomass: A short review. *Nano Struct. Nano Objects* 24, 100601. https://doi.org/10.1016/j.nanoso.2020.100601.

Hoffmann, C., Christoffi, N. (2001). Testing the toxicity of influents to activated sludge plants with the *Vibrio fischeri* bioassay utilizing a sludge matrix. *Environ. Toxicol.* 16(5), 422–427.

Huang, P.-J., Chang, K.-L., Hsieh, J.-F., Chen, S.-T. (2015). Catalysis of rice straw hydrolysis by the combination of immobilized cellulase from *Aspergillus niger* on β-cyclodextrin-Fe3O4 nanoparticles and ionic liquid. *BioMed Res. Int.* 2015, 1–9. https://doi.org/10.1155/2015/409103.

ISO/TS 27687:2008 Nanotechnologies. (2008). *Terminology and Definitions for Nano-Objects — Nanoparticle, Nanofibre and Nanoplate.* Available at: https://www.iso.org/standard/44278.html.

Kamperidou, V., Terzopoulou, P. (2021). Anaerobic digestion of lignocellulosic waste materials. *Sustainability* 13(22), 12810. https://doi.org/10.3390/su132212810.

Klasen, H.J. (2000). Historical review of the use of silver in the treatment of burns.I. *Earlyuses. Burns* 26(2), 117–130.

Komemoto, K., Lim, Y.G., Nagao, N., Onoue, Y., Niwa, C., Toda,T. (2009). Effect of temperature on VFA's and biogas production in anaerobic solubilization of food waste. *Waste Manag.* 29(12), 2950–2955.

Kumar, P., Barrett, D.M., Delwiche, M.J., Stroeve, P. (2009). Methods for pretreatment of lignocellulosic biomass for efficient hydrolysis and biofuel production. *Ind. Eng. Chem. Res.* 48(8), 3713–3729.

Lima, J.S., Araújo, P.H.H., Sayer, C., Souza, A.A.U., Viegas, A.C., de Oliveira, D. (2017). Cellulase immobilization on magnetic nanoparticles encapsulated in polymer nanospheres. *Bioproc. Bioprocess Biosyst. Eng.* 40(4), 511–518. https://doi.org/10.1007/s00449-016-1716-4.

Liu, G., Wang, D., Wang, J., Mendoza, C. (2011). Effect of ZnO particles on activated sludge: Role of particle dissolution. *Sci. Total Environ.* 409(14), 2852–2857.

Montag, D., Schink, B. (2016). Biogas process parameters–energetics and kinetics of secondary fermentations in methanogenic biomass degradation. *Appl. Microbiol. Biotechnol.* 100(2), 1019–1026.

Mosier, N., Wyman, C., Dale, B., Elander, R., Lee, Y., Holtzapple, M., Ladisch, M. (2005). Features of promising technologies for pretreatment of lignocellulosic biomass. *Bioresour. Technol.* 96(6), 673–686.

Qu, X., Alvarez, P.J.J., Qilin, L. (2013). Applications of nanotechnology in water and wastewater treatment. *Water Res.* 47(12), 3931–3946.

Rathee, K., Dhull, V., Dhull, R., Singh, S. (2016). Biosensors based on electrochemical lactate detection: A comprehensive review. *Biochem. Biophys. Rep.* 5, 35–54.

Ribeiro, F., Gallego-Urrea, J.A., Jurkschat, K., Crossley, A., Hassellöv, M., Taylor, C., Soares, A.M., Loureiro, S. (2014). Silver nanoparticles and silver nitrate induce high toxicity to *Pseudokirchneriella subcapitata, Daphnia magna* and *Daniorerio. Sci. Total Environ.* 466, 232–241.

Sankaran, R., Markandan, K., Khoo, S.K., Cheng, K.C., Ashokkumar, V., Deepanraj, B., Show, L.P. (2021). The expansion of lignocellulose biomass conversion Into bioenergy via nanobiotechnology. *Front. Nanotechnol.* https://doi.org/10.3389/fnano.2021.793528.

Sanusi, I.A., Suinyuy, T.N., Kana, G.E.B. (2021). Impact of nanoparticle inclusion on bioethanol production process kinetic and inhibitor profile. *Biotechnol. Rep. (Amst)* 29, e00585. https://doi.org/10.1016/j.btre.2021.e00585.

Sode, K., Tsugawa, W., Aoyagi, M., Rajashekhara, E., Watanabe, K. (2008). Propionate sensor using coenzyme-A transferase and acyl-CoA oxidase. *Protein Pept. Lett.* 15(8), 779–781.

Srivastava, N., Srivastava, M., Mishra, P.K., Ramteke, P.W. (2016). Application of ZnO nanoparticles for improving the thermal and pH stability of crude cellulase obtained from *Aspergillus fumigatus* AA001. *Front. Microbiol.* 7, 514–519. https://doi.org/10.3389/fmicb.2016.00514.

Sun, Y., Cheng, J. (2002). Hydrolysis of lignocellulosic materials for ethanol production: A review. *Bioresour. Technol.* 83(1), 1–11. https://doi.org/10.1016/s0960-8524(01)00212-7.

Tahir, N., Tahir, M.N., Alam, M., Yi, W., Zhang, Q. (2020). Exploring the prospective of weeds (*Cannabis sativa* L., *Parthenium hysterophorus* L.) for biofuel production through nanocatalytic (Co, Ni) gasification. *Biotechnol. Biofuels* 13(1), 1–10. https://doi.org/10.1186/s13068-020-01785-x.

Theivasanthi, T., Alagar, M. (2011). Studies of copper nanoparticles effects on micro-organisms. *Ann. Biol. Res.* 2, 368–373.

Victor, F.P., Ferrer, A.S. (2011). *Enhancement of Biogas Production in Anaerobic Digesters Using Iron Oxide Nanoparticles*, Catalan Institute of Nanoscience and Nanotechnology (ICN2), University Autonma Barcelona, Barcelona, Spain.

Wang, H., Covarrubias, J., Prock, H., Wu, X., Wang, D., Bossmann, S.H. (2015). Acid-functionalized magnetic nanoparticle as heterogeneous catalysts for biodiesel synthesis. *J. Phys. Chem. C* 119(46), 26020–26028. https://doi.org/10.1021/acs.jpcc.5b08743.

WBA Global Bioenergy Statistics 2017. www.worldbioenergy.org. Accessed 18 May 2021.

Weiland, P. (2006). Biomass digestion in agriculture: A successful pathway for the energy production and waste treatment in Germany. *Eng. Life Sci.* 6(3), 302–309. https://doi .org/10.1002/elsc.200620128.

WHO. (2003). *Silver in Drinking-Water, Background Document for Development of WHO Guidelines for Drinking-Water Quality,* World Health Organization, Geneva, Switzerland, Volume 2.

Wu, D., Fan, W., Kishen, A., Gutmann, J.L., Fan, B. (2014). Evaluation of the antibacterial efficacy of silver nanoparticles against *Enterococcus faecalis* biofilm. *J. Endod.* 40(2), 285–290.

Yang, L., Xu, F., Ge, X., Li, Y. (2015). Challenges and strategies for solid-state anaerobic digestion of lignocellulosic biomass. *Renew. Sustain. Energ. Rev.* 44, 824–834.

Zada, B., Mahmood, T., Malik, S.A. (2013). Effect of iron nanoparticles on hyacinth's fermentation. *Int. J. Sci.* 2, 78–92.

Zang, Y., Li, C.Z., Chueh, C.C., Williams, S.T., Jiang, W., Wang, Z.H., Yu, J.S., Jen, A.K.Y. (2014). Integrated molecular, interfacial, and device engineering towards high performance non fullerene based organic solar cells. *Adv. Mater.* 26(32), 5708–5714.

Zanuso, E., Gomes, D.G., Ruiz, H.A., Teixeira, J.A., Domingues, L. (2021). Enzyme immobilization as a strategy towards efficient and sustainable lignocellulosic biomass conversion into chemicals and biofuels: Current status and perspectives. *Sustain. Energ. Fuels* 5(17), 4233–4247. https://doi.org/10.1039/d1se00747e.

Zhang, X., Yan, S., Tyagi, R.D., Surampalli, R.Y. (2011). Synthesis of nanoparticles by microorganisms and their application in enhancing microbiological reaction rates. *Chemosphere* 82(4), 489–494. https://doi.org/10.1016/j.chemosphere.2010.10.023.

Zhang, Y., Feng, Y., Quan, X. (2015). Zero-valent iron enhanced methanogenic activity in anaerobic digestion of waste activated sludge after heat and alkali pretreatment. *Waste Manag.* 38, 297–302.

Zhu, J.Y., Agarwal, U.P., Ciesielski, P.N., Himmel, E.M., Gao, R., Deng, Y., Mortis, M., Osterberg, M. (2021). Towards sustainable production and utilization of plant-biomass-based nanomaterials: A review and analysis of recent developments. *Biotechnol. Biofuels* 14(1), 114. https://doi.org/10.1186/s13068-021-01963-5.

13 Bio-nanotechnology in Waste to Energy Conversion in a Circular Economy Approach for Better Sustainability

Neha Saxena

CONTENTS

13.1 INTRODUCTION

What does nano-bio mean exactly? Even among nano-experts, this is a tough question to answer since the definition of nanotechnology is still not clear, and it varies depending on the expert view. For example, the chemical background of nanobiotechnology involves the development of biomolecules to develop new materials; for a physicist it may be defined as the study of biological interactions at the molecular level; and for a biotechnologist it may be defined as the application of nanotools to diagnose and cure diseases. Until recently, nano and biotechnologies combined

DOI: 10.1201/9781003316374-13

were called nano-bio, based on the notion "bio is nano" as they are generated from the bottom-up. As a result, the approach is distinct from conventional technologies, which are considered to use a variety of raw materials to build anything at the nanoscale. Shaping the World Atom by Atom was the subtitle of the 1999 pamphlet Nanotechnology, and atomic accuracy became known as "nanoscale precision" (Sen et al., 2013). The capacity to create things by combining molecules instead of atoms was one of the most hoped-for benefits of nanotechnology (Baird et al., 2004). One-billionth of a meter equals ten hydrogen atoms side by side in a nanometer. Processes regulating the physical and chemical characteristics at molecular-scale level are used to create nanotechnology (Nembhard et al., 2007). As a result, terms like "molecule by molecule" or "atom by atom" are important indications for both nanotechnology and nano-bio. As a result, we must return to the 19th-century understanding of atoms and molecules.

Over the next decade, predictions suggest that many scientists and engineers will work in the field of nanotechnology and bio-nanotechnology. Nanoscale developments in biological science and technology, in fact, offer applications in practically every field, with revolutionary socio-economic consequences. Major advances are expected in fields as varied as nanocomposite materials for solar power production and nanoscale devices with high precision functions for medical services, for example. The pharmaceutical business is still in its infancy and is still stuck in a Neanderthal mindset, focusing mostly on relatively basic chemicals to treat ailments. According to the new discipline of biological nanotechnology, in future we might construct medical weapons of comparable complexity to those used by "the adversary" in the future. We ought to develop complicated and intelligent machines at molecular level that can tackle viruses and bacteria that use nuanced infection methods on a more level playing field.

Penicillin is a wonder medicine that launched the antibiotics revolution, yet compared to its foes, it's a pretty basic system. Fullerenes are carbon hollow cage-ball like molecules, and their other extended forms like carbon nano tubes (CNTs) and stable allotropic forms like graphene, diamond, and graphite, offer exciting potential as fundamental materials for innovative nanoscale applications using biomolecules. The morphological structure of materials is a mesmerizing field, where structure-related features are of interest in nanoscale engineering, where potential nanoscale devices with superior performance and sustainable, eco-friendly applications are developed.

The carbon-based systems are expected to be used in medical circumstances and have chemical synthesis capabilities to develop and design molecular devices with sophisticated functions. Fullerene cages have the potential to be zero-toxic carriers of elements that are radioactive in chemotherapeutic applications, and drug delivery is just one of them. The possibility of paradigm-shifting advancements in medical techniques is particularly enticing. One of the earliest collections of papers on this intriguing and complex topic may be found in this volume. If any of these stimulating advancements are to be realized, the next generation of scientists will require a solid education in the field of nanoscale science and technology, which should be integrated into undergraduate and graduate curricula in biology and medicine.

Bio-nanotechnology is a vital 21st-century functional technology that is attracting attention all around the world. The ability to use biomolecule structures and processes for new applications in materials, biosensors, bioelectronics, and medicine has spawned the fast-expanding field of nanobiotechnology. Atoms exhibit extraordinary diversity and individuality at the nanoscale. Bio-nanotechnology is a word that refers to the functional uses of biomolecules in nanotechnology. It is a merger of bioscience and nanotechnology and is based on the principles and chemical processes of living beings. Because the development of nanotechnology techniques is guided by studying the structure and function of natural nanosized molecules found in living cells, it incorporates the study, creation, and illumination of connections between nutrition, food science, structural molecular biology, and nanotechnology.

13.2 APPLICATION OF BIO-NANOTECHNOLOGY

13.2.1 APPLICATION OF BIO-NANOTECHNOLOGY IN THE ENERGY SECTOR: CONVERSION OF BIOMASS TO VALUE ADDED FUELS

The nanoscience and bio-nanotechnology applications in the areas of energy, the environment, and life sciences will be the main focus in this chapter of the book. It has been found that there will be approximately a 36% rise in the world's primary energy consumption from 2008 to 2035 which will be met by electricity according to the sources from the International Energy Agency (IEA) (Moniz, 2010). As a result, renewable energy will play a critical role in moving the globe toward a more stable, secure, and sustainable energy route. Modern researchers are developing techniques to find economical and eco-friendly energy sources, and one of the potential sources for this purpose is biomass. Figure 13.1 depicts the various biomass resources available that can be utilized for conversion into value added products.

Green energy is a clean energy source that has proved to be eco-friendly in nature and is sustainable for future applications. Many researchers have proposed methods to decrease the use of fossil fuels and prevent the destructive impacts to the environment (Bhowmik et al., 2017). Carbon capture and coal processing are some of the methods that focus on reducing the emission of greenhouse gases. Another such method is the conversion of biomass to value added energy fuel which comes with the advantage of being renewable and eco-friendly in nature (Kim et al., 2020). In recent decades, the application of biomass in the form of bio waste and used cooking oil (UCO) has been enhanced and is employed in producing green and eco-friendly products like biofuels, biochemicals, biomedical products, etc. (Patel et al., 2021). Biomass waste is increasing at an alarming rate, from 2000 million tons in 2015 to approximately 4000 million tons in 2025 in the US and Europe. This large amount of waste cannot be dumped in landfill, so there is a need to develop a system for the conversion of biomass to green and sustainable products like biofuels and chemicals. Various biomass conversion processes have been studied by many researchers and are depicted together in Figure 13.2.

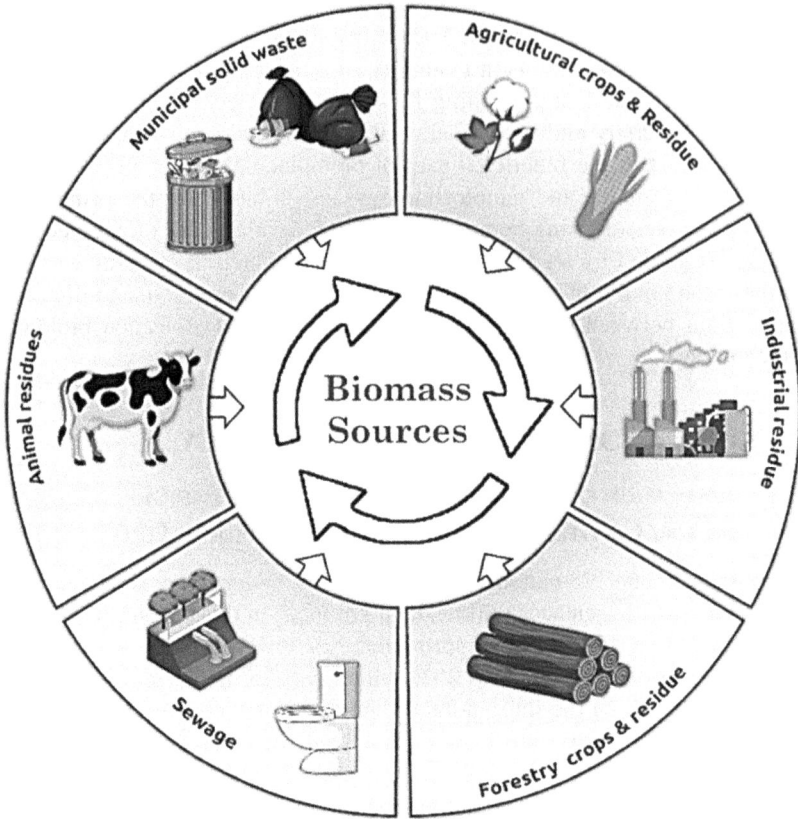

FIGURE 13.1   Various biomass resources.

Biomass conversion in the 21st century is assumed to be the sustainable choice for vast applications including energy sources, biofuels, and biochemicals (Hanchate et al., 2021). The conversion of the biomass is important to provide a sustainable environment to future generations.

13.2.2   APPLICATION OF BIO-NANOTECHNOLOGY IN THE ENVIRONMENTAL SECTOR

The existing environmental problems and advanced solutions to them are addressed by environmental nanosciences. Bio-nanotechnology plays a crucial role in providing effective solutions to a wide range of environmental science and engineering problems. Some of the major contributions of these biological nanosciences in the environment include emission control from a wide range of sources, green technology development for minimizing undesirable by-products, and the remediation of water in the environment and polluted water bodies as shown in Figure 13.3.

The growth of environmental nanosciences hinges on the remediation of pollutants, the creation of renewable energy, and the efficacy of monitoring systems.

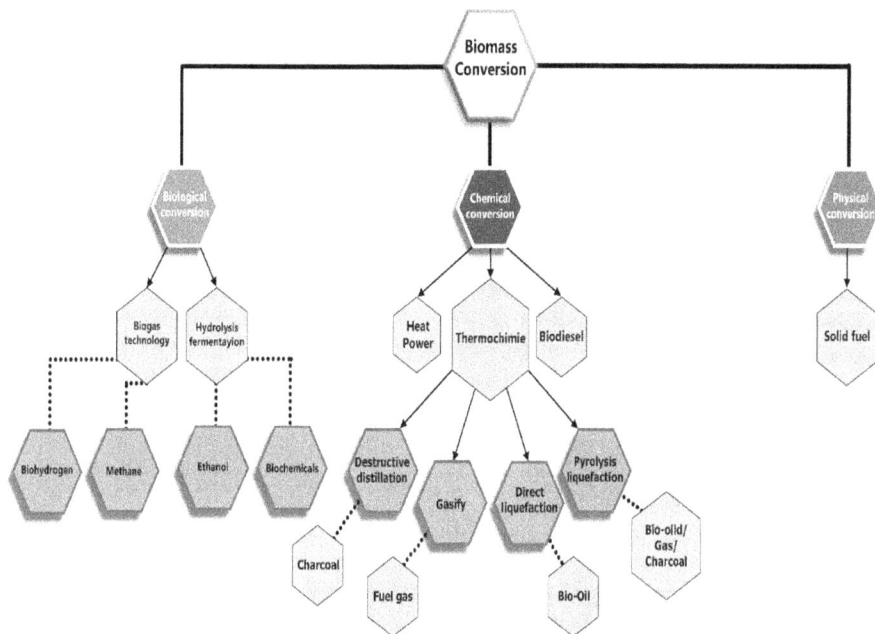

FIGURE 13.2 Different biomass conversion processes that employ biotechnology in the energy sector.

Some of the considerations in nanoscience technology and bio-nanotechnology are the advanced availability of technological devices, including their cost and size in terms of environmental implications; minimizing the effect of pollutants from air, water, and soil sources; the prospect of advanced sensors in environmental science; rapid advances in nanoscience in health care management coupled with the environment; the artificial nanoparticle and its impact on the ecosystem; and nanotechnology (Mansoori et al., 2008; Mansoori et al., 2019; Stander et al., 2011). Air pollution, water contamination, and soil contamination are all examples of environmental issues in the nanotechnology (Figure 13.4).

Silver nanoclusters are employed as nano-catalysts in the production of propylene dioxide, a common chemical used in paints, detergents, and polymers, to reduce by-product pollution (Bhawana et al., 2012). Nanoscience catalysts such as nanofibers in combination with manganese oxide aid in the removal of volatile organic compounds from industrial chimneys (Gomes et al., 2015). Carbon nanotubes absorb gases from large-scale power plants and industrial facilities 100 times quicker than conventional techniques, and nanostructured membranes with microscopic holes are employed to remove methane and CO_2 from exhaust.

Zero valent iron nanoparticles (nZVI) are used to remediate polychlorinated biphenyls (PCBs) and a variety of other pollutants in soil. Soil washing, thermal desorption, and landfill disposal are the most prevalent forms of traditional technology now in use. The idea of electrokinetic remediation, which employs direct

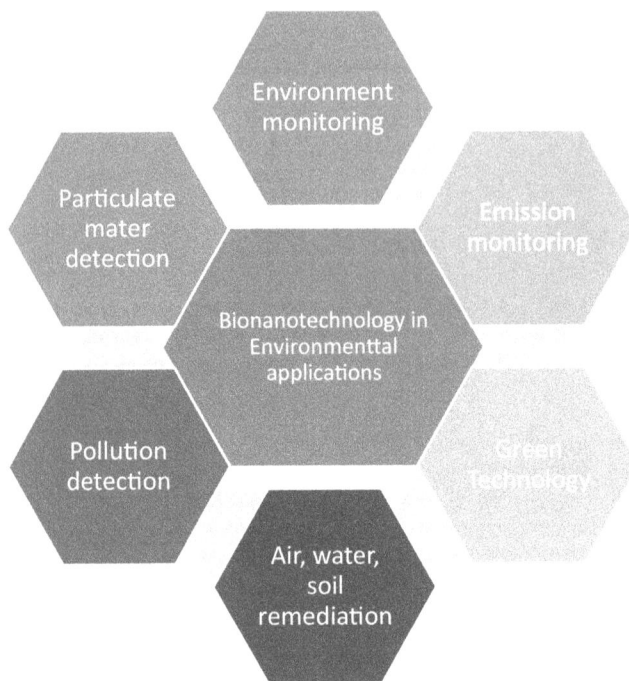

FIGURE 13.3   Contributions of bionanotechnology in environmental applications.

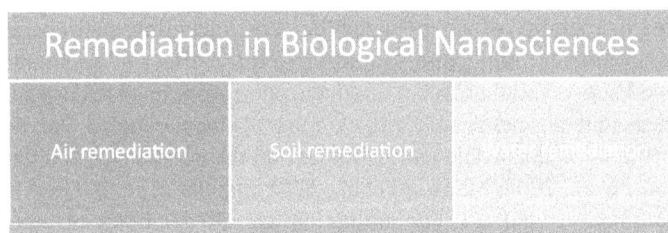

FIGURE 13.4   Different types of environmental remediation in biological nanosciences.

current, was applied to improve nanoparticle mobility. The use of a direct kinetic current to introduce nZVI into the soil would improve the destruction of pollutants in the soil and the in-situ transformation process rather than transferring contaminants using traditional techniques (Gillham, 1996). The other frequent soil pollutant, Malathion, was similarly remediated by utilizing nZVI. Three decades ago, zero valence nanoparticles were used to cure or decontaminate water. Using nanoscience technology, halogenated group pollutants in water were decontaminated by zero valence iron nanoparticles in permeable reactive barriers (PRB). The decontamination of trichloroethene, a harmful pollutant found in industrial effluent, is one example of this strategy (Gillham, 1996, Nutt et al., 2005). The zero valent iron

(ZVI or Fe0) nanoparticles remedial treatment was employed for all pollutants in various studies, and these materials are also very well suited for diverse organic and inorganic contaminants in groundwater.

The water purification procedure based on nanoscience technology is simpler and less expensive, as it involves injecting nanoparticle catalysts into the groundwater system rather than pumping water. Deionization is made cheaper and more consistent by using nanosized fibers for the electrode. Iron nanoparticles, for example, might be used to clean contaminated groundwater by removing organic solvents. Ion exchange resins, which have a substrate made up of an organic polymer with nanosized holes, are employed in separation, purification, and decontamination procedures in which ions are trapped on the surface and then exchanged. The drawbacks of nanoscience technology in remediation must be addressed, as nanoparticles used for commercial and industrial purposes have not been well investigated, and environmental and health risks must be considered.

Apart from the minor drawbacks, bio-nanotechnology has the potential to play an important role in pollution management by precisely manipulating and developing nanomaterials at the nanoscale level, hence boosting the pollutants' affinity, capacity, and selectivity. This strategy reduces the amount of hazardous waste in the atmosphere and aquatic ecosystem, laying the groundwork for environmental protection agencies (Zad et al., 2018; Ahmadpour et al., 2003; Shahsavand et al., 2004). The benefits of bio-nanotechnology include solutions for energy consumption, pollution management, and greenhouse gas emissions, all of which have significant economic, social, and environmental implications. Other advantages include more efficient industrial processes, waste reduction through high-precision manufacturing, and improvements in air, water, and soil quality through the detection and elimination of pollutants, as well as the remediation of environmental damages, which reduces the need for large-scale industrial plants.

13.2.3 APPLICATION OF BIO-NANOTECHNOLOGY IN LIFE SCIENCE

Bio-nanotechnology is the production and use of materials, technologies, and systems at the nanoscale, i.e., at the atomic, molecular, and supramolecular level. Nanotechnology is derived from the Greek term nano, which means "dwarf." For advances in biological and medicinal systems, it is critical to first understand nanoparticles and their characteristics. However, a new area of research known as "nanomedicine" has been recognized as a diverse field in the last five years, and it has expanded quickly since then. Richard P. Feynman, the late Nobel physicist, had the brilliant notion of designing, manufacturing, and introducing small nanorobots into the human body to accomplish repairs at the molecular and cellular level. In his foresightful 1959 address, "There is Plenty of Room at the Bottom," he advocated that machine tools be used to build minor machine tools, which might then be used to make even smaller machine tools that work at the atomic level (Feynman, 2018). Feynman was well aware of the new technology's potential medicinal uses. According to Feynman, nanomedicine is a multidisciplinary area that includes chemistry, biology, physics, material science, and engineering and plays a significant role in improving human health. Nanomedicine illustrates the use of nanotechnological techniques in the areas of biomedical research. The inclusive objective of nanomedicine is to diagnose as correctly and as early as possible, to cure as effectively as possible without any side effects, and to noninvasively assess treatment efficacy. Biological nanotechnology's potential is multifarious, giving not just enhancements to existing procedures but also whole new tools and capabilities. The fundamental characteristics and bioactivity of pharmaceuticals and other materials can be changed by changing them at the nanoscale scale. Controlling drug or agent properties such as solubility, blood pool retention times, controlled release over short or extended time periods, environmentally driven controlled release, or very precise site-targeted delivery may all be done using these technologies. Furthermore, the enhanced efficient surface area per unit volume obtained by utilizing nanometer-sized particles may be used in a variety of ways. This section highlights some of the most recent achievements in the biomedical field with nanomaterials and tools. It also provides researchers with a comprehensive summary of the current state of nanomaterials research and recommends future avenues for using nanomaterials to achieve yet-to-be-achieved biological goals. Nanomaterials' distinct optical, magnetic, and electrical characteristics make them potential platforms for biological applications such as biosensing, imaging, and drug delivery (Han et al., 2001; Kamat 2002; Katz et al., 2004). Because all attributes of nanomaterials are influenced by their size and form, one of the key research fields is the investigation of techniques for their synthesis. Traditionally, there have been two types of synthetic methods for nanomaterials: top-down and bottom-up. A common "top-down" process, also known as a physical method, is mechanically grinding bulk material and then stabilizing the resultant nanosized particles using colloidal protective agents (Gaffet et al., 1996; Amulyavichus et al., 1998). A "bottom-up" approach tries to develop nanomaterials and devices one molecule/atom at a time, similar to how live creatures make macromolecules.

Biotechnology, medicines, gene therapy, medication delivery, tissue engineering, medical devices, and human diagnostics have all benefited from breakthroughs in nanoscience technology and bio-nanotechnology (Nakamura et al., 2011; Nanjwade et al., 2011; Pandurangappa et al., 2011; Elgindy et al., 2011; Patil et al., 2011) (Figure 13.5).

Drug delivery systems have long been a crucial component of nanoscience-based therapy approaches (Figure 13.6). The green synthesis of silver and other nanoparticles has shown to be a viable strategy since it is clean, eco-friendly, and low-toxicity (Manam et al., 2020; Manam et al., 2021). Due to the extensive research, huge surface area, and antibacterial, anti-diabetic, antioxidant, and cytotoxic effects, silver nanoparticles have been widely used as a drug delivery method (Manam et al., 2013; Manam et al., 2014b; Murugesan, 2014a; Murugesan, 2014b; Manam et al., 2014c).

The benefit of silver and other nanoparticles in drug delivery is that they can be utilized to transport medications, particularly chemotherapeutic agents used in cancer therapy, to the target region with less accumulation in healthy cells, reducing harm to them. In cancer therapy, the enhanced permeability and retention effect (EPR) in nanoscale materials is significant. Lack of oxygen and malfunctioning lymphatic systems prevent therapeutic intervention during the formation of tumors;

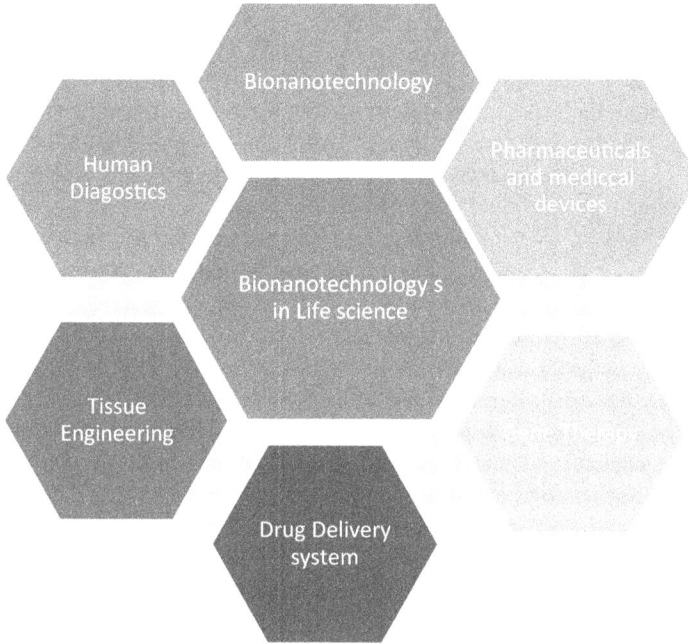

FIGURE 13.5 The application of bio-nanotechnology in life sciences.

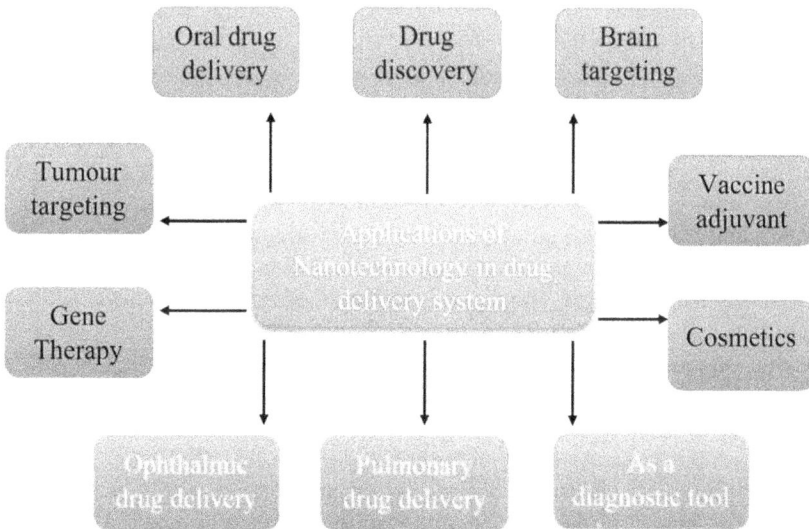

FIGURE 13.6 The application of bio-nanotechnology in drug delivery systems.

however, nanoparticles serve as drug delivery vehicles and are kept at the target site via blood vessel pores.

Radio and magnetic signals, as well as the induction of chemical nanoparticles, are used to target the sick region. Where elements such as infrared heat or a combination of certain chemicals are present, the nanoparticles are programmed to release the associated medicine into the host. Nanoscale particles such as silver nanoparticles, nanoshells, dendrimers with an organic basis, and polymers with hollow capsules are used to encapsulate the target medication, and the negative effects of the medicine may be managed more accurately in inpatient treatments by dosage reduction. Chemotherapy, gene therapy, radiation, imaging, orthopedics, and wound management are all examples of uses where the medicine is designed to convey the therapeutic material (Langer et al., 1999; Nimesh et al., 2006; Soppimath et al., 2001; Jung et al., 2000). Poor bioavailability of medications that cannot be taken orally can be improved by applying nanoscience technology in drug delivery systems. Pharmaceuticals subjected to denaturation owing to high pH and drugs with a limited shelf life can also benefit from the adhesion process in drug delivery systems (El-Shahbouri et al., 2002: Hu et al., 2004; Arangoa et al., 2001; Arbos et al., 2004). Recent encapsulation research and the creation of appropriate animal models have improved the immunization efficacy of nanoparticles used in antigen delivery for vaccination systems (Diwan et al., 2003; Sánche and Alonso et al., 2005; Lutsiak et al., 2002).

Polymer nanoparticles, quantum dots, and nano-fabricated structures are some of the different types of drug delivery methods that use nanoscience technology. When compared to standard therapeutic medications such as paclitaxel and gemcitabine, the usage of liposomal nanoparticles with PEG-WHI-P131 formulation in patients with chest/breast cancer showed improved therapeutic potential in vivo (Dibirdik et al., 2010). Similarly, for tagging biomolecules in diagnostics, chitosan-CdS quantum nanodots are recommended (Knight et al., 2010). During the medication delivery process, the improved permeable layer activity of nano-fabricated structures prevents superfluous molecules from passing through (Mehrotra et al., 2010). Cardiovascular disorders are a big worry nowadays, and nanoscience technology, with tools like quantum dots, nanocrystals, and nano-barcodes, is being utilized to recognize and ignore signals in the erythrogenic process in heart difficulties. These technologies are used extensively in the commercial scientific imaging and diagnostic businesses, as well as in the engineering of sick tissues (Wickline et al., 2006).

Tissue engineering, a significant biological nanotechnological subject, continues to expand economically by serving as a bridge between the biomedical and pharmaceutical sectors (Rippel et al., 2011; Kim et al., 2011; Shih et al., 2015). Internal tissue implants, which provide organ replacements via xenotransplants, are employed in producing scaffold nanomaterial tissues and organs, whereas nanostructured materials are used to create the optical profiles of contact lenses (Sun et al., 2013). The separation and identification of diverse cells by recognizing sparse calls, which are physiologically distinct from somatic cells such as cancer, fetal, and HIV-infected T cells, is another major subject in nanoscience. Cells are sorted based on surface charge and biocompatible surfaces with precise nanopores in this nanoscience technique, which employs electrodes incorporated into microchannels (Fakruddin et al., 2012). Because of their capacity to pass through cell membranes and lipid bilayers, liposomes, a lipid bilayer, have been widely employed in targeted medication administration (Wang et al., 2018; Ewert et al., 2005). Linking nanoparticle (PEG) treated liposomes with reporter expression from rhesus monkey brains to human insulin reporter monoclonal antibodies has been reported (Zhang et al., 2003). As a result, both developed and developing countries benefit from the broad-spectrum benefits of nanoscience and bio-nanotechnology. The numerous advantages include eco and environmentally clean energy, improved health care, clean drinking water using nano-filtration to trap organisms and toxins, advanced diagnostic measures, the treatment of dreadful diseases, potent drug delivery systems, nanoscale product development, advanced solutions over conventional methods, ease of life for humans and the environment, and so on (Macoubrie, 2004).

13.2.4 Applications of Bio-Nanotechnology as Regenerative Medicine

When artificial organs are transplanted in place of natural organs, the regenerative regeneration of tissue offers a novel method for treating any disease or injury. Regeneration is the revolution of the millennium in the field of medicine. R. Langer, and a surgeon, J.P. Vacanti (Langer et al., 1999), were the ones who proposed tissue engineering in the 1980s. The key technology employed is biodegradable polymer scaffolds, that preform and target the tissue shape, allowing the rebuilding of cartilage tissues for the growth of human tissues. Preformed biodegradable polymer scaffolds and several other types of particular cells have been combined to rebuild bone, cartilage, and blood arteries for therapeutic purposes. Tissue engineering has been employed for tissue engineering with in vitro development. The initial clinical results have reported that, for the expansion of the urinary bladder, a conventional biodegradable scaffold is not beneficial in the development of functional tissue due to a lack of nutrients and oxygen supply inside the tissue morphology (Chung, 2006). Thus, the assembly of various functional activities for the development tissue architecture, like blood-capillary mimicking vascular lumina, is important for maintenance and tissue growth. Bio-nanotechnology refers to technology and science at nanoscale in a biological system. Nanotechnology presents great potential for benefits in its application and research that also attract investment from businesses and governments throughout the world. Nanomaterials, including semiconductor devices (Fukumori et al., 2009), carbon nanotubes (Wilder et al., 1998), fullerene (Fukumori et al., 2009), graphenes zeolites (Yamamoto et al., 2003), inorganic nanoparticles (Xiao et al., 2003), and nano-sensors including scanning probe microscopes and scanning electron microscopes (Goldstein et al., 2003), have been invented and found application. In recognition of the enormous scientific and commercial potential of nanotechnology, President Clinton established the National Nanotechnology Initiative (NNI) in 2000 (Mnyusiwalla et al., 2003). The NNI confines the activities of several federal agencies with a proposed budget of 1 billion dollars in 2004. As per Moore's law (Moore, 1965), top-down methodologies have been developed and implemented over the years, and recently DNA chips have been claimed to be an important methodology developed for application in medical fields. On the other side, for the application of bottom-up methodologies in the field of medical science, quantum dots have been used as imaging devices in the body (Choi et al., 2010) and drug delivery systems as nanocarriers for the treatment of tumor patients (Safra et al., 2000). The clinical diagnosis shows the progress made in this field and highlights significant achievements and holes at the theoretical and applied level. In spite of the fact that nanotechnology techniques have opened the way for new innovations, applications like tissue engineering in combination with nanotechnology are becoming some of the most significant areas in the medical field.

13.2.5 Application of Bio-Nanotechnology in the Production of Functional Food

A number of functional foods have been introduced to the market as a result of the generally recognized relationship between nutrition and health. Food materials that are dosed with bioactive substances or that include dietary supplements to give specific health advantages are referred to as functional foods. Functional foods, on the other hand, do not have a single, widely acknowledged definition. As a result, the word "functional food" should be regarded as a concept rather than as a collection of precise food items. In 2000, the global market for functional foods was valued at \$47.6 billion, with the United States accounting for the greatest proportion, followed by Europe and Japan (Sloan, 2002). In recent years, the spectrum of functional foods has expanded dramatically, currently including infant foods, bakery products and cereals, confectionary dairy foods, ready to cook meals and snacks, soft drinks, and meat products. Drinks are expected to continue to be the most popular category of functional foods, and next are breads and cereals (Sloan, 2002). Nanomaterials are generally 60,000 times smaller in size than viruses (Sekhon, 2010); they are believed to be so tiny that bacteria would require a microscope to view them (Pray and Yaktine, 2009). Figure 13.7 provides a classification of food packaging based on the use of nanotechnology and bio-based nanotechnology (Robertson, 2008).

Food nanotechnology, as the name implies, is the application of nanotechnology to the manufacture of food. It covers numerous applications, such as altering the characteristics of meals by adding nano-additives and nano-ingredients, improving food delivery, quality, and safety, and designing packaging with improved qualities (Buzby, 2010).

Food nanotechnology has enormous potential, with suggested applications spanning the whole food sector from packaging to developed products to their analysis now being researched. However, at the time of writing, the most common applications of this new technology are food packaging and food items that contain nano-sized ingredients (Sekhon, 2010). The high degree of interest in nanotechnology's possible use in food production arises from the assumption that nanotechnology plays a significant role in food systems by providing an innovative and more efficient approach to transporting nutrients and bioactive substances to the body's critical areas (Maynard et al., 2006).

```
              ┌─────────────────────┐
              │   Third-generation  │
              │  biobased polymers  │
              └─────────────────────┘
```

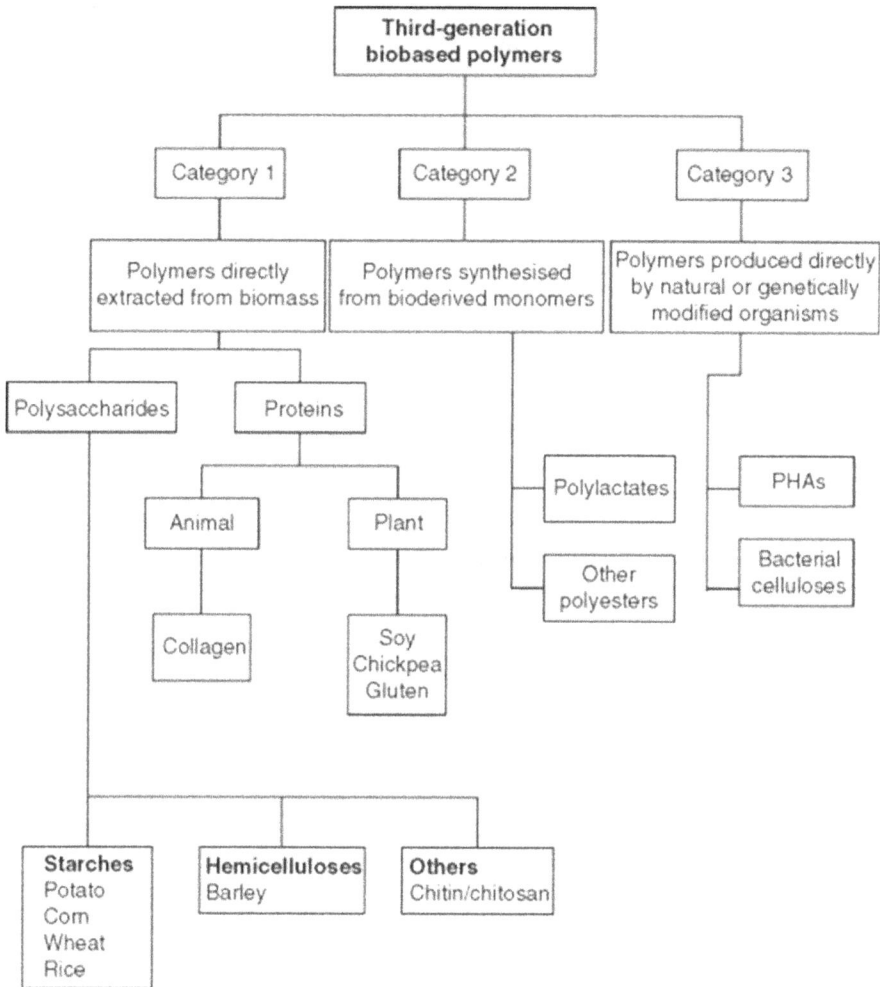

FIGURE 13.7   The classification of food packaging.

In comparison to the usage of nanotechnology in medication delivery and pharmaceuticals, where these have been studied over the last two decades, nanotechnology in the food sector is relatively new (Sozer and Kokini, 2009). Surprisingly, the existence of nanomaterials in food is not new; several traditional cuisines produce nanosized particles. Whey proteins are permitted to collect in the form of protein nanoparticles in the production of ricotta cheese, for instance. However, nanomaterials have been used in the manufacturing of these traditional foods without any knowledge of the alterations occurring at the sub-microscopic level.

The purposeful fabrication of nanoparticles for use in the manufacture of food is referred to as food nanotechnology. Nanomaterials are created using either a "top–down" or a "bottom-up" technique. Grinding, milling, etching, and lithography are used to physically reduce materials to nanoscale size. This "top-down" method is used to create the majority of nanomaterials on the market today. However, as nanotechnology advances, it is expected that the adoption of the "bottom-up" method, in which nanomaterials are formed from individual atoms and molecules with the ability to self-assemble, will increase (Sanguansri and Augustin, 2006). Reduced salt content and allergenicity, as well as higher bioavailability of vitamins, phytonutrients, PUFA, and probiotics, are some of the health-promoting benefits that innovative technologies offered in the current market may provide. These technologies enable the manufacture of beneficial meals while still preserving their freshness. However, to get the necessary outcomes, technologies like pulsed electric field (PEF) and high pressure processing (HPP) must frequently be paired with traditional methods like heating and enzymatic hydrolysis, which risks producing goods that lack the sensory attributes that consumers like. Hurdle technologies that combine these unique processing methods might be a feasible solution for creating goods with specified characteristics. A combination of HPP with nano-encapsulation, for example, might provide a more effective way to boost probiotic viability without compromising the functional food's organoleptic qualities. The commercial availability of a large number of functional meals or components made with these unique technologies attests to the advantages that these technologies may provide once developed. However, more work needs to be done in order to fully exploit the capabilities of these technologies. If customers are to accept nanotechnology products, concerns about safety must be addressed. If HPP and PEF are to be widely used, new and less expensive designs that minimize the initial high cost of installation may be required.

13.3 BIO-NANOTECHNOLOGY AND A CIRCULAR ECONOMY APPROACH FOR BETTER SUSTAINABILITY

To achieve a sustainable society, it is pre-requisite to carry out deep studies on issues like waste management, product cycle assessment, and the circular economy. Waste management in the present scenario goes hand in hand with the circular economy. The circular economy, often called "circularity," is an economic system that deals with global challenges like pollution, global warming, biodiversity loss, biowaste, etc. The idea of the circular economy aims at transforming traditional patterns of production, consumption, and economic growth into the circular flow and utilization of these resources to prevent pollution and promote waste management as shown in Figure 13.8 (Wang et al., 2020).

The circular economy is the newest strategy that have been developed and conceptualized throughout the world. The major focus has been laid on the 4R rule (reduce, reuse, recycle, and recover). Figure 13.9 depicts the 4R rule for the waste management of food items from various sources. This 4R rule further leads to R-Hierarchy (rethink, redesign, repair, redo, redistribute, and recover) that is combined with the concept of sustainable development (Jabbour et al., 2019). The circular economy is proposed to preserve natural resources and optimize the use of fossil fuels so as to provide a clean and green future.

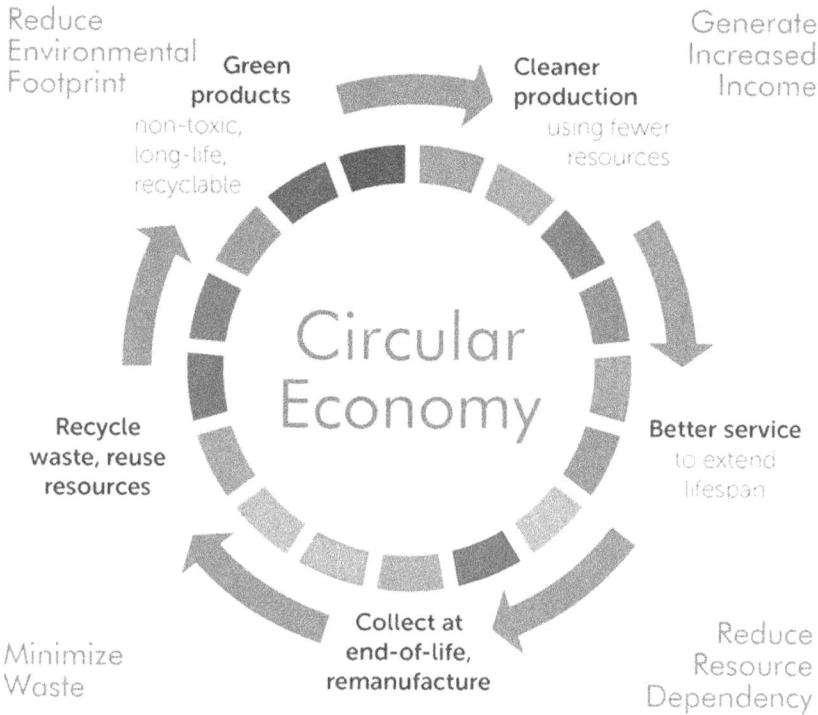

FIGURE 13.8 The image depicts the flow of the circular economy.

Many researchers across the world describe the circular economy as a regenerative system where the flow of resources and speed are lowered aiming for the maintenance of durable products, their repair, reuse, and recycling (Gupta et al., 2019; Kurdve and Bellgran, 2021). The circular economy aims for potential sustainable development and economic growth by proper waste management. It also focuses on the development of eco-friendly, cost-effective solutions and a sustainable environment for future generations.

13.4 FUTURE CHALLENGES IN BIO-NANOTECHNOLOGY

The combination of the traditional areas of biology with nanotechnology has given rise to the topic of bio-nanotechnology, which is a catch-all word covering a variety of related technologies such as nanobiology, nanotoxicology, DNA nanotechnology, and other complex elements of nanoscience. Bio-nanotechnology is essentially a fusion of biological study in various fields of nanotechnology, such as food, nutrition, and health. Nanobiology, including nanoscale devices, nanoparticles, and other notions that come within the purview of nanotechnology, has spawned a slew of modern concepts. This one-of-a-kind technique drawn from biology will enable scientists to develop nanoscience that may be employed in cutting-edge biological

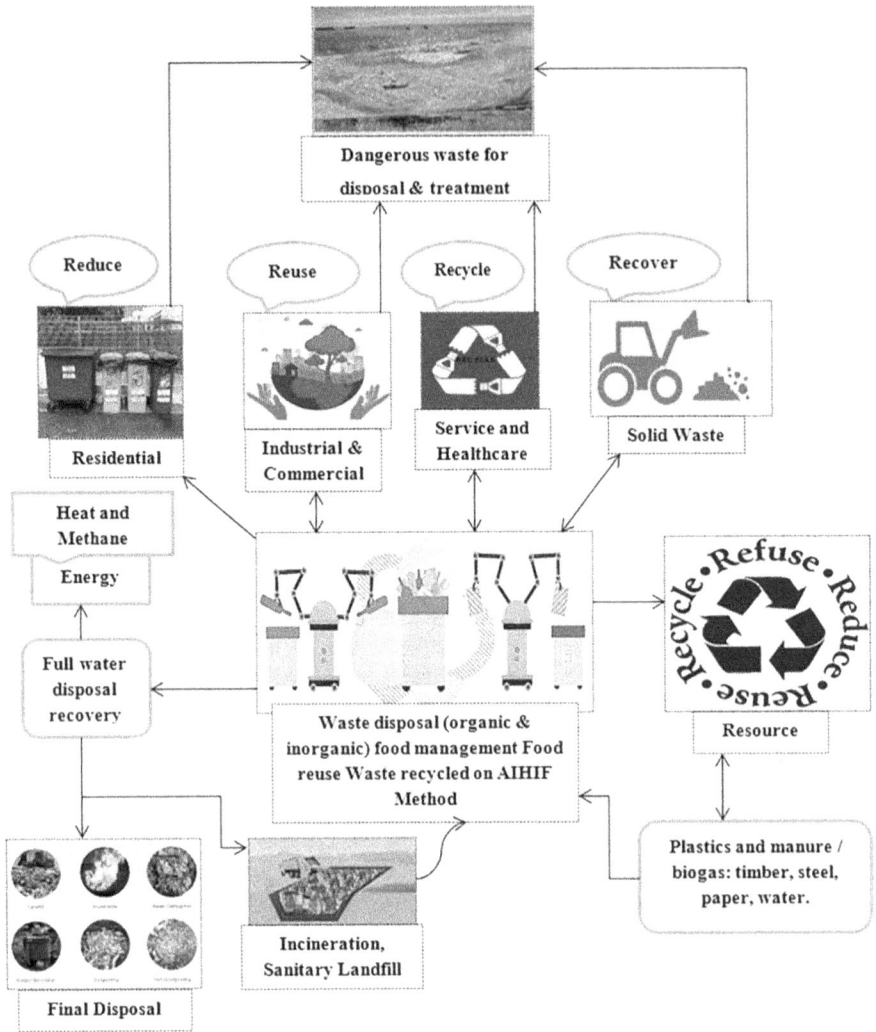

FIGURE 13.9   The 4R rule for the waste management of food items.

and medical studies in the future. Bio-nanotechnology can teach us how to develop systems, potentially leading to a number of beautiful systems that can be used in a variety of fields (Subbiah et al., 2010). Nanoscience is expected to have a bright future, with applications in all aspects of life. There's also a chance that bio-nano-technology will have the potential to change the globe in the not-too-distant future. Molecular manufacturing, which is the capacity to transform materials into life by developing molecules to ordinary systems, is one of the main priorities for the future of nanoscience. Nanobiology techniques have the potential to enhance the health of people suffering from significant diseases or accidents. Medical practitioners are able to treat disease pathophysiology at the cellular and molecular level via nanosurgery. Another key goal in the field of nanotechnology is the development of unique, inventive instruments for the biological, medical, and surgical fields. New nanotools are frequently created by repurposing existing nanotool apps. Nanobiology will also entail the application of nanotools to pertinent medical and biological difficulties, as well as the refinement of such applications. Novel nanotool-based approaches might, for example, be used to remove cancer, as surgical procedures could be conducted at the cellular level in the organs where cancer cells spread (Kudgus et al., 2011).

Food and pharmaceutical businesses might potentially benefit from nanodelivery technologies (de la Rica et al., 2012). Because of variations in the efficacy of the chemicals of interest, toxicological considerations must also be adequately explored. Bio-nanotechnology has the potential to tackle some of the world's challenges. It has the potential to end world hunger as well as a number of other potential issues, such as pollution. Furthermore, advances in bio-nanotechnology may pave the way for the molecular manufacturing of healthy body cells, tissues, and various organs, including the utilization of nanorobots, that have ability to execute activities currently performed by humans. Nanorobots might identify individual molecules of toxins in water, keeping the environment greener and cleaner (Douglas et al., 2012). Nanorobots might thus be an important tool for cleaning contaminated surroundings, removing toxins from water bodies, and perhaps cleaning up oil spills. Bio-nanotechnology may be able to assist in the rebuilding of the depleted ozone layer in the not-too-distant future, and so may have a beneficial impact on the environment. In the medical field, bio-nanotechnology will have a significant influence. Patients might consume fluids containing nanorobots that are programmed to fight cancerous cells and viruses and remodel the DNA molecular structure of cells during medical therapy. Cancerous cells might be identified and eliminated, and the procedure could be aided by the surgical implantation of healthy cells. Furthermore, nanosurgery might be used to maintain or heal over the years, ranging from natural aging to diabetes to bone spurs, leaving nothing that nanosurgery couldn't fix. The field of bio-nanotechnology broadens to incorporate heterogeneous molecular and cellular nanosystem networks, where molecules and supramolecular structures function as separate devices. Computers and robotics might be shrunk to absurdly tiny dimensions. The development of effective genetic therapies and anti-aging medicines will be among the medical uses. Inside cells, new medicinal pathways will produce proteins that work collectively to shield cells from a wide range of temperature impacts and allow them to perform in a variety of conditions much more quickly. As a result, nanotechnology should improve every industrial area over time, including healthcare and food. It should also benefit the environment by making more efficient use of resources and improving pollution control systems. It is necessary to help the public to visualize nanotechnology in a larger context where human values and quality of life are still relevant and critical if this strong new science is to realize its full potential. Because nanotechnology is such a young science, its future is an unknown and difficult territory. Nanoscience will provide people the power to alter matter at the atomic or molecular level in order to create something valuable at the nanoscale.

13.5 CONCLUSIONS

In the present chapter, the application of bio-nanotechnology in the energy, environment, life sciences, food packaging, and biomedical sectors has been discussed, where the conversion of biomass to value added products has been the focus. The author has aimed to study the use of green and clean energy sources from bio-nanotechnology that are pre-requisite for sustainable development. The study of the circular economy has also been discussed, including the application of the 4R rule. The use of green energy and strategies of the circular economy will help us to develop a sustainable environment for future generations.

REFERENCES

Ahmadpour, A., Shahsavand, A., & Shahverdi, M. R. (2003, February). Current application of nanotechnology in environment. In: *Proceedings of the 4th Biennial Conference of Environmental Specialists Association*, Tehran.

Ali Mansoori, G., Bastami, T. R., Ahmadpour, A., & Eshaghi, Z. (2008). Environmental application of nanotechnology. *Ann. Rev Nano Res.*, 439–493. https://doi.org/10.1142/9789812790248_0010

Amulyavichus, A., Daugvila, A., Davidonis, R., & Sipavichus, C. (1998). Study of chemical composition of nanostructural materials prepared by laser cutting of metals. *Fiz. Met. Metalloved.*, 85(1), 111–117.

Arangoa, M. A., Campanero, M. A., Renedo, M. J., Ponchel, G., & Irache, J. M. (2001). Gliadin nanoparticles as carriers for the oral administration of lipophilic drugs. Relationships between bioadhesion and pharmacokinetics. *Pharm. Res.*, 18(11), 1521–1527.

Arbós, P., Campanero, M. A., Arangoa, M. A., & Irache, J. M. (2004). Nanoparticles with specific bioadhesive properties to circumvent the pre-systemic degradation of fluorinated pyrimidines. *J. Control. Release*, 96(1), 55–65.

Baird, D., & Shew, A. (2004). Probing the history of scanning tunneling microscopy. *Discov. Nano.*, 2, 145–156.

Bhawana, P., & Fulekar, M. (2012). Nanotechnology: Remediation technologies to clean up the environmental pollutants. *Res. J. Chem. Sci.*, 2, 90–96.

Bhowmik, C., Bhowmik, S., Ray, A., & Pandey, K. M. (2017). Optimal green energy planning for sustainable development: A review. *Renew. Sustain. Energy Rev.*, 71, 796–813.

Buzby, J. C. (2010). Nanotechnology for food applications: More questions than answers. *J. Consum. Aff.*, 44(3), 528–545.

Choi, H. S., Liu, W., Liu, F., Nasr, K., Misra, P., Bawendi, M. G., & Frangioni, J. V. (2010). Design considerations for tumour-targeted nanoparticles. *Nat. Nanotechnol.*, 5(1), 42–47.

Chung, S. Y. (2006). Bladder tissue-engineering: A new practical solution? *Lancet*, 9518(367), 1215–1216.

De La Rica, R., Aili, D., & Stevens, M. M. (2012). Enzyme-responsive nanoparticles for drug release and diagnostics. *Adv. Drug Deliv. Rev.*, 64(11), 967–978.

Dibirdik, I., Yiv, S., Qazi, S., & Uckun, F. (2010). *In vivo* anti-cancer activity of a liposomal nanoparticle construct of multifunctional tyrosine kinase inhibitor 4-(4'-hydroxyphenyl)-amino-6, 7-dimethoxyquinazoline. *J. Nanomed. Nanotechnol.*, 1, 101.

Diwan, M., Elamanchili, P., Lane, H., Gainer, A., & Samuel, J. (2003). Biodegradable nanoparticle mediated antigen delivery to human cord blood derived dendritic cells for induction of primary T cell responses. *J. Drug Target.*, 11(8–10), 495–507.

Douglas, S. M., Bachelet, I., & Church, G. M. (2012). A logic-gated nanorobot for targeted transport of molecular payloads. *Science*, 335(6070), 831–834.

Elgindy, N., Elkhodairy, K., Molokhia, A., & ElZoghby, A. (2011). Biopolymeric nanoparticles for oral protein delivery: Design and in vitro evaluation. *J. Nanomed.Nanotechnol.*, 2(3), 110.

El-Shabouri, M. H. (2002). Positively charged nanoparticles for improving the oral bioavailability of cyclosporin-A. *Int. J. Pharm.*, 249(1–2), 101–108.

Ewert, K., Evans, H. M., Ahmad, A., Slack, N. L., Lin, A. J., Martin Herranz, A., & Safinya, C. R. (2005). Lipoplex structures and their distinct cellular pathways. *Adv. Genet.*, 53, 119–155.

Fakruddin, M., Hossain, Z., & Afroz, H. (2012). Prospects and applications of nanobiotechnology: A medical perspective. *J. Nanobiotechnol.*, 10(1), 1–8.

Feynman, R. (2018). *There's Plenty of Room at the Bottom* (pp. 63–76). CRC Press.

Fukumori, K., Akiyama, Y., Yamato, M., Kobayashi, J., Sakai, K., & Okano, T. (2009). Temperature-responsive glass coverslips with an ultrathin poly (N-isopropylacrylamide) layer. *Acta Biomater.*, 5(1), 470–476.

Gaffet, E., Tachikart, M., El Kedim, O., & Rahouadj, R. (1996). Nanostructural materials formation by mechanical alloying: Morphologic analysis based on transmission and scanning electron microscopic observations. *Mater. Charact.*, 36(4–5), 185–190.

Gillham, R. W. (1996). In situ treatment of groundwater: Metal-enhanced degradation of chlorinated organic contaminants. In *Advances in Groundwater Pollution Control and Remediation* (pp. 249–274). Springer.

Goldstein, J. I., Newbury, D. E., Echlin, P., Joy, D. C., Lyman, C. E., Lifshin, E., ... & Michael, J. R. (2003). Specimen preparation of hard materials: Metals, ceramics, rocks, minerals, microelectronic and packaged devices, particles, and fibers. In *Scanning Electron Microscopy and X-ray Microanalysis* (pp. 537–564). Springer.

Gomes, H. I., Dias-Ferreira, C., Ottosen, L. M., & Ribeiro, A. B. (2015). Electroremediation of PCB contaminated soil combined with iron nanoparticles: Effect of the soil type. *Chemosphere*, 131, 157–163.

Gupta, S., Chen, H., Hazen, B. T., Kaur, S., & Gonzalez, E. D. S. (2019). Circular economy and big data analytics: A stakeholder perspective. *Technol. Forecast. Soc. Change*, 144, 466–474.

Han, M., Gao, X., Su, J. Z., & Nie, S. (2001). Quantum-dot-tagged microbeads for multiplexed optical coding of biomolecules. *Nat. Biotechnol.*, 19(7), 631–635.

Hanchate, N., Ramani, S., Mathpati, C. S., & Dalvi, V. H. (2021). Biomass gasification using dual fluidized bed gasification systems: A review. *J. Clean. Prod.*, 280, 123148.

Hu, L., Tang, X., & Cui, F. (2004). Solid lipid nanoparticles (SLNs) to improve oral bioavailability of poorly soluble drugs. *J. Pharm. Pharmacol.*, 56(12), 1527–1535.

Jabbour, C. J. C., de Sousa Jabbour, A. B. L., Sarkis, J., & Godinho Filho, M. (2019). Unlocking the circular economy through new business models based on large-scale data: An integrative framework and research agenda. *Technol. Forecast. Soc. Change*, 144, 546–552.

Jung, T., Kamm, W., Breitenbach, A., Kaiserling, E., Xiao, J. X., & Kissel, T. (2000). Biodegradable nanoparticles for oral delivery of peptides: Is there a role for polymers to affect mucosal uptake. *Eur. J. Pharm. Biopharm.*, 50(1), 147–160.

Kamat, P. V. (2002). Photophysical, photochemical and photocatalytic aspects of metal nanoparticles. *J. Phys. Chem. B*, 106(32), 7729–7744.

Katz, E., & Willner, I. (2004). Integrated nanoparticle–biomolecule hybrid systems: Synthesis, properties, and applications. *Angew. Chem. Int. Ed.*, 43(45), 6042–6108.

Kim, H., Choi, J., Park, J., & Won, W. (2020). Production of a sustainable and renewable biomass-derived monomer: Conceptual process design and techno-economic analysis. *Green Chem.*, 22(20), 7070–7079.

Kim, S., Khang, G., & Lee, D. (2011). Application of nanomedicine in cardiovascular diseases and stroke. *Curr. Pharm. Des.*, 17(18), 1825–1833.

Knight, L. C., Romano, J. E., Krynska, B., Faro, S., Mohamed, F. B., & Gordon, J. (2010). Binding and internalization of iron oxide nanoparticles targeted to nuclear oncoprotein. *J. Mol. Biomark. Diagn.*, 1(1). 10.4172/2155-9929.1000102

Köping-Höggård, M., Sánchez, A., & Alonso, M. J. (2005). Nanoparticles as carriers for nasal vaccine delivery. *Expert Rev. Vaccines*, 4(2), 185–196.

Kudgus, R. A., Bhattacharya, R., & Mukherjee, P. (2011). Cancer nanotechnology: Emerging role of gold nanoconjugates. *Curr. Med. Chem. Anti Cancer Agents*, 11(10), 965–973.

Kurdve, M., & Bellgran, M. (2021). Green lean operationalisation of the circular economy concept on production shop floor level. *J. Clean. Prod.*, 278, 123223.

Langer, R. S., & Vacanti, J. P. (1999). Tissue engineering: The challenges ahead. *Sci. Am.*, 280(4), 86–89.

Lutsiak, M. E., Robinson, D. R., Coester, C., Kwon, G. S., & Samuel, J. (2002). Analysis of poly (D, L-lactic-co-glycolic acid) nanosphere uptake by human dendritic cells and macrophages in vitro. *Pharm. Res.*, 19(10), 1480–1487.

Macoubrie, J. (2004). Public perceptions about nanotechnology: Risks, benefits and trust. *J. Nanopart. Res.*, 6(4), 395–405.

Manam, D., & Kiran, V. (2021). Role of nanoscience and bionanotechnology in energy, environment, and life sciences. *Environ. Technol. Innov.* https://papers.ssrn.com/sol3/papers.cfm?abstract_id=3922022

Manam, D., Kiran, V., & Murugesan, S. (2013). Biogenic silver nanoparticles by *Halymenia poryphyroides* and its in vitro anti-diabetic efficacy. *J. Chem. Pharm.*, 5(12), 1001–1008.

Manam, D., Kiran, V., & Murugesan, S. (2014a). Biological synthesis of silver nanoparticles from marine alga *Colpomenia sinuosa* and its in vitro anti-diabetic activity. *Am. J. Biochem. Biotechnol.*, 3(1), 1–7.

Manam, D., Kiran, V., & Murugesan, S. (2014b). Bio-synthesis of silver nano particles from marine alga *Halymenia poryphyroides* and its antibacterial efficacy. *Int. J. Curr. Microbiol. Appl. Sci.*, 3(4), 96–103.

Manam, D., Kiran, V., & Murugesan, S. (2014c). Biosynthesis of silver nanoparticles from marine alga *Colpomenia sinuosa* and its antibacterial efficacy. *Int. J. Curr. Microbiol. Appl. Sci.*, 3(4), 887–893.

Manam, D., Kiran, V., & Subbaiah, D. M. (2020). Biosynthesis and characterization of silver nanoparticles from marine macroscopic brown seaweed *Colpomenia sinuosa* and its anti-fungal efficacy against dermatophytic and non dermatophyticfungi. *Int. J. Pharm. Sci.*, 11(2), 59–68.

Mansoori, G. A., & Keshavarzi, T. E. (2019). Recent developments in phase transitions in small/nano systems. In *Advances in Phytonanotechnology* (pp. 1–16). Academic Press.

Maynard, A. D., Aitken, R. J., Butz, T., Colvin, V., Donaldson, K., Oberdörster, G., Warheit, D. B., Ryan, J., Seaton, A., Stone, V., Tinkle, S. S., Tran, L., Walker, N. J., & Warheit, D. B. (2006). Safe handling of nanotechnology. *Nature*, 444(7117), 267–269.

Mehrotra, A., Nagarwal, R. C., & Pandit, J. K. (2010). Fabrication of lomustine loaded chitosan nanoparticles by spray drying and in vitro cytostatic activity on human lung cancer cell line L132. *J. Nanomedic. Nanotechnol.*, 1, 103.

Mnyusiwalla, A., Daar, A. S., & Singer, P. A. (2003). 'Mind the gap': Science and ethics in nanotechnology. *Nanotechnology*, 14(3), R9.

Moniz, E. J. (2010). *Nanotechnology for the Energy Challenge*. John Wiley & Sons.

Moore, G. E. (1965). Cramming more components onto integrated circuits. *Electronics.*, 86(1), 114–117.

Murugesan, S. (2014b). In vitro cytotoxic activity of silver nano particle biosynthesized from *Colpomenia sinuosa* and *Halymenia poryphyroides* using DLA and EAC cell lines. *World J. Pharm. Res.*, 2(9), 926–930.

Murugesan, S. (2014a). Invitro antioxidant activity of silver nano-particles from *Colpomenia sinuosa* and *Halymenia poryphyroides*. *World J. Pharm. Res.*, 2(9), 817–820.

Nakamura, J., Nakajima, N., Matsumura, K., & Hyon, S. H. (2011). In vivo cancer targeting of water-soluble taxol by folic acid immobilization. *J. Nanomed. Nanotechnol.*, 2, 106.

Nanjwade, B. K., Derkar, G. K., Bechra, H. M., Nanjwade, V. K., & Manvi, F. V. (2011). Design and characterization of nanocrystals of lovastatin for solubility and dissolution enhancement. *J. Nanomed. Nanotechnol.*, 2(2), 14–22.

Nembhard, H. B. (2007). Nanotechnology: A big little frontier for quality. *Qual. Prog.*, 40(7), 23.

Nimesh, S., Manchanda, R., Kumar, R., Saxena, A., Chaudhary, P., Yadav, V., Mozumdar, S., & Chandra, R. (2006). Preparation, characterization and in vitro drug release studies of novel polymeric nanoparticles. *Int. J. Pharm.*, 323(1–2), 146–152.

Nutt, M. O., Hughes, J. B., & Wong, M. S. (2005). Designing Pd-on-Au bimetallic nanoparticle catalysts for trichloroethene hydrodechlorination. *Environ. Sci. Technol.*, 39(5), 1346–1353.

Pandurangappa, C., & Lakshminarasappa, B. N. (2011). Optical absorption and photoluminescence studies in gamma-irradiated nanocrystalline CaF2. *J. Nanomed. Nanotechnol.*, 2(2), 592–595.

Patel, S. K., Gupta, R. K., Das, D., Lee, J. K., & Kalia, V. C. (2021). Continuous biohydrogen production from poplar biomass hydrolysate by a defined bacterial mixture immobilized on lignocellulosic materials under non-sterile conditions. *J. Clean. Prod.*, 287, 125037.

Patil, A., Chirmade, U. N., Trivedi, V., Lamprou, D. A., Urquart, A., & Douroumis, D. (2011). Encapsulation of water insoluble drugs in mesoporous silica nanoparticles using supercritical carbon dioxide. *J. Nanomed. Nanotechnol.*, 2(3). DOI: 10.4172/2157-7439.1000111.

Rippel, R. A., & Seifalian, A. M. (2011). Gold revolution—Gold nanoparticles for modern medicine and surgery. *J. Nanosci. Nanotechnol.*, 11(5), 3740–3748.

Robertson, G. (2008). State-of-the-art biobased food packaging materials. In *Environmentally Compatible Food Packaging* (pp. 3–28). Woodhead Publishing.

Safra, T., Muggia, F., Jeffers, S., Tsao-Wei, D. D., Groshen, S., Lyass, O., Henderson, R., Berry, G., & Gabizon, A. (2000). Pegylated liposomal doxorubicin (doxil): Reduced clinical cardiotoxicity in patients reaching or exceeding cumulative doses of 500 mg/m². *Ann. Oncol.*, 11(8), 1029–1034.

Sanguansri, P., & Augustin, M. A. (2006). Nanoscale materials development–A food industry perspective. *Trends Food Sci. Technol.*, 17(10), 547–556.

Sekhon, B. S. (2010). Food nanotechnology–An overview. *Nanotechnol. Sci. Appl.*, 3, 1.

Sen, S., Banerjee, A., & Acharjee, A. (2013). Nanotechnology: Shaping the world atom by atom. *Int. J. Modr Eng. Res. Technol*, 3(4), 2219–2225.

Shahsavand, A., & Ahmadpour, A. (2004, June). The role of nanotechnology in environmental culture development. In *Proceedings of the First International Seminar on the Methods for Environmental Culture Development*, Tehran.

Shih, M. F., Wu, C. H., & Cherng, J. Y. (2015). Bioeffects of transient and low-intensity ultrasound on nanoparticles for a safe and efficient DNA delivery. *J. Nanomed. Nanotechnol.*, 6(2), 1.

Sloan, A. E. (2002). The top 10 functional food trends: The next generation. *Food Technol.*, 56, 32–57.

Soppimath, K. S., Aminabhavi, T. M., Kulkarni, A. R., & Rudzinski, W. E. (2001). Biodegradable polymeric nanoparticles as drug delivery devices. *J. Control. Release*, 70(1–2), 1–20.

Sozer, N., & Kokini, J. L. (2009). Nanotechnology and its applications in the food sector. *Trends Biotechnol.*, 27(2), 82–89.

Stander, L., & Theodore, L. (2011). Environmental implications of nanotechnology—An update. *Int. J. Environ. Resour.*, 8(2), 470–479.

Subbiah, R., Veerapandian, M., & Yun, S. K. (2010). Nanoparticles: Functionalization and multifunctional applications in biomedical sciences. *Curr. Med. Chem.*, 17(36), 4559–4577.

Sun, Y. L., Yang, Y. W., Chen, D. X., Wang, G., Zhou, Y., Wang, C. Y., & Stoddart, J. F. (2013). Mechanized silica nanoparticles based on pillar arenes for on command cargo release. *Small*, 9(19), 3224–3229.

Wang, H., Liu, S., Jia, L., Chu, F., Zhou, Y., He, Z., Guo, M., Chen, C., & Xu, L. (2018). Nanostructured lipid carriers for microRNA delivery in tumor gene therapy. *Cancer Cell Int.*, 18(1), 1–6.

Wang, H., Schandl, H., Wang, X., Ma, F., Yue, Q., Wang, G., & Zheng, R. (2020). Measuring progress of China's circular economy. *Resour. Conserv. Recycl.*, 163, 105070.

Wickline, S. A., Neubauer, A. M., Winter, P., Caruthers, S., & Lanza, G. (2006). Applications of nanotechnology to atherosclerosis, thrombosis, and vascular biology. *Arterioscler. Thromb. Vasc. Biol.*, 26(3), 435–441.

Wilder, J. W., Venema, L. C., Rinzler, A. G., Smalley, R. E., & Dekker, C. (1998). Electronic structure of atomically resolved carbon nanotubes. *Nature*, 391(6662), 59–62.

Xiao, Y., Patolsky, F., Katz, E., Hainfeld, J. F., & Willner, I. (2003). "Plugging into enzymes": Nanowiring of redox enzymes by a gold nanoparticle. *Science*, 299(5614), 1877–1881.

Yaktine, A., & Pray, L. (Eds.). (2009). *Nanotechnology in Food Products: Workshop Summary*. National Academies Press.

Yamamoto, K., Sakata, Y., Nohara, Y., Takahashi, Y., & Tatsumi, T. (2003). Organic-inorganic hybrid zeolites containing organic frameworks. *Science*, 300(5618), 470–472.

Zad, T. J., Astuti, M. P., & Padhye, L. P. (2018). Fate of environmental pollutants. *Water Environ. Res.*, 90(10), 1104–1170.

Zhang, Y., Schlachetzki, F., Li, J. Y., Boado, R. J., & Pardridge, W. M. (2003). Organ-specific gene expression in the rhesus monkey eye following intravenous non-viral gene transfer. *Mol. Vis.*, 9, 465–472.

14 Enhancement of Biohydrogen Production by the Application of Microbial Nanotechnology

Tirtharaj Datta, Afan Ahmed, Saurabh Kumar Jha,
Rashi Srivastava, Pallavi Singh, and Bindu Naik

CONTENTS

DOI: 10.1201/9781003316374-14

14.1 INTRODUCTION

Lignocellulosic Biomass (LCB) is composed of energy-dense biopolymers with a layered structure, such as polysaccharides (cellulose (38–50%) and hemicelluloses (17–32%)) and an aromatic polymer, lignin (15–30%), as well as minor quantities of proteins, minerals, pectin, and other components. It is a great substrate for materials that is renewable, recyclable, biodegradable, and chemically accessible. It contains macrostructures (tree branches, grass stalks, and so on) that have been subjected to mechanical comminution operations in order to lower particle size before being converted. The defensive inner structure of lignocellulosic biomass has contributed to the hydrolytic stability and structural robustness of plant cell walls, as well as their resistance to microbial destruction (Singhvi et al., 2019). At the microscale level, the permeable structure of the biomass is the most prominent characteristic, which is extremely species dependant and aids in the transfer of water and nutrients (Zhang et al., 2019). The cell walls, which make up the biomass's mesostructure, are made up of three ultra-structural domains: the primary cell wall (PCW), middle lamella, and secondary cell wall (SCW) (Lee et al., 2014). The primary cell wall is formed during cell growth, and when cell growth is halted, a thick cell wall, known as the secondary cell wall, forms in layers (Singh et al., 2021). Biodiesel is a long-term, ecologically benign alternative to petroleum-based diesel (Singh et al., 2021). Biofuel is an ester comprised of a carbon chain (C14–C24) synthesized from a range of lipids, such as animal fats, vegetable oils, and waste oil (Khoobbakht et al., 2016). Biodiesels have been touted as a more environmentally friendly alternative to diesel, with lower emissions and health risks. According to data published in 2014 by the Renewable Energy Policy Network for the 21st Century, bioethanol and biodiesel are the most commonly utilized biofuels and are responsible for transportation globally. Because of its recyclability, renewability, and involvement in reducing pollutant emissions in recent years, biodiesel has been regarded as one of the most promising renewable fuels (Gaurav et al., 2019). As it contains less sulphur and aromatic chemicals than petroleum-derived fuels, biodiesel is considered environmentally friendly (Christophe et al., 2012). Biofuel is aerobic, non-toxic, low-polluting, and biodegradable, with no engine changes required (Khan et al., 2019). Several significant breakthroughs in nano-catalyst technology have recently been made, exhibiting a symmetrical relationship with cost-competitive biodiesel production. The emergence of nanotechnology in recent years, with revolutionary implications in a variety of sectors, has been promising for humankind. The creation, production, and use of materials with atomic or molecular accuracy at dimensions of 1–100 nm are common definitions (Ingle et al., 2020). Nanomaterials are the foundations of nanotechnology and contain a variety of unique qualities. The surface area of metal nanoparticles, for example, is hundreds of times that of macroscale materials of the same mass. Nanocatalysts are known for their high activation energy, regulated reaction rate, and selective reactivity, as well as their ease of recuperation (Thangaraj et al., 2019). Nanocatalysts that are highly porous, magnetic, carbon-based, and metal oxide-based are all accessible. Carbon-based nanocatalysts include fullerene, graphite, inorganic nanotubes, and carbon black, whereas metal-based nanocatalysts include metal oxides. Nanocatalysts include iron oxide, cerium oxide, iron, silver, titanium oxide, silver, aluminium, cobalt, calcium oxide, and zinc oxide (Naylor & Higgins, 2017), as well as hydrotalcites, zeolites, sulphated oxides, zirconia, and other materials (Himmel, 2009). Nanocatalysts can achieve excellent yields with relatively mild reaction conditions and shorter reaction times. These catalysts' reusability is also exceptional, since they continue to operate well after 11 cycles.

14.1.1 Components of Lignocellulosic Biomass

14.1.1.1 Cellulose

Cellulose is composed of glucose units linked together by 1,4-glycosidic connections, with amorphous areas including large holes and other irregularities, and dense and crystalline sections (Rahardjo et al., 2021). Because of its structure, denaturing cellulose into simple sugars takes time. A non-crystalline matrix, comprised of lignins, hemicelluloses, proteins, and pectins, surrounds and links cellulose microfibrils (cellulose organized in orthogonal layers) and functions as glue between the microfibrils (Karimi, 2015).

14.1.1.2 Hemicellulose

Hemicellulose is a branching uncrystallized polymer with a complicated structure. The monomeric units of hemicellulose include monosaccharides such as galactose, arabinose, xylose (highly prevalent), glucose, and mannose. Hemicellulose is a matrix substance in lignocellulosic biomass that is non-covalently connected to cellulose fibres. Moreover, the absence of crystal lattice makes its denaturation easier than cellulose (Rahardjo et al., 2021). Hemicellulose will be degraded into its monosaccharides during the pretreatment process. Hemicellulose monosaccharides produce furfural, 5-hydromethyl-2-furaldehyde (HMF), levulinic acid, and formic acid (Rasmussen). In comparison to cellulose, hemicellulose has a porous structure that is easily hydrolyzed and has a low degree of polymerization (100–200) (Nanda et al., 2014).

14.1.1.3 Lignin

Lignin is a three-dimensional macromolecule composed up of phenyl propane units (p-hydroxyphenyl (H), guaiacyl (G), and syringyl (S)). The cell walls of ferns, vascular plants, and club mosses all contain lignin. Lignin, along with hemicellulose, is present in the primary wall, middle lamella, secondary wall, and the cellulose microfibrils' voids. The lignin polymer, also known as "liquid crystal", is composed of hemicellulose over cellulose with diverge networking cluster (Rasmussen et al., 2014). The lignin monomer breaks down the connection between lignin and lignocellulose, which impedes biological processes, and makes up the alcohol with the assistance of the pretreatment process (Antunes et al., 2019).

14.1.1.4 Biomass Recalcitrance

Recalcitrance refers to a cellulose biomass's inherent resistance to infections, enzymes, and chemicals (Lynd et al., 1999), and it is assumed to be mostly impacted by lignin, along with its amount, location, and type (syringyl/guacyl ratio) (Studer et al., 2011; Ding et al., 2012). Plant recalcitrance is hypothesized to be influenced by the quantity of cellulose, hemicellulose, pectin, and other organic matter. Biomasses are inherently resistant due to the crystallinity of cellulose and the presence of inert lignin. Pre-processing is required in order to make cellulose available for lysis by enzymes, which aids in the collection of sugars like maltose and sucrose, needed as precursors in biofuel production. Methods for overcoming the recalcitrance of cellulosic biomass include acid hydrolysis, gasification, and pretreatment (Vincent, 1999).

14.2 PRETREATMENT OF LIGNOCELLULOSIC BIOMASS

Pretreatment is a process in the preparation of lignocellulosic biomass to improve its suitability for enzymatic conversion consequently resulting in efficient production, which would otherwise be constrained by high production costs and low quantity of production (Sannigrahi et al., 2010). Pretreatment lowers cellulose crystallinity by eliminating lignin and hemicelluloses, increasing the permeability of lignocellulosic materials, and removing lignin and hemicelluloses. During the anaerobic digestion (AD) process, several treatments are used to minimize biomass crystallinity and increase solubility.

14.2.1 Methods

Ionic liquid pretreatment disrupts hydrogen bonds between cellulose microfibers (Wyman et al., 2013), resulting in cellulose breakdown, decreased crystallinity, and enhanced porosity, all of which improve biomass digestibility (Feng & Chen, 2008). Ionic liquids (ILs) containing pyridinium, imidazolium, phosphonium-based cations, and ammonium, as well as alkyl or allyl side chains connected to anions such acetate, phosphonate, and chloride (Zhu, 2008), have been utilized in lignocellulosic biomass pretreatment (Wang et al., 2017). Mechanical pretreatment can increase the available reactive area of organic matter by changing the composition of particles. This alteration in composition and hence size can increase organic matter availability and responsiveness to microbe and enzyme action, as well as boosting biomass digestion and altering cellulose crystallinity during AD. The mechanical pretreatment of LCB includes methods such as milding (Neshat et al., 2017) and irradiation (Abraham et al., 2020). The humidity of biomass and particle size reduction is used in milling processes. When rice straw is pulverized, for example, it produces 3% more methane (Zhang & Zhang, 1999). Microwave radiation is directly supplied to the biomass in the irradiation method, allowing for rapid heating with a low-temperature gradient, resulting in polar orientation variations in polar molecules, enhancing the solubility of lignocellulosic biomass (Pellera & Gidarakos, 2017). Whenever sodium hydroxide and microwave treatment are coupled, for example, paddy straw (Kaur & Phutela, 2016) produces 55% more biogas.

Chemical pretreatments are classified as acid, alkaline, or oxidative depending on the type of chemical used. The hemicellulose, lignin, and cellulose will be solubilized/hydrolyzed depending on the chemical pretreatment we utilize. Acid will shatter van der Waals bonds, covalent bonds, and hydrogen bonds, that are joined together in biomass, such as wheat straw (16% rise in methane following sulfuric acid treatment) (Taherdanak et al., 2016) and water hyacinth (161% biogas is generated) (Sarto et al., 2019). Alkaline pretreatment induces increased surface area, and porosity, with a reduction in the monomeric constituents present in lignocellulosic biomass, allowing for delignification (Chandra et al., 2012).

Rice straw, for example, sees a 50% increase in biogas yield when treated with NaOH and KOH. The oxidative pretreatment causes electrophilic substitutions to occur, aromatic nuclei to break, side chains to dislocate, and alkyl aryl ether linkages to cleave, disrupting the lignin and hemicellulose relationship (Paudel et al., 2017).

The direct impact of microbial activity on lignocellulosic substrates is referred to as biological pretreatment (Zheng et al., 2014). Biological pretreatment procedures are more consistent with Alzheimer's disease since no hazardous by-products are produced throughout the process that might worsen the disease.

Pretreatment includes microbial and enzyme treatments. The microbial preparation of lignocellulosic biomass uses a fungal approach (based on lignin/cellulose content degradation). The two types of fungus are brown rot fungi (Wagner et al., 2018) and soft rot fungi. For microbial growth, moisture content, pH, oxygen concentration, and nutrient concentration are all crucial. For example, BRFM985 *Polyporus brumalis* (52% of biogas produced during AD). The success of enzyme pretreatment is determined by the enzyme's hobby, substrate specificity, the quantity of enzyme employed for treatment, the enzyme's tolerance to diverse inhibitors, enzymatic interest, the anaerobic fermentation device, enzymatic stability at various temperatures, and pH. Enzymes are categorized as cellulose/hemicellulose-degrading or lignin-degrading primarily on the basis of hobby toward lignocellulosic biomass (Koupaie et al., 2019), for example corn stover (36.9% increase in biogas technology through hydrolysis of the substrate using cellulase enzyme) and manure fibre (34% increase in methane production by means of combining laccase enzyme with other pretreatment approaches).

14.2.2 Role of Nanotechnology in the Pretreatment of Lignocellulosic Biomass

A novel methodology is established to solve this issue, taking into account the numerous difficulties connected with traditional pretreatment approaches. Recent breakthroughs in nanotechnology are expected to result in a plethora of different nanomaterials that can be utilized to immobilize enzymes. Immobilizing enzymes on nanoscale substances improves their catalytic efficacy (cellulase, hemicellulase, and laccase) (Gaikwad et al., 2019). The phrase "nanobiocatalyst" refers to enzymes immobilized in nanoparticles. Nanoparticles including silica nanoparticles, carbon nanotubes, nickel nanoparticles, oxide nanoparticles, and others are combined with enzymes to generate a nanobiocatalyst (Gaikwad et al., 2019). Nanomaterials with a wide surface area for enzyme attachment can increase the effectiveness of immobilized enzymes by allowing for high concentrations of enzymes relative to particle concentration (Rai et al., 2019), as well as helping in the hydrolysis of various lignocellulosic biomass components.

14.2.3 NICKEL NANOPARTICLES

A metal like nickel can be used in the hydrogenation process, which involves sorbitol formation from glucose (Grewal et al., 2017). Nickel, in the form of a nanomaterial catalyst, can help break down lignocellulosic material into hemicellulose (Chandel et al., 2011). Nickel has the properties of efficient catalysis, surface reactiveness, and adsorption capability (Kobayashi et al., 2014), and thus catalyses biofuel production from lignocellulosic material. The stability of the cellulase enzyme at various temperatures is reduced with nickel-cobaltatite nanoparticles; at 1 mM concentration, nanoparticles illustrated enzyme balance at 80°C for eight hours, while enzyme stability was decreased with the absence of nanoparticles, indicating that nanoparticles can be a useful resource in the manufacturing of cellulase enzyme (Srivastava et al., 2017).

14.2.4 SILVER NANOPARTICLES

Cellulase and other lignocellulolytic enzymes may be effectively immobilized using silica nanoparticles. A review study (Lupoi et al., 2011) stated silica nanoparticles with immobilized cellulose can carry out catalysis in concurrent SSF procedures for additional biofuel production technology from lignocellulosic substances. Silica nanoparticles have been found to have a higher affinity for the adsorption of cellulase. Another study employed physical adsorption and chemical binding to immobilize two mesoporous silica nanoparticles with changing particle size, pore size, and surface area. Cellulase immobilized on mesoporous silica nanoparticles with large pore diameters by chemical binding was shown to provide high yield and stable glucose from cellulose (Chang et al., 2011).

14.2.5 CARBON NANOTUBES

The electrical, thermal, and mechanical properties of carbon nanoparticles are good. Single walled carbon nanotubes (SWCNTs) and multi-walled carbon nanotubes (MWCNTs) are two types of carbon nanotubes. Because of its structural arrangement, MWCNT is favoured for enzyme immobilization due to its increased electrical characteristics, a high edge to basal plane ratio, and quick electrode kinetics (Ahmad et al., 2018). Ninety-eight percent of cellulase was produced by covalently connecting MWCNTs with carboxylic acid as functional groups, acting upon sludge obtained from a wastewater treatment plant using microbe *Trichoderma reesei* RUT C-30.

14.3 ROLE OF NANOPARTICLES IN THE PRODUCTION OF BIODIESEL

Nanotechnology is the science which studies particles up to the size of 100 nm. Richard Feynman used the word "nanotechnology" in his address "Plenty of Room at the Bottom" (1959). Nano is a unit of measurement, which basically stands for one-billionth of a meter (1.0×10^{-9}), which the human eye is unable to see. The fast decline of fossil fuel supplies makes coordinating future energy needs extremely difficult. Limited carbon biofuel has been identified as a potential alternative for reducing reliance on fossil fuels. Because of their high robustness, lipid content, and absence of need to compete with food crops, biofuels derived from algae are the ideal replacement. However, the biodiesel produced as a by-product of biofuel is confronted with issues of energy-intensiveness and increased manufacturing costs. A novel and fast emerging branch of nanotechnology has provided an alternative for developing durable nano-biocatalytic systems with long-term stability and cheap capture costs. According to preliminary research, adding nanomaterials to an algal growth system increased microalgal extension and accelerated lipid aggregation. Furthermore, the lipid regulation may be heightened as a result of the petition of nanoparticles (Mandotra et al., 2018).

Further, nanoparticles have the potential to improve lipid parentage management and possibly cause microalgal loss without causing harm. CaO and MgO nanoparticles, in particular, have been found to be useful as enzyme carriers or as a heterogeneous stimulant in the conversion of biodiesel from oil (Zhang et al., 2013).

The common usage of red meat and vegetable oil is food industry and biofuel production, consequently results in feedstock loss and hence competition between the industries. As a result, substitute substrate is required in biofuel production. Heterotrophic microalgae are well-established and able to store large amounts of lipid (Bano et al., 2020). Biodiesel, as a prospective fuel, is witnessing high usage and research at a rapid rate, owing to the administration and diminishing benefits of fossil fuels and other non-renewable resources. The study is primarily focused on the transformation of biomass into liquid fuels (Nizami et al., 2018). Due to the squat and solidity of biomass, it is extremely difficult to manipulate, freeze, and store, posing a major barrier in the production of sweeping biodiesel (Balat, 2011). Currently, vast swaths of research into expanding biodiesel production have risen to prominence, owing to massive increases in fuel costs, the requirement of massive amounts of energy, and issues relating to global warming due to the release of greenhouse gases from currently used sources of energy (Borah et al., 2019).

Biofuel is often manufactured by mixing homogeneous and heterogeneous stimulants (Ibrahim et al., 2020). However, in the production of biodiesel, the title role and aggregating the nanocatalysts are important. This led to the conclusion that riposte expansion is crucially associated with nanocatalysts which also remain unchanged during the reactions. It also remarkably reduces the response profiles' invigoration energy. Furthermore, a censorious framework for adjudicating the capitulation of biodiesel production is the aggregation of nanocatalyst is seen to function (Anbessie et al., 2019). Another researcher has reviewed the unique properties of nanocatalysts used in biofuel production. With no stimulant crowding in the process of primed reactive condition variables, no biofuel was produced, implying the importance of the appearance and aggregation of the nanocatalyst for biodiesel yield (Show et al., 2019). Biofuel production can make an important contribution in reducing the use of energy derived from fossil fuels. In contrast to non-renewable energy resources, the paperback scheming of nanocatalysts, the fabrication of microbial sensing, and the growing escalation methodology are breviloquently obliging in growing biofuel production on a wide scale.

With the advent of nanotechnology, nanocatalysts have become more important in enhancing the capacity and yield quality of biofuel while also reducing reaction cadence.

14.4 METHODOLOGIES FOR THE PRODUCTION OF BIOHYDROGEN USING NANOPARTICLES

Because of its strong energy explosion and attractive environmental properties, biohydrogen is recognized as a fuel with great future possibilities. The generation of biohydrogen does not require fossil fuels, unlike current commercial hydrogen production systems. It's a renewable biofuel that's also a great way to cut down on fossil-fuel-related pollution. While numerous biohydrogen conversion technologies have been developed, microbial conversion is still in its infancy. This chapter focuses on biohydrogen metabolisms, variables affecting production processes, and bioprocess and bioreactor design considerations in order to increase biohydrogen production in bioreactor and bioprocess design (Lee et al., 2000).

14.4.1 Carbon Nanotubes

Graphite, graphene, CNTs, and fullerene are just a few of the amazing structures that may be created through sp2 hybridization. CNTs are nanometre-diameter cylindrical structures which are hollow inside and rolled with graphite plates. The length/diameter ratios are huge with an absence of pores and opening, as defined by Iijima in. Carbon atoms in graphite sheets have a honeycomb structure that is similar to the carbon network in shells. The two varieties of carbon nanotubes are single-walled and multi-walled carbon nanotubes, SWNTs and MWNTs, respectively. A continuous graphene sheet is wrapped into a cylinder with a diameter of a few nanometres and a length of several microns to make SWNT (Endo et al., 2005).

14.4.2 Application of CNT-based Microbial Immobilization to Optimize Biohydrogen Generation

Low-cost catalysts that exhibit negligible deactivation throughout the process have also been used to decompose methane into hydrogen and carbon molecules. Catalysts made of carbon are low in cost hence avoiding the requirement of regeneration. Further, there is the probability of the contamination of released hydrogen gas due to the presence of CO or CO_2 emissions. This problem is solved by removing the oxygenate groups prior to achieving the methane decomposition temperature. Carbon black, coal char, or activated carbons are some of the thoroughly researched commercially available carbons utilized as catalysts. The high carbon emission deactivates these catalysts, posing a major issue in their usage. For short reaction times, activated carbon is the most active catalyst, although it deactivates fast. Carbon blacks, on the other hand, have a high rate of methane breakdown irrespective of the reaction times. Moreover, carbon nanotubes show low activity despite the fact that they contain high metal concentrations. The role of a catalyst in catalysis is crucial. CNTs are presently being employed as a catalytic support due to their unique shape and characteristic features.

Modification of the structure of CNTs is required for their use as a catalyst. This is achieved through chemical reduction, impregnation, hydrothermal technique, and precipitation (Liu et al., 2011). Savva and co-workers published a key study on the use of CNTs for the single step breakdown of ethylene with Ni as catalyst. This study sheds light on the impacts of the catalyst synthesis process, Ni metal concentration on catalytic stability and activity in the context of ethylene breakdown and support chemical composition, followed by carbon removal by catalyst regeneration. The Ni/CNTs catalyst's catalytic performance was shown to be highly influenced by the Ni metal loading. For both the ethylene and methane breakdown processes, the Ni/CNTs catalyst combination yielded the greatest carbon capacity and hydrogen production. In comparison with a 0.3 wt.% Ni and SiO_2 catalyst tested under identical experimental circumstances, hydrogen production was 50 times higher, evident by 0.5 wt.% Ni/CNTs catalyst. Furthermore, in the 0.15 wt.% to 10 wt.% ranges investigated, 0.5 wt.% Ni was shown to be an approximately optimal metal loading. The geometric and electrical properties of the support may influence the catalysing power of supported Ni catalysts. The geometric factor refers to the development of nickel aggregates of a specific size on the support surface, whereas the electric factor refers to the change of Ni's electric characteristics as a result of electrical interactions with the support. A CNTs-supported Ni catalyst, described by Hou et al., with excellent catalytic activity was effectively used for low-temperature organic compound reformation in bio-oil using the homogeneous deposition–precipitation technique. With the best Ni loading (15 wt.% Ni–CNTs), the H_2 output was around 92.5% above the catalyst.

By reducing Ni–Cu alloys with formaldehyde, Liao and Yang created a novel catalyst combination mounted on CNTs, which they employed in methanol steam reforming. Sulfuric acid and nitric acid were used to produce surface flaws in CNTs and ethanol was used to increase lipophobicity. The mounting process of Ni and Cu alloys on CNTs included the simultaneous reduction of Cu- and Ni- precursors with tetra-n-methylammonium hydroxide to prevent Ni–Cu particles from aggregating. At the same time, the activated carbon and CNT mounted Ni and Cu catalysts (Ni and Cu/C)/(Ni and Cu/CNTs) were examined. At a temperature of 360°C and 20 wt.% concentration Ni20 and Cu80/CNTs, the hydrogen production in fossil fuel reforming of methanol approached 100%. Ni20 and Cu80/CNTs showed much greater catalytic activity than Ni20 and Cu80/C and Ni20 and Cu80/CNTs. Utilizing CNTs as supports resulted in better metallic dispersion and less metal particle agglomeration in comparison to activated carbon used for same purpose. Despite the enormous contact area of activated carbon (C), the Ni and Cu alloys moved along with accumulation to form large diameter particles, lowering the activity of Ni and Cu/C substantially.

Eswaramoorthi employed carbon nanotubes as a support for Cu and Zn catalysts in the incomplete oxidation of methanol aiming to produce hydrogen. CVD was used to make high-quality CNTs utilizing an anodic aluminium oxide template in addition to acetylene, used as the carbon source (Eswaramoorthi et al., 2008). A simultaneous precipitation technique was used to make the Cu and Zn/CNTs catalysts. Cu deposition on the surface of CNTs resulted in the formation of strong proton donor sites. Up to the threshold value, the increase in temperature of reaction and copper loading increases the methanol transformation velocity and hydrogen selectivity (12 wt.%). Catalyst Cu–Zn 12 wt.% and 9 wt.% respectively supported on CNTs were highly stable in time-on-stream experiments, with inhibition in the generation of CO. The outcomes showed that CuO is a key component responsible for hydrogen synthesis with decreased CO creation, but copper ions hinder methanol from yielding hydrogen. Yang and Liao used a wet impregnation and chemical reduction approach to create a novel Cu and ZnO–CNTs combination as a nanocatalyst for fossil fuel reforming methanol reactions. Roughness and reactive groups on the surface were introduced by pretreatment with sulphuric acid and nitric acid at a temperature of 60°C for a time period of 24 hours, and then lipophobicity was increased by ethanol. Cu nanoparticles with a particle size of around 10 nm were dispersed well on the surface of CNTs. The hydrogen production from hydrocarbon was efficient at a temperature of 280°C with a range of 83% and approached efficiency of 100% at high temperatures, specifically temperatures greater than 320°C with a concentration ratio of water to methanol being 1.5 molar using a 23 wt.% catalyst combination of ZnO_20 and Cu80 as CNTs (Zeng et al., 2006).

Wang et al. produced Pd and Pt catalysts for the incomplete removal of hydrogen from methylcyclohexane and cyclohexane to create uncontaminated residual hydrogen using CNTs, in the form of stacked-cone, as the fixation medium. The scientists found that a 0.25 wt.% Pt and SC as CNT catalyst had equal catalysing selectivity and reactivity for producing hydrogen gas and aromatic compounds as an industrial 1 wt.% Pt and Al_2O_3 catalyst. For the removal of hydrogen from cyclohexane and its methyl derivative, a concentration of 0.1 wt.% Pt and SC as CNT exhibits strong reaction activity, and this catalyst combination shows the greatest number of turnovers for the generation of hydrogen. For the dehydrogenation of both cyclohexane and methylcyclohexane, Pd supported on SC-CNT lacked in effectiveness in comparison to a combination of Pt/SC-CNT. The same researchers made a Pt catalyst by fixating as cone to form CNTs for the partial hydrogen removal of decalin as well as tetralin, thus yielding purified naphthalene and hydrogen in a more recent article. They showed that 1.0 wt.% concentration of Pt and SC as CNT shows greater efficiency than a concentration of 1.0 wt.% of catalyst Pt fixed on carbon black obtained from soot of acetylene or granular carbon for removal of hydrogen from tetralin at a temperature of 240°C. Also, it was more efficient than a concentration of 1.0 wt.% Rh, Pd, or Pt metal fixed on aluminium oxide for the dehydrogenation of tetralin at 240°C. Pt was better disseminated on SC as CNT than on the other fixating medium, according to HRTEM analysis of the utilized catalysts.

Currently, one of the most significant barriers to CNT adoption is its exorbitant prices, as well as the large scale of manufacturing obtained by CVD. CNTs as a catalyst support, on the other hand, appear to be successful and appealing in terms of their widespread use (Wang et al., 2006, Shen et al., 2008).

14.5 STRATEGIES FOR THE PRODUCTION OF BIOHYDROGEN FROM WASTEWATER

At high temperature and pressure, hydrogen may be produced using methods such as fossil-based hydrocarbon reformation, coal gasification, and partial oxidation. Biohydrogen production is divided into four categories: (i) photodecomposition of organic compounds (by photosynthetic bacteria), (ii) bio-photolysis of water (by algae/Cyanophyta), (iii) dark fermentative hydrogen (by facultative bacteria and anaerobes), and (iv) microbial fuel cells (MFC). Each biological production process comes with its own set of upsides and downsides. During algal (green) and cyano-bacterial photosynthesis the water molecule is broken down into gas (H_2) and liquid (H_2O); however, the slow growth of algal cells and the suppression of the hydrogenase enzyme in the presence of oxygen residues largely limit its use. In photosynthetic organisms and dark fermentation bacteria, the decomposition of the organic compounds for energy and the release of energy are similar. Organic acids are utilized as a substrate by photosynthetic bacteria, but they are hostile to ammonia and oxygen, making them unsuitable for commercial hydrogen production. Dark fermentation, on the other hand, breaks down a wide spectrum of organic wastes, including complex lignocellulose, food waste, and industrial effluent, into simpler monomers (sucrose, glucose).

The dark fermentation, on the other hand, has a lower COD removal effectiveness (33%) since it requires further treatment before being discharged into the system. Furthermore, the dark fermentation's biomass growth rate and hydrogen generation rate are both higher than those of conventional hydrogen production techniques, making it an appealing contender for industrial and commercial biohydrogen production.

14.5.1 Bio-Photolysis of Water Using Algae/Cyanophyta

Photosynthesis and hydrogen generation catalysed by hydrogenases are the two processes in bio-photolysis or photo-biological hydrogen production.

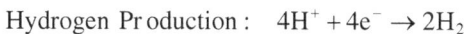

$$\text{Photosynthesis}: \quad 2H_2O \rightarrow 4H^+ + 4e^- + O_2$$

$$\text{Hydrogen Pr oduction}: \quad 4H^+ + 4e^- \rightarrow 2H_2$$

Biophotolysis and photofermentation are light-catalysed biological processes that use water to create hydrogen. In biological systems, they are water-splitting mechanisms that occur in the presence of light. Water, light, and photosensitive microorganisms are therefore the three main components of biophotolysis. The molecules H_2 and O_2 are created by light energy (Hay et al., 2013). In a closed system, photosensitive microorganisms such as microalgae are grown in a photobioreactor for the generation of hydrogen. Biophotolysis-based hydrogen generation has been divided into three categories: direct, indirect, and photofermentation (Ghiasian, 2019). In biophotolysis and photofermentation, the entire hydrogen generation process may be described as

$$6H_2O + 6CO_2 \xrightarrow{\ h(\upsilon)\ } C_6H_{12}O_6 + 6O_2$$

$$C_6H_{12}O_6 + 6H_2O \xrightarrow{\ h(\upsilon)\ } 6CO_2 + 12H_2$$

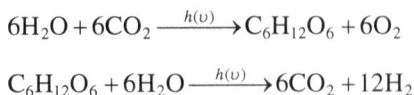

The microalgae *Chlamydomonas reinhardtii* has been the best investigated for direct biophotolysis. Photosystems (both PSI and PSII) and hydrogenase are required. PSII (680 nm) and/or PSI (700 nm) absorb light in the form of photons, producing a powerful oxidant capable of oxidizing water into protons, electrons/reducing equivalents, and O_2. Electrons decrease protons to generate H_2 (Brentner et al., 2010).

Direct biophotolysis is the process of using algae/cyanobacteria to produce H_2 through a series of processes. The coordinated action of the two photosystems utilized in plant-type photosynthesis mediates the overall response (PSI and PSII). Photon absorption divides water throughout the process, resulting in a drop in ferredoxin, which is used to enhance H_2 production via proton reduction. The light-driven breakdown of water into H_2 and O_2 involves the nitrogenase and hydrogenase enzymatic activities (Pinto et al., 2002). *Chlamydomonas reinhardtii* and *Anabaena cylindrica* (Prabakar et al., 2018) are green algae or cyanobacteria that mediate direct biophotolysis without or with respiratory O_2 absorption. In both procedures, water is split into O_2 and H_2. Direct biophotolysis with respiratory O_2 uptake is a two-stage process involving PSII and PSI. Under aerobic conditions, electron equivalent from water splitting are used in the first step of direct biophotolysis to fix CO_2. In the second stage, water is split into H_2 and O_2, with O_2 being separated and routed to the first phase. Several species and strains from at least 14 genera have been tested for H_2 production under a variety of growth conditions (Ramprakash et al., 2014). The reported H_2 generation rates in these investigations are varied depending on the species and environment. The capacity to create hydrogen from water in mild settings, such as at moderate temperatures and pressures, is a benefit of biophotolysis.

14.5.2 DISADVANTAGES

Low photochemical efficiency and significant by-product (oxygen, O_2) inhibition of hydrogenase (H_2ase) are severe issues (Ramprakash et al., 2015). H_2 is generated by nitrogenase in the absence of ammonium ions (NH_4) during photofermentation by anoxygenic photosynthetic bacteria. Because no O_2 is released in this scenario, O_2 inhibition is not a concern (Sivagurunathan et al., 2020). However, nitrogenase's poor H_2 generation activity, NH_4-induced inhibition of nitrogenase expression, and low photochemical efficiency are all significant disadvantages.

14.6 APPLICATION OF MICROBIAL NANOTECHNOLOGY FOR THE BIOREMEDIATION OF WASTEWATER AND PRODUCTION OF BIOHYDROGEN

The nanotechnological approach is viewed as a single, comprehensive package that incorporates various procedural aspects of traditional approaches while lowering the costs of wastewater treatment plants. Nanotechnology is now being used in a variety of fields, including food toxin detection, nanosensors, nanonutraceuticals, targeted medication delivery, imaging, theranostics, and photodynamic microorganism inactivation (Sivagurunathan et al., 2014, Vatsala et al., 2008). The usage of nanotechnology is more sustainable and eco-friendly when nanomaterials are biofabricated and microorganisms are used simultaneously. Chemically manufactured nanoparticles may have drawbacks in terms of chemical use and aqueous solution self-agglomeration. As a result, green nanoparticle manufacturing using plant extracts, fungal enzymes, and bacterial enzymes might be a viable option. They form metallic nanoparticles by acting as reductive agents for the metal complex salt. They develop higher firmness in an aqueous system due to co-precipitation or the inclusion of proteinaceous and bioactive substances in the nanoparticles' outer face. *Aspergillus tubingensis* (STSP 25) prefabricated iron oxide nanoparticles were obtained from the rhizosphere of *Avicennia officinalis* in the Sundarbans, India (Veeravalli et al., 2019). Synthesized nanoparticles were able to eliminate more than 90% of toxic metals [Pb (II), Ni (II), Cu (II), and Zn (II)] from wastewater after five rounds of regeneration. Metal ions were chemically adsorbed on the nanoparticles' surfaces during endothermic processes. Another experiment used exopolysaccharides (EPS) from *Chlorella vulgaris* to co-precipitate iron oxide nanoparticles. Fourier-transform infrared (FTIR) spectroscopy was used to show that EPS functional groups may effectively modify nanoparticles. The nanocomposite also removed 91% of $PO4\ 3-$ and 85% of $NH4+$ (Yang et al., 2007).

The use of microbes to synthesize nanoparticles has been shown to be a cost-effective and environmentally beneficial technique. *Escherichia* sp. SINT7, a copper-resistant bacterium, was used to make copper nanoparticles. It was discovered that biogenic nanoparticles degraded azo dye and textile wastewater. Reactive black-5, congo red, direct blue-1, and malachite green were decreased by 83.61, 97.07, 88.42, and 90.55%, respectively, at a lower concentration of 25 mg/L, whereas this was reduced to 76.84, 83.90, 62.32, and 31.08%, respectively, at a higher concentration of 100 mg/L. Industrial wastewater was also treated, and the suspended particles, chloride, and phosphate ions in treated samples were reduced. The performance of biogenic nanoparticles like these boosts industry's cost-effective and long-term manufacturing (Zhang et al., 2020). The extracellular transfer of electrons allowed these nanoparticles to degrade Napthol Green B dye. The utilization of *Pseudoalteromonas* sp. CF10-13 in nanoparticle manufacturing gives an environmentally acceptable biodegradation approach. The generation of hazardous gases and metal complexes was suppressed by the endogenous creation of nanoparticles. The utilization of biogenic particles in the clean-up of industrial effluents is a better method. Apart from directly producing nanoparticles from microbes, there are various additional ways in which microorganisms might aid in the advancement of nanotechnology. Microorganisms, for example, might supply catalytic enzymes that, in combination with nanoparticles, aid in wastewater clean-up.

Using technology to convert waste materials into usable goods is drawing the interest of researchers all around the world. We can decrease trash while also producing valuable things with this method. Adsorbents, clinker, biogas, biohydrogen, biomolecules, and a range of other products are all made using this approach (Laurinavichene et al., 2018). Nanotechnology has contributed in the enhancement of waste-to-resource conversion production rates. Kumar and colleagues released a study in 2019 on employing nanoparticles to promote biohydrogen production and improve dark fermentation processes (Mahanty et al., 2020). Nanoparticles added to fermentative bacteria have brought unprecedented possibilities for producing biohydrogen from wastewater. Elreedy et al. (Mishra et al., 2014) generated biohydrogen using mixed culture bacteria and single, dual, and multiple nanoparticles. They observed that biohydrogen production was at its highest when a large number of nanoparticles were utilized (14% higher than when single nanoparticles were used). It was discovered that nanoparticles boosted hydrogenase and dehydrogenase activity, resulting in increased biohydrogen production. Similarly, combining nickel oxide and hematite nanoparticles increased biohydrogen production by 1.2–4.5 times as compared to employing just nanoparticles. The combination of nanoparticles produced the maximum hydrogen output of 8.83 mmol/g COD. The enhanced activity of hydrogenase and ferredoxin oxidoreductase enzymes is responsible for this rise (Noman et al., 2020, Oh et al., 2013). As a result, nanotechnology may be employed to provide green energy for long-term industrial growth and environmentally responsible manufacturing.

14.7 CONCLUSION

The application of nanotechnology for the smart and successful generation of biohydrogen and related biofuels is clearly demonstrated in the study. The demand for renewable energy sources has risen in recent decades, necessitating the construction of newer and better facilities to meet it. We may also conclude that the use of nanoparticle delivery systems is a milestone for the production of energy resources. Nanotechnology has attracted researchers' interest due to its numerous advantages, including a high available surface area, the capacity to serve several purposes, stability in difficult environments, quick and efficient material manipulations, greater interactivity, and so on. The combination of microbes and enzymes with nanotechnology has resulted in a more environmentally friendly approach to industrial wastewater control and biofuel production. The employment of microorganisms can reduce the danger associated with chemically generated nanoparticles. The residues that remain are either biocompatible or readily separated using basic filtration/precipitation methods. The commercialization of these nanotechnological capabilities is the greater hurdle. As of yet, just 1% of these nanotechnological features have been commercialized. As a result, companies will use these simple and effective microorganism-assisted nanotechnology procedures on a broad scale as a stepping stone. This needs ongoing support and confirmation from academics, as well as government financing, in order to cultivate nanotechnology's potential for sustainable and cost-effective manufacturing in industry. Long-term basic and applied research is required in this field, but if successful, it will result in a long-term solution for sustainable hydrogen generation. Understanding the natural mechanisms and genetic rules that govern H_2 generation is critical. In bigger bioreactors, metabolic and genetic engineering might be employed to show the process. Another alternative is to use artificial photosynthesis to duplicate the two phases.

REFERENCES

Abraham, A., Mathew, A. K., Park, H., Choi, O., Sindhu, R., Parameswaran, B., … & Sang, B.-I. (2020). Pretreatment strategies for enhanced biogas production from lignocellulosic biomass. *Bioresource Technology*, 122725. https://doi.org/10.1016/j.biortech.2019.12272.

Ahmad, R., & Khare, S. K. (2018). Immobilization of *Aspergillus niger* cellulase on multiwall carbon nanotubes for cellulose hydrolysis. *Bioresource Technology*, 252, 72–75.

Anbessie, T., Mamo, T. T., & Mekonnen, Y. S. (2019). Optimized biodiesel production from waste cooking oil (WCO) using calcium oxide (CaO) nano-catalyst. *Scientific Reports*, 9(1), 18982. https://doi.org/10.1038/s41598-019-55403-4.

Antunes, F. A. F., Chandel, A. K., Terán-Hilares, R., Ingle, A. P., Rai, M., dos Santos Milessi, T. S., … & Dos Santos, J. C. (2019). Overcoming challenges in lignocellulosic biomass pretreatment for second-generation (2G) sugar production: Emerging role of nano, bio-technological and promising approaches. *3 Biotech*, 9(6), 1–17.

Balat, M. (2011). Potential alternatives to edible oils for biodiesel production–A review of current work. *Energy Conversion and Management*, 52(2), 1479–1492. https://doi.org/10.1016/j.enconman.2010.10.011.

Bano, S., Ganie, A. S., Sultana, S., Sabir, S., & Khan, M. Z. (2020). Fabrication and optimization of nanocatalyst for biodiesel production: An overview. *Frontiers in Energy Research*, 350.

Borah, J. M., Das, A., Das, V., Bhuyan, N., & Deka, D. (2019). Transesterification of waste cooking oil for biodiesel production catalyzed by Zn substituted waste egg shell derived CaO nanocatalyst. *Fuel*, 242, 345–354. https://doi.org/10.1016/j.fuel.2019.01.060.

Brentner, L. B., Peccia, J., & Zimmerman, J. B. (2010). Challenges in developing biohydrogen as a sustainable energy source: Implications for a research agenda. *Environmental Science and Technology*, 44(7), 2243–2254.

Chandel, A., Da Silva, S. S., & Singh, O. V. (2011). Detoxification of lignocellulosic hydrolysates for improved bioethanol production. In *Biofuel Production-Recent Developments and Prospects* (pp. 225–246). Intech.

Chandra, R., Takeuchi, H., Hasegawa, T., & Kumar, R. (2012). Improving biodegradability and biogas production of wheat straw substrates using sodium hydroxide and hydrothermal pretreatments. *Energy*, 43(1), 273–282.

Chang, R. H. Y., Jang, J., & Wu, K. C. W. (2011). Cellulase immobilized mesoporous silica nanocatalysts for efficient cellulose-to-glucose conversion. *Green Chemistry*, 13(10), 2844–2850.

Christophe, G., Kumar, V., Nouaille, R., Gaudet, G., Fontanille, P., Pandey, A.; Soccol, C. R.; Larroche, C. (2012). Recent developments in microbial oils production: A possible alternative to vegetable oils for biodiesel without competition with human food? *Brazilian Archives of Biology and Technology*, 55(1), 29–46.

Ding, S. Y., Liu, Y. S., Zeng, Y., Himmel, M. E., Baker, J. O., & Bayer, E. A. (2012). How does plant cell wall nanoscale architecture correlate with enzymatic digestibility? *Science*, 338(6110), 1055–1060.

Endo, M., Muramatsu, H., Hayashi, T., Kim, Y. A., Terrones, M., & Dresselhaus, M. S. (2005). 'Buckypaper' from coaxial nanotubes. *Nature*, 433(7025), 476–476.

Eswaramoorthi, I., Sundaramurthy, V., Das, N., Dalai, A. K., & Adjaye, J. (2008). Application of multi-walled carbon nanotubes as efficient support to NiMo hydrotreating catalyst. *Applied Catalysis A: General*, 339(2), 187–195.

Feng, L., & Chen, Z. L. (2008). Research progress on dissolution and functional modification of cellulose in ionic liquids. *Journal of Molecular Liquids*, 142(1–3), 1–5.

Gaikwad, S., Ingle, A. P., da Silva, S. S., & Rai, M. (2019). Immobilized nanoparticles-mediated enzymatic hydrolysis of cellulose for clean sugar production: A novel approach. *Current Nanoscience*, 15(3), 296–303.

Gaurav, A., Dumas, S., Mai, C. T. Q., & Ng, T. T. (2019). A kinetic model for a single step biodiesel production from a high free fatty acid (FFA) biodiesel feedstock over a solid heteropolyacid catalyst. *Green Energy and Environment*, 4(3), 328–341.

Ghiasian, M. (2019). Biophotolysis-based hydrogen production by cyanobacteria. In *Prospects of Renewable Bioprocessing in Future Energy Systems* (pp. 161–184). Springer.

Grewal, J., Ahmad, R., & Khare, S. K. (2017). Development of cellulase-nanoconjugates with enhanced ionic liquid and thermal stability for in situ lignocellulose saccharification. *Bioresource Technology*, 242, 236–243.

Hay, J. X. W., Wu, T. Y., Juan, J. C., & Md. Jahim, J. (2013). Biohydrogen production through photo fermentation or dark fermentation using waste as a substrate: Overview, economics, and future prospects of hydrogen usage. *Biofuels, Bioproducts and Biorefining*, 7(3), 334–352.

Himmel, M. E. (2009). *Biomass Recalcitrance: Deconstructing the Plant Cell Wall for Bioenergy*. Wiley-Blackwell.

Ibrahim, M. L., Adlina Nik Abdul Khalil, N. N., Islam, A., Rashid, U., Ibrahim, S. F., Mashuri, S. I. S., & Taufiq-Yap, Y. H. (2020). Preparation of Na2O supported CNTs nanocatalyst for efficient biodiesel production from waste-oil. *Energy Conversion and Management*, 205, 112445. https://doi.org/10.1016/j.enconman.2019.112445.

Ingle, A. P., Chandel, A. K., Philippini, R., Martiniano, S. E., & da Silva, S. S. (2020). Advances in nanocatalysts mediated biodiesel production: A critical appraisal. *Symmetry*, 12(2), 256.

Karimi, K. (2015). *Biofuel and Biorefinery technologies, Vol. 1. Lignocellulose Based Bioproducts*. Berlin: Springer

Kaur, K., & Phutela, U. G. (2016). Enhancement of paddy straw digestibility and biogas production by sodium hydroxide-microwave pretreatment. *Renewable Energy*, 92, 178–184.

Khan, I., Saeed, K., & Khan, I. (2019). Nanoparticles: Properties, applications and toxicities. *Arabian Journal of Chemistry*, 12(7), 908–931.

Khoobbakht, G., Akram, A., Karimi, M., & Najafi, G. (2016). Exergy and energy analysis of combustion of blended levels of biodiesel, ethanol and diesel fuel in a DI diesel engine. *Applied Thermal Engineering*, 99, 720–729.

Kobayashi, H., Hosaka, Y., Hara, K., Feng, B., Hirosakia, Y., & Fukuoka, A. (2014). Control of selectivity, activity and durability of simple supported nickel catalysts for hydrolytic hydrogenation of cellulose. *Green Chemistry*, 16(2), 637. https://doi.org/10.1039/c3gc41357h.

Koupaie, E. H., Dahadha, S., BazyarLakeh, A. A., Azizi, A., & Elbeshbishy, E. (2019). Enzymatic pretreatment of lignocellulosic biomass for enhanced biomethane production – A review. *Journal of Environment Management*, 233, 774–784.

Laurinavichene, T., Tekucheva, D., Laurinavichius, K., & Tsygankov, A. (2018). Utilization of distillery wastewater for hydrogen production in one-stage and two-stage processes involving photofermentation. *Enzyme and Microbial Technology*, 110, 1–7.

Lee, H. V., Hamid, S. B. A., & Zain, S. K. (2014). Conversion of lignocellulosic biomass to nanocellulose: Structure and chemical process. *The Scientific World Journal*.

Lee, S. M., & Lee, Y. H. (2000). Hydrogen storage in single-walled carbon nanotubes. *Applied Physics Letters*, 76(20), 2877–2879.

Liu, R., Lin, Y., Chou, L. Y., Sheehan, S. W., He, W., Zhang, F., … & Wang, D. (2011). Water splitting by tungsten oxide prepared by atomic layer deposition and decorated with an oxygen-evolving catalyst. *Angewandtechemie International Edition*, 50(2), 499–502.

Lupoi, J. S., & Smith, E. A. (2011). Evaluation of nanoparticle-immobilized cellulase for improved yield in simultaneous saccharification and fermentation reactions. *Biotechnology and Bioengineering*, 108(12), 2835–2843.

Lynd, L. R., Wyman, C. E., & Gerngross, T. U. (1999). Biocommodity engineering. *Biotechnology Progress*, 15(5), 777–793.

Mahanty, S., Chatterjee, S., Ghosh, S., Tudu, P., Gaine, T., Bakshi, M., … & Chaudhuri, P. (2020). Synergistic approach towards the sustainable management of heavy metals in wastewater using mycosynthesized iron oxide nanoparticles: Biofabrication, adsorptive dynamics and chemometric modeling study. *Journal of Water Process Engineering*, 37, 101426. https://doi.org/10.1016/j.jwpe.2020.10142.

Mandotra, S. K., Kumar, R., Upadhyay, S. K., & Ramteke, P. W. (2018). Nanotechnology: A new tool for biofuel production. In *Green Nanotechnology for Biofuel Production* (pp. 17–28). Springer.

Mishra, P., & Das, D. (2014). Biohydrogen production from *Enterobacter cloacae* IIT-BT 08 using distillery effluent. *International Journal of Hydrogen Energy*, 39(14), 7496–7507.

Nanda, S., Mohammad, J., Reddy, S. N., Kozinski, J. A., & Dalai, A. K. (2014). Pathways of lignocellulosic biomass conversion to renewable fuels. *Biomass Conversion and Biorefinery*, 4(2), 157–191.

Naylor, R. L., & Higgins, M. M. (2017). The political economy of biodiesel in an era of low oil prices. *Renewable and Sustainable Energy Reviews*, 77, 695–705.

Neshat, S. A., Mohammadi, M., Najafpour, G. D., & Lahijani, P. (2017). Anaerobic co-digestion of animal manures and lignocellulosic residues as a potent approach for sustainable biogas production. *Renewable and Sustainable Energy Reviews*, 79, 308–322.

Nizami, A., & Rehan, M. (2018). Towards nanotechnology-based biofuel industry. *Biofuel Research Journal*, 18(2), 798–799. https://doi.org/10.18331/BRJ2018.5.2.2.

Noman, M., Shahid, M., Ahmed, T., Niazi, M. B. K., Hussain, S., Song, F., & Manzoora, I. (2020). Use of biogenic copper nanoparticles synthesized from a native Escherichia sp. as photocatalysts for azo dye degradation and treatment of textile effluents. *Environment and Pollution*, 257, 113514. https://doi.org/10.1016/j.envpol.2019.113514.

Oh, Y. K., Raj, S. M., Jung, G. Y., & Park, S. (2013). Metabolic engineering of microorganisms for biohydrogen production. In *Biohydrogen* (pp. 45–65). Elsevier.

Paudel, S. R., Banjara, S. P., Choi, O. K., Park, K. Y., Kim, Y. M., & Lee, J. W. (2017). Pretreatment of agricultural biomass for anaerobic digestion: Current state and challenges. *Bioresource Technology*, 245(A), 1194–1205.

Pellera, F. M., & Gidarakos, E. (2017). Microwave pretreatment of lignocellulosic agroindustrial waste for methane production. *Journal of Environmental Chemical Engineering*, 5(1), 352–365.

Pinto, F. A. L., Troshina, O., & Lindblad, P. (2002). A brief look at three decades of research on cyanobacterial hydrogen evolution. *International Journal of Hydrogen Energy*, 27(11–12), 1209–1215.

Prabakar, D., Manimudi, V. T., Sampath, S., Mahapatra, D. M., Rajendran, K., & Pugazhendhi, A. (2018). Advanced biohydrogen production using pretreated industrial waste: Outlook and prospects. *Renewable and Sustainable Energy Reviews*, 96, 306–324.

Rahardjo, A. H., Azmi, R. M., Muharja, M., Aparamarta, H. W., & Widjaja, A. (2021, February). Pretreatment of tropical lignocellulosic biomass for industrial biofuel production: A review. *IOP Conference Series: Materials Science and Engineering*, 1053(1), 012097.

Rai, M., Ingle, A. P., Pandit, R., Paralikar, P., Biswas, J. K., & da Silva, S. S. (2019). Emerging role of nanobiocatalysts in hydrolysis of lignocellulosic biomass leading to sustainable bioethanol production. *Catalysis Reviews*, 61(1), 1–26.

Ramprakash, B., & Muthukumar, K. (2014). Comparative study on the production of biohydrogen from rice mill wastewater. *International Journal of Hydrogen Energy*, 39(27), 14613–14621.

Ramprakash, B., & Muthukumar, K. (2015). Comparative study on the performance of various pretreatment and hydrolysis methods for the production of biohydrogen using

Enterobacter aerogenes RM 08 from rice mill wastewater. *International Journal of Hydrogen Energy*, 40(30), 9106–9112.

Rasmussen, H., Sørensen, H. R., & Meyer, A. S. (2014). Formation of degradation compounds from lignocellulosic biomass in the biorefinery: Sugar reaction mechanisms. *Carbohydrate Research*, 385, 45–57.

Sannigrahi, P., Pu, Y., & Ragauskas, A. (2010). Cellulosic biorefineries—Unleashing lignin opportunities. *Current Opinion in Environmental Sustainability*, 2(5–6), 383–393.

Sarto, S., Hildayati, R., & Syaichurrozi, I. (2019). Effect of chemical pretreatment using sulfuric acid on biogas production from water hyacinth and kinetics. *Renewable Energy*, 132, 335–350.

Shen, W., Huggins, F. E., Shah, N., Jacobs, G., Wang, Y., Shi, X., & Huffman, G. P. (2008). Novel Fe–Ni nanoparticle catalyst for the production of CO-and CO2-free H2 and carbon nanotubes by dehydrogenation of methane. *Applied Catalysis A: General*, 351(1), 102–110.

Show, K. Y., Yan, Y. G., & Lee, D. J. (2019). Biohydrogen production: Status and perspectives. In *Biofuels: Alternative Feedstocks and Conversion Processes for the Production of Liquid and Gaseous Biofuels* (pp. 693–713). Academic Press.

Singh, R., Arora, A., & Singh, V. (2021). Biodiesel from oil produced in vegetative tissues of biomass–A review. *Bioresource Technology*, 326, 124772.

Singhvi, M. S., & Gokhale, D. V. (2019). Lignocellulosic biomass: Hurdles and challenges in its valorization. *Applied Microbiology and Biotechnology*, 103(23), 9305–9320.

Sivagurunathan, P., & Lin, C. Y. (2020). Biohydrogen production from beverage wastewater using selectively enriched mixed culture. *Waste and Biomass Valorization*, 11(3), 1049–1058.

Sivagurunathan, P., Sen, B., & Lin, C. Y. (2014). Batch fermentative hydrogen production by enriched mixed culture: Combination strategy and their microbial composition. *Journal of Bioscience and Bioengineering*, 117(2), 222–228.

Srivastava, N., Srivastava, M., Manikanta, A., Singh, P., Ramteke, P. W., & Mishra, P. K. (2017). Mishra nanomaterials for biofuel production using lignocellulosic waste. *Environmental Chemistry Letters*, 15(2), 179–184.

Studer, M. H., DeMartini, J. D., Davis, M. F., Sykes, R. W., Davison, B., Keller, M., … & Wyman, C. E. (2011). Lignin content in natural *Populus* variants affects sugar release. *Proceedings of the National Academy of Sciences of the United States of America*, 108(15), 6300–6305.

Taherdanak, M., Zilouei, H., & Karimi, K. (2016). The influence of dilute sulfuric acid pretreatment on biogas production from wheat plant. *International Journal of Green Energy*, 13(11), 1129–1134.

Thangaraj, B., Solomon, P. R., Muniyandi, B., Ranganathan, S., & Lin, L. (2019). Catalysis in biodiesel production—A review. *Clean Energy*, 3(1), 2–23.

Vatsala, T. M., Raj, S. M., & Manimaran, A. (2008). A pilot-scale study of biohydrogen production from distillery effluent using defined bacterial co-culture. *International Journal of Hydrogen Energy*, 33(20), 5404–5415.

Veeravalli, S. S., Shanmugam, S. R., Ray, S., Lalman, J. A., & Biswas, N. (2019). Biohydrogen production from renewable resources. In *Advanced Bioprocessing for Alternative Fuels, Biobased Chemicals, and Bioproducts* (pp. 289–312). Woodhead Publishing.

Vincent, J. F. V. (1999). From cellulose to cell. *Journal of Experimental Biology*, 202(23), 3263–3268.

Wagner, A. O., Lackner, N., Mutschlechner, M., Prem, E. M., Markt, R., & Illmer, P. (2018). Biological pretreatment strategies for second-generation lignocellulosic resources to enhance biogas production. *Energies*, 11(7), 1797.

Wang, F. L., Li, S., Sun, Y. X., Han, H. Y., Zhang, B. X., Hu, B. Z., … & Hu, X. M. (2017). Ionic liquids as efficient pretreatment solvents for lignocellulosic biomass. *RSC Advances*, 7(76), 47990–47998.

Wang, Y., Shah, N., Huggins, F. E., & Huffman, G. P. (2006). Hydrogen production by catalytic dehydrogenation of tetralin and decalin over stacked cone carbon nanotube-supported Pt catalysts. *Energy and Fuels*, 20(6), 2612–2615.

Wyman, C. E., Dale, B. E., Balan, V., Elander, R. T., Holtzapple, M. T., Ramirez, R. S., Ladisch, M. R., Mosier, N. S., Lee, Y. Y., Gupta, R., Thomas, S. R., Hames, B. R., Warner, R., Kumar, R. (2013). Comparative performance of leading pretreatment technologies for biological conversion of corn Stover, poplar wood, and switchgrass to sugars. In *Aqueous Pretreatment of Plant Biomass for Biological and Chemical Conversion to Fuels and Chemicals* (pp. 239–259). John Wiley and Sons, Ltd.

Yang, P., Zhang, R., McGarvey, J. A., & Benemann, J. R. (2007). Biohydrogen production from cheese processing wastewater by anaerobic fermentation using mixed microbial communities. *International Journal of Hydrogen Energy*, 32(18), 4761–4771.

Zeng, Q., Li, Z., & Zhou, Y. (2006). Synthesis and application of carbon nanotubes. *Journal of Natural Gas Chemistry*, 15(3), 235–246.

Zhang, L., Peng, X., Zhong, L., Chua, W., Xiang, Z., & Sun, R. (2019). Lignocellulosic biomass derived functional materials: Synthesis and applications in biomedical engineering. *Current Medicinal Chemistry*, 26(14), 2456–2474.

Zhang, R., & Zhang, Z. (1999). Biogasification of rice straw with an anaerobic-phased solids digester system. *Bioresource Technology*, 68(3), 235–245.

Zhang, X. L., Yan, S., Tyagi, R. D., & Surampalli, R. Y. (2013). Biodiesel production from heterotrophic microalgae through transesterification and nanotechnology application in the production. *Renewable and Sustainable Energy Reviews*, 26, 216–223.

Zhang, Z., Zhang, H., Li, Y., Lu, C., Zhu, S., He, C., … & Zhang, Q. (2020). Investigation of the interaction between lighting and mixing applied during the photo-fermentation biohydrogen production process from agricultural waste. *Bioresource Technology*, 312, 123570.

Zheng, Y., Zhao, J., Xu, F., & Li, Y. (2014). Pretreatment of lignocellulosic biomass for enhanced biogas production. *Progress in Energy and Combustion Science*, 42, 35–53.

Zhu, S. (2008). Use of ionic liquids for the efficient utilization of lignocellulosic materials. *Journal of Chemical Technology & Biotechnology: International Research in Process, Environmental & Clean Technology*, 83(6), 777–779.

15 Case Studies

An Insight into Green Energy Resources, Infrastructure, Economies, Energy Schemes, and Sustainability in India

Akshay Raj and Vaibhav Sharma

CONTENTS

DOI: 10.1201/9781003316374-15

15.1   INTRODUCTION

Electricity generation sources such as coal, oil, and natural gas have contributed to one-third of worldwide greenhouse gas emissions. It is critical to increase living standards by providing cleaner and more dependable power. India's energy consumption is rising in order to meet the country's economic growth ambitions. The availability of growing quantities of energy is a critical prerequisite for a country's economic progress. The Ministry of Power (MoP) structured the National Energy Plan (NEP), which was a ten-year detailed action plan with the goal of providing electricity across the nation, and has prepared a further plan to guarantee that power is given to residents effectively and at a fair cost. According to the World Resource Institute Report 2017, India is responsible for about 6.65% of total world carbon emissions, ranking fourth after China (26.83%), the United States (14.36%), and the European Union (9.66%). Climate change has the potential to disrupt the world's natural equilibrium. The United Nations

Framework Convention on Climate Change (UNFCCC) and the Paris Agreement have set Intended Nationally Determined Contributions (INDCs). The latter are intended to meet the objective of keeping the global temperature rise well below 2 degrees Celsius. Global power demand will peak in 2030, according to the World Energy Council. India is one of the world's major users of coal and imports expensive fossil fuel.. Coal and oil supply over 74% of the world's energy requirements. According to the Center for Monitoring the Indian Economy, India imported 171 million tons of coal in 2013–2014, 215 million tons in 2014–2015, 207 million tons in 2015–2016, 195 million tons in 2016–2017, and 213 million tons in 2017–2018. Therefore, there is an urgent need to develop alternative energy sources for generating power.

As a result, the country will see a quick and worldwide shift to renewable energy technology, allowing it to achieve sustainable growth while avoiding catastrophic climate change. Renewable energy sources are critical to ensuring long-term energy security with lower emissions. Renewable energy technologies are already acknowledged to have the potential to considerably cover power demand while reducing emissions. In recent years, the country has created a sustainable energy supply strategy. Energy conservation has been pushed among residents in order to boost the use of solar, wind, biomass, waste, and hydroelectric energies. Clean energy is undeniably less hazardous and, in many cases, less expensive. By 2022, India hopes to have 175 GW of renewable energy, including 100 GW from solar energy, 10 GW from bio-power, 60 GW from wind power, and 5 GW from small hydropower facilities. Investors have vowed to achieve more than 270 GW, far exceeding the lofty objectives. The following are the promises: foreign firms will contribute 58 GW, private corporations will contribute 191 GW, the private sector will contribute 18 GW, and Indian Railways will contribute 5 GW. According to recent predictions, solar potential will exceed 750 GW in 2047, while wind potential will be 410 GW. To meet the ambitious aim of producing 175 GW of renewable energy by 2022, the government must generate 330,000 new employment and livelihood prospects.

A combination of push and pull policies, complemented by specific initiatives, should encourage the development of renewable energy technology. Technology advancement, proper regulatory policies, tax deduction, and attempts at efficiency enhancement due to research and development (R&D) are some of the pathways to energy and environmental conservation that should ensure that renewable resource bases are used in a cost-effective and timely manner. As a result, measures for increasing investment prospects in the renewable energy sector, as well as jobs for unskilled employees, technicians, and contractors, are reviewed. This chapter also discusses the government's technological and financial endeavors, policy and regulatory framework, and training and educational programs for the growth and development of renewable energy sources. The advancement of renewable technology has faced obvious impediments, necessitating a discussion of these impediments. Furthermore, it is critical to identify potential ways to overcome these hurdles, and as a result, appropriate suggestions for the steady expansion of renewable energy have been made. Given the country's tremendous renewables potential, consistent legislative measures and an investor-friendly administration may be the major catalysts for India to become a worldwide leader in clean and green energy.

15.2 STATUS OF RENEWABLE ENERGY IN INDIA

The origins of renewable energy development in India may be traced back to the late 1980s' worldwide oil crisis. Since then, the Indian government has worked tirelessly to promote the renewable energy sector through a variety of strategic policy and regulatory initiatives. Given the constitutional status of energy as a concurrent item (number 38 on the concurrent list), major policy measures to enhance the renewable energy industry are developed from time to time by both the federal and provincial governments. The most recent policy impetus to transition to a greener energy regime, however, is expressed in the Government of India's transformational energy strategy, which aims to create 175 GW of renewable energy by 2022. With the precise policy proclamation of the Jawaharlal Nehru National Solar Mission (JNNSM), solar power has been given pride of place in the renewable basket. Similar legislative measures, such as ensuring 24/7 electricity availability across the country by 2019, are visible manifestations of the emphasis placed on renewable energy. This point is reinforced by India's global climate promises to the UNFCCC in the form of INDCs. The global climate commitment to getting 40% of its energy from renewable sources by 2040 is a strong indication of renewable energy policy goals.

Not only that, but the present energy balance reflects a shifting governmental focus on energy generation, with a growing percentage of renewables in the country's energy basket.

However, mapping primary energy use does not indicate such a transformation, owing to an overreliance on biofuels and oil products. This clearly demonstrates that rural India is still reliant on fossil fuels for primary energy usage. This also demonstrates that India has a long way to go in terms of total energy transformation.

On the other hand, renewable energy as a source of power is rapidly displacing traditional energy sources in the country. This may be seen in the country's varied renewable energy growth estimates. For example, the National Action Plan on Climate Change (NAPCC) targets 15% renewable energy consumption by 2020.

15.3 INDIA'S RENEWABLE ENERGY FINANCING STRUCTURE AND PATTERN

Renewable energy system technological artifacts differ greatly from traditional power systems. As a result, the cost factors differ for both industries. In comparison to traditional energy systems, the cost features of renewable energy projects are such that they are relatively capital-intensive in nature, with zero fuel cost (Bhattacharyya and Maheshwari, 2010; Hirth and Steckel, 2016). These zero recurrent expenditures have a favorable impact on average power rates, which are less variable. Given the technological features of the industry, it has been estimated that the investment required to reach the Government of India's ambitious objective of establishing 175 GW would be roughly $189 billion (CPI 2016).

It's worth delving more into the market structure of renewable energy in India. Unlike the traditional energy sector, the private sector is responsible for creating the renewable energy industry. The sector's challenge thus is to mobilize private capital at a rate and pace consistent with policy aims and goals. This is necessary owing to the restricted availability of public funds. Whatever meager public money is available, it is mostly utilized to attract private capital to flow into the industry. It is based on the assumption that the private sector has the potential to fund the industry; nevertheless, the government must build an enabling environment.

According to a deconstructed financial mapping for the industry, the renewable energy sector is driven by private investors, with a dependence on banking institutions to mobilize the required money. However, the financial community has been hesitant to finance renewable energy projects, owing to the risks and uncertainties connected with these initiatives. This is obvious in a recent study, which clearly shows that a significant amount of money from government-owned banks and other financial organizations has been directed into coal projects rather than renewables (CFA 2018). This financing pattern is also quite similar to debt-equity financing, in which 70% of funds are derived from debt and 30% are mobilized as equity money. It is also claimed that the loan costs of renewable energy projects in India are 24 to 32% more than those in the United States and Europe, based on the leveled cost of energy (CPI 2012). On the other hand, the equity component of funding, which was typically given by project developers, has undergone a transformation. According to recent financing patterns, equity money is given by outside parties such as private equity (PE) investors. This is obvious when equity investment in renewable energy in India shifts from balance-sheet to project-based financing (Bhattacharyya and Maheshwari, 2010; Hirth and Steckel, 2016). Though the financing structure of renewable energy in India is dominated by bank finances, a range of investors have recently emerged in the Indian renewable energy sector. Commercial banks, private equity investors, institutional investors, and development banks are among them. International banks, on the other hand, are apparent by their absence. A thorough study of the bank and non-bank funding of renewable energy projects reveals that numerous banks have invested around $2,570 million; IDFC has committed around 20% of the total sum. The table also clearly shows that the majority of the commitments are made by non-banking financial firms. Renewable energy in India is dominated by bank financing; nevertheless, a range of investors have recently emerged in the Indian renewable energy sector. Commercial banks, private equity investors, institutional investors, and development banks are among them. Various types of investors are present in the renewable energy sector; the venture capital type of investor is the most recent form of investor in the area to support the equity component under the project-based financing method. Foreign banks, on the other hand, are notable for their absence.

15.4 SOLAR ENERGY AND ITS CURRENT STATUS IN INDIA

Since ancient times, our planet has been regarded as the source of all life. The industrial era taught us about sunlight as a source of energy. India has a large potential for solar energy. Approximately 5,000 trillion kWh of energy is incident over India's geographical surface each year, with the majority of areas receiving 4–7 kWh per square meter each day. In India, solar photovoltaic electricity can be successfully harvested, allowing for massive scalability. Solar also permits dispersed power generation and quick capacity expansion with short lead periods. Off-grid, decentral-ized, and low-temperature applications will benefit rural electrification while also addressing other energy demands for electricity, heating, and cooling in both rural and urban locations. Solar is the most secure of all energy sources since it is plenti-ful. In theory, a small proportion of total incoming solar energy can cover the entire country's electricity needs (if caught efficiently).

Solar energy has had a noticeable influence on the Indian energy landscape in recent years. Decentralized and distributed solar energy applications have benefited millions of people in Indian communities by addressing their cooking, lighting, and other energy demands in an environmentally benign way. The social and economic benefits include reduced drudgery among rural women and girls engaged in long-distance fuel wood collection and cooking in smoky kitchens, reduced risks of con-tracting lung and eye diseases, job creation at the village level, and, ultimately, an improvement in the standard of living and the creation of opportunities for economic activities at the village level. Furthermore, in recent years, India's solar energy sector has emerged as a prominent participant in grid-connected power generation capacity. It supports the government's vision of sustainable growth while emerging as a vital actor in meeting the nation's energy demands and ensuring energy security.

The National Institute of Solar Energy estimated the country's solar potential to be 748 GW, assuming that solar PV modules cover 3% of the waste land area. Solar energy has assumed a prominent role in India's National Action Plan on Climate Change, with the National Solar Mission serving as one of the primary missions. On January 11, 2010, the National Solar Mission (NSM) was launched. The National Sustainable Movement (NSM) is a significant project of the Government of India, with strong participation from states, to promote ecologically sustainable growth while addressing India's energy security issues. It would also represent a significant contribution by India to the global effort to address climate change problems. The Mission's goal is to position India as a global leader in solar energy by establishing the regulatory conditions for the rapid adoption of solar technology across the coun-try. By 2022, the Mission intends to have installed 100 GW of grid-connected solar power plants. This is consistent with India's INDCs target of achieving approxi-mately 40% of cumulative electric power installed capacity from non-fossil fuel-based energy resources by 2030, as well as reducing the emission intensity of its GDP by 33 to 35% from 2005 levels.

To achieve the aforementioned target, the Government of India has launched a number of schemes to encourage the generation of solar power in the country, including the Solar Park Scheme, VGF Schemes, CPSU Schemes, Defense Scheme, Canal Bank & Canal Top Scheme, Bundling Scheme, Grid Connected Solar Rooftop Scheme, and so on.

Among the policy measures implemented were the establishment of a trajectory for Renewable Purchase Obligation (RPO), which included solar, a waiver of Inter State Transmission System (ISTS) costs and losses for inter-state sales of solar and wind power for projects to be completed by March 2022, must-run status, guidelines for solar power procurement through a tariff-based competitive bidding procedure, solar photovoltaic system and device deployment standards, rooftop solar panel installation and smart city development guidelines, amendments to building codes to require the deployment of rooftop solar for new construction or buildings with a greater floor area ratio, infrastructure status for solar projects, raising tax-free solar bonds, obtaining long-term financing from multilateral lending institutions, and so on.

India has just surpassed Italy to take fifth place in the world in solar power deployment. In the previous five years, solar power capacity has expanded more than 11 times, from 2.6 GW in March 2014 to 30 GW in July 2019. Solar tariffs in India are now quite competitive, and the country has reached grid parity.

15.5 ENERGY SCENARIO IN INDIA

Energy is one of the most important inputs for every country's economic success. The energy industry is particularly important in emerging nations, given the ever-increasing energy demands that need large expenditures to satisfy.

The following characteristics may be used to categorize energy into different types:

- Energy, both primary and secondary
- Energy, both commercial and non-commercial
- Energy sources: renewable and non-renewable

Nature's primary energy sources are those that are discovered or stored there. Coal, oil, natural gas, and biomass are some of the most common primary energy sources (such as wood). Nuclear energy from radioactive chemicals, thermal energy stored in the earth's interior, and potential energy owing to gravity are some of the other basic energy sources available.

Some energy sources have non-energy applications, such as the use of coal or natural gas as a feedstock in fertilizer factories.

There are two types of energy: commercial and non-commercial.

15.5.1 ENERGY FOR BUSINESS

Commercial energy refers to energy sources that are offered on the market for a set price. Electricity, coal, and refined petroleum products are by far the most significant kinds of economic energy. In today's world, commercial energy is at the heart of industrial, agricultural, transportation, and commercial growth. Commercialized fuels are the primary source of energy in industrialized nations, not just for economic production but also for numerous home functions.

Electricity, lignite, coal, oil, natural gas, and so on are some examples.

15.5.2 ENERGY THAT ISN'T FOR PROFIT

Non-commercial energy refers to energy sources that are not accessible for a fee on the commercial market. Fuels such as firewood, cow dung, and agricultural wastes, which are traditionally harvested and not purchased at a price, are examples of non-commercial energy sources utilized mostly in rural communities. Traditional fuels are another name for them. In energy accounting, non-commercial energy is often overlooked.

Non-commercial energy sources include, for instance, in rural areas, firewood and agricultural waste; solar energy for water heating, electricity generation, and drying grain, fish, and fruits; animal power for transportation, threshing, lifting water for irrigation, and crushing sugarcane; and wind energy for lifting water and electricity generation.

15.5.3 Energy, Both Renewable and Non-Renewable

Renewable energy is energy derived from almost limitless sources. Wind power, solar power, geothermal energy, tidal power, and hydroelectric power are examples of renewable resources. Renewable energy's most essential advantage is that it may be utilized without releasing damaging pollutants.

Non-renewable energy refers to traditional fossil fuels like coal, oil, and gas, which are expected to run out over time.

15.6 RESERVES OF PRIMARY ENERGY ON THE PLANET

15.6.1 Coal

By the end of 2003, the proven worldwide coal resource was projected to be 984,453 million tons. The United States held the highest percentage of the world reserve (25.4%), followed by Russia (15.9%), and China (15.9%). With 8.6%, India ranked fourth on the list.

15.6.2 Oil

By the end of 2003, the worldwide proven oil resource was projected to be 1147 billion barrels. Saudi Arabia had the greatest part of the reserve, accounting for over 23%.

(A barrel of oil holds around 160 liters.)

15.6.3 Gas

By the end of 2003, the worldwide proven gas resource was projected to be 176 trillion cubic meters. With over 27% of the reserve, the Russian Federation was the greatest contributor.

(Ref: BP World Energy Statistical Review, June 2004.)

The world's oil and gas reserves are only expected to last 45 and 65 years, respectively. Coal is expected to last around 200 years.

15.6.4 Consumption of Primary Energy on a Global Scale

At the end of 2003, worldwide primary energy consumption was 9741 million tons of oil equivalents (Ministry of Renewable Energy).

15.7   PRIMARY ENERGY CONSUMPTION BY FUEL

15.7.1   DISTRIBUTION OF ENERGY AMONG DEVELOPED
AND DEVELOPING COUNTRIES

Despite the fact that emerging nations account for 80% of the world's population (a fourfold growth in the last 25 years), their energy consumption accounts for just 40% of overall global energy consumption. The high levels of energy use in industrialized nations are responsible for their high living standards.

In comparison to highly industrialized countries, population growth in underdeveloped countries has kept per capita energy consumption low.

Energy distribution in developed and developing countries has been studied. The global average for energy usage per person is 2.2 tons of coal. Individuals in developed nations consume four to five times as much as the global average, and nine times as much as people in poor countries. An American consumes 32 times as much commercial energy as an Indian.

In India, coal dominates the energy mix, accounting for 55% of total primary energy production. Natural gas's proportion in primary energy output has risen steadily over the years, from 10% in 1994 to 13% in 1999. During the same time span, oil's proportion in primary energy output fell from 20% to 17%.

15.8   INDIA'S ENERGY CONSUMPTION BY SECTOR

Energy is necessary for economic development. However, the link between growing energy consumption and economic development is not always obvious. For example, 6% growth in India's gross domestic product (GDP) would result in a 9% increase in demand for the country's energy industry under current circumstances.

The ratio of energy consumption to GDP is a helpful metric in this regard. A high ratio indicates energy reliance and a significant impact of energy on GDP growth. The industrialized nations keep their energy to GDP ratios below one by concentrating on energy efficiency and less energy-intensive pathways. For underdeveloped nations, the ratios are significantly larger.

15.8.1   INDIA'S ENERGY REQUIREMENTS

15.8.1.1   Energy Consumption Per Capita

In comparison to developed nations, India's per capita energy consumption is excessively low. It accounts for just 4% of the United States' GDP and 20% of the global average. India's per capita consumption is expected to rise as the economy grows, boosting energy demand.

15.8.1.2   Intensity of Energy

The term "energy intensity" refers to the amount of energy used per unit of GDP. A country's development stage is indicated by its energy intensity. India's energy intensity is 3.7 times that of Japan, 1.55 times that of the United States, 1.47 times Asia's average, and 1.5 times the global average.

15.9 INDIA'S LONG-TERM ENERGY OUTLOOK

In India, coal is the most common energy source for power generation, accounting for over 70% of total domestic electricity. India's energy consumption is predicted to rise over the next 10–15 years, and despite plans for new oil and gas facilities, coal is likely to remain the leading power production source. Despite large gains in total installed capacity over the previous decade, the gap between supply and demand for electricity has continued to widen. The ensuing gap has harmed industrial production and slowed economic development. However, in order to fulfill future demand, domestic coal supply will need to be considerably increased. Production is now approximately 290 million tons per year, but by 2030, demand for coal is predicted to more than quadruple. Indian coal is normally of low quality, and as a result, it will need to be beneficiated to enhance its quality; coal imports will also need to dramatically grow to meet industrial and power production needs.

15.9.1 OIL

According to forecasts in the Tenth Five-Year Plan, India's consumption for petroleum products would climb from 97.7 million tonnes in 2001–2002 to about 139.95 million tonnes in 2006–2007. During the plan period, the compound annual growth rate (CAGR) was set at 3.6%, according to the plan document. Domestic crude oil output was expected to increase slightly from 32.03 million tonnes in 2001–2002 to 33.97 million tonnes by the conclusion of the tenth plan period (2006–2007). India's oil self-sufficiency has steadily fallen, from 60% in the 1950s to 30% now.

By 2020, the unemployment rate is expected to drop to 8%. Imports will have to provide roughly 92% of India's total oil consumption by 2020, as indicated in Figure 1.8.

15.9.2 NATURAL GAS

India's natural gas output was expected to increase from 86.56 million cubic meters per day in 2002–2003 to 103.08 million cubic meters per day (cmpd) in 2006–2007. It is based mostly on the prediction of a more than doubling of private operators' output to 38.25 mm cmpd.

India is now experiencing a peak demand shortfall of roughly 14% and an energy deficit of 8.4%. With this in mind, and in order to sustain GDP growth of 8% to 10%, India's government has set an extremely conservative goal of 215,804 MW power generation.

15.9.3 ELECTRICITY

In India, electricity tariffs are structured in a relatively straightforward manner. While high-tension consumers are taxed for both demand (kVA) and energy (kWh), low-tension (LT) consumers are simply charged for the energy used (kWh) under most electricity boards' tariff systems. The price per kWh varies a lot across states and even between client categories within a state. In India, tariffs have been adjusted to take into account the time of use and the voltage level of supply. Some State Electricity Boards have extra revenue from consumers in the form of fuel surcharges, electricity fees, and taxes in addition to the basic prices. For example, demand rates for an industrial customer may range from Rs. 150 to Rs. 300 per kVA, while energy prices may range from Rs. 2 to Rs. 5 per kWh. When it comes to the tariff adjustment system, even if some states have regulatory commissions for tariff review, the choices to make adjustments are still political, and there is no automated adjustment mechanism in place to guarantee that the power boards recoup their expenses.

15.10   REFORMS IN THE ENERGY SECTOR

Since the beginning of India's economic reforms in 1991, there has been a growing understanding of the necessity to expand these changes in numerous areas of the economy that had been essentially in the hands of the government for decades. It is now widely recognized that if macroeconomic policy change is to have any credibility, it must be founded on reforms that affect the operation of many vital sectors of the economy, the most important of which are the infrastructure and energy sectors.

15.10.1   COAL

The government has acknowledged the necessity for new coal policy initiatives as well as a rationalization of the legislative and regulatory framework that would control the industry's future growth. One of the most significant reforms is the government's approval of coal imports to meet our needs. The private sector has been given permission to harvest coal for its own purposes solely. Further revisions are being considered, which would require amendment to the Coal Mines Nationalization Act, which is now awaiting Parliament's approval.

The ultimate goal of some of the current policies, as well as those under discussion, is to ensure that a competitive environment is established for the different companies in this business to operate in. This would not only result in increased efficiency but also reduced costs, ensuring a bigger supply of coal at cheaper prices. Competition would also have the positive consequence of bringing in new technology, which is desperately needed due to stagnation in the coal industry's technological progress.

15.10.2   NATURAL GAS AND OIL

Since 1993, private investors have been able to freely import and sell liquefied petroleum gas (LPG) and kerosene; private investment in lubricants, which are not subject to price regulations, has also been permitted. The cost of naphtha and a few other fuels has been reduced. The government created the New Exploration Licensing Policy (NELP) in 1997 to encourage investment in domestic oil and gas exploration and production. Furthermore, the refining sector has been opened to private and foreign investors in order to reduce refined product imports and encourage downstream pipeline investment. For the building of liquefied natural gas (LNG) import facilities, investors are being given attractive conditions.

15.10.3 ELECTRICITY

The Central Electricity Regulatory Commission (CERC) was established after the passing of the Electricity Regulatory Commission Legislation, with the primary goal of regulating central power production utilities. Tariff-setting and competition-promoting regulatory bodies have also been established at the state level. Private power generating ventures were also permitted. Separate generation, transmission, and distribution firms were requested of the state SEBs. All SEB networks would be linked to establish a single national electricity grid, according to plans.

15.10.4 2003 ELECTRICITY ACT

The government passed the Electrical Act in 2003, which aims to improve the quality of the electricity industry. The Act aims to provide a free-market development environment for the electricity industry by removing the government from the regulatory process. It repeals the Indian Electricity Act of 1910, the Electricity (Supply) Act of 1948, and the Electricity Regulatory Commissions Act of 1998, and replaces them.

> To consolidate the laws relating to generation, transmission, distribution, trading, and use of electricity, and generally for taking measures conducive to the development of the electricity industry, promoting competition therein, protecting consumers' interests, and ensuring the supply of electricity to all areas, rationalization of electricity tariffs, ensuring transparent policies regarding subsidies, and promotion of efficient and environmentally benign policing,

according to the Act's objectives.

The following are the main characteristics of the Electricity Act of 2003:

15.10.4.1 Environment and Energy

The use of energy resources in industry pollutes the atmosphere, causing environmental harm. Sulfur dioxide (SO_2), nitrous oxide (NOX), and carbon monoxide (CO) emissions from boilers and furnaces, as well as chlorofluorocarbon (CFC) emissions from refrigerant usage, are only a few instances of air pollution. Toxic gasses are emitted in the chemical and fertilizer industries. Particulate matter is emitted by cement and electricity plants.

15.10.4.2 Changes in Climate

The blanket of greenhouse gasses (water vapor, carbon dioxide, methane, ozone, and so on) around the earth has grown thicker as a result of human activities, particularly the combustion of fossil fuels. The consequent rise in global temperature is affecting the intricate network of processes that enable life on earth to flourish, including rain-fall, wind patterns, ocean currents, and plant and animal distribution.

It is possible to see light. The outer atmosphere scatters around 30% of the sun-light back into space, while the remaining 70% reaches the earth's surface, where it is reflected as infrared radiation. Greenhouse gasses delay the escape of slow-moving infrared radiation. More infrared radiation is trapped by a thicker layer of greenhouse gasses, raising the earth's temperature

Greenhouse gasses make up just 1% of the atmosphere, yet they operate as a blanket over the world, or like the glass roof of a greenhouse, keeping it 30 degrees warmer than it would be otherwise – the earth would be too cold to exist without them. Emissions of carbon dioxide from the burning of coal, oil, and natural gas; extra methane and nitrous oxide from agricultural operations and changes in land use; and other man-made gasses with a long life in the atmosphere are all responsible for thickening the greenhouse layer. The rate of increase in greenhouse gas emis-sions is alarming. If greenhouse gas emissions continue at their current rates, carbon dioxide levels in the atmosphere will almost certainly increase twice or three times from pre-industrial levels during the 21st century.

Changes in climate, such as cloud cover, precipitation, wind patterns, and season length, will accompany even a small increase in the earth's temperature. Millions of people rely on weather patterns like monsoon rains to continue as they have in the past in an already overcrowded and stressed world. Even little modifications will be difficult and disruptive.

The "enhanced greenhouse effect" is caused by carbon dioxide, which accounts for 60% of the total. Humans are using coal, oil, and natural gas at a significantly greater pace than they are being generated. This releases carbon from the fuels into the atmosphere, disrupting the carbon cycle (a precise, well-balanced system in which carbon is exchanged between plants and animals).

Over millions of years, the air, seas, and land plants have changed. Currently, carbon dioxide levels in the air are growing by over 10% every 20 years.

15.11   CLIMATE CHANGE EVIDENCE CURRENTLY AVAILABLE

Cyclones, storms, and hurricanes are becoming more frequent, while floods and droughts are becoming more severe. Extreme weather events are becoming more common, and they can't be explained away as random occurrences.

Computer models indicate a trend toward more violent storms and hotter, longer dry spells. Warmer temperatures result in more evaporation, and a warmer atmo-sphere can contain more moisture, thus there is more water aloft that may fall as rain. Dry places, on the other hand, are more likely to lose moisture as the temperature warms, resulting in more severe droughts and desertification.

15.12   FUTURE CONSEQUENCES

Even the smallest climate changes predicted for the 21st century are likely to be significant and disruptive. Future climate change predictions cover a wide range of topics. The global temperature could rise by 1.4 to 5.8 degrees Celsius, and the sea level could rise by 9 to 88 cm. As a result, sea level rises this century are expected to be significant to catastrophic. This uncertainty reflects the natural processes that make up the climate's complexity, interconnectedness, and sensitivity.

15.12.1 FLOODING AND SEVERE STORMS

The minimum warming predicted for the next 100 years is more than double the 0.6°C increase seen since 1900, which is already having significant consequences. Extreme weather events are occurring more often, as anticipated by computer models, and are projected to worsen and become even more common. More catastrophic storms and floods are predicted in the future along the world's increasingly populous coasts.

15.12.2 FOOD SCARCITY

Although regional and local effects may vary, most tropical and subtropical regions can expect a general decrease in potential crop yields. Mid-continental regions, like the "grain belt" of the United States and enormous swaths of Asia, are expected to dry up. Even with a small rise in temperature in Sub-Saharan Africa, where dry-land agriculture depends primarily on rain, production would plummet. In a world already plagued by food shortages and famines, such shifts might trigger food supply disruptions.

15.12.3 FRESHWATER SUPPLIES ARE DWINDLING.

The quality and quantity of freshwater sources will be harmed as a result of saltwater intrusion caused by increasing sea levels. This is a big worry, since billions of people currently lack access to clean water on the planet. In many regions of the globe, rising ocean levels are already poisoning subsurface water supplies.

15.12.4 BIODIVERSITY IS DISAPPEARING.

The majority of the world's endangered species (about 25% of mammals and 12% of birds) may become extinct in the next several decades as rising temperatures change the forests, marshes, and oceans.

They rely on grasslands and rangelands, and human development prevents them from moving elsewhere.

15.12.5 DISEASES ARE BECOMING MORE PREVALENT.

Higher temperatures are expected to expand the range of some dangerous "vector-borne" diseases, such as malaria, which kills one million people each year, the majority of whom are children.

15.13 A WORLD IN TURMOIL DUE TO EXCESSIVE USE OF NATURAL RESOURCES

As a result of ongoing ecologically destructive activities such as overgrazing, deforestation, and depleted agricultural soils, nature will be more sensitive to climate change than in the past.

Similarly, the huge human population of the globe, much of which is impoverished, is susceptible to climate change. Millions of people live in unsafe areas such as floodplains or slums in the developing world's major cities. There is often nowhere else for them to go. In the distant past, man and his predecessors traveled in response to changes in habitat. In the future, there will be significantly less space for migration.

Climate change will most probably be unjust. North America and Western Europe, as well as other countries such as Japan, are responsible for the vast majority of past and current greenhouse gas emissions. These emissions are incurred as a result of the high living conditions enjoyed by the citizens of such nations.

The developing world, on the other hand, will be the worst hit by climate change. Storms, floods, droughts, disease outbreaks, and disruptions in food and water supplies are all exacerbated by their lack of resources. They are hungry for economic progress for themselves, but climate change may make this already tough process much more onerous. The world's poorest countries have done very little to create global warming, yet they are the most vulnerable to its impacts.

15.14 ENERGY SAFETY

The primary goal of energy security for a country is to lessen a country's reliance on foreign energy sources for economic development.

Throughout the projected period, India will continue to face an energy supply shortage. Since 1985, when the nation became a net coal importer, the deficit has worsened. In the 1990s, India was unable to significantly increase its oil output. Oil demand is increasing at a rate of about 10% per year, resulting in significant oil import expenses. Furthermore, the government subsidizes refined oil product prices, exacerbating the government's total financial loss.

During the years 1991–1999, oil and coal imports increased at rates of 7% and 16% per year, respectively. The country's reliance on imported energy is expected to grow in the future. In 2006, oil imports were expected to cover 75% of total oil consumption needs and coal imports would meet 22% of total coal consumption requirements, according to estimates. Gas and LNG imports are expected to rise in the future. This reliance on imported energy exposes the nation to foreign price shocks and supply changes, jeopardizing the country's energy security.

Increasing reliance on oil imports necessitates reliance on Middle Eastern imports, an area prone to interruptions in oil supply as a result of unrest. This begs for diversity of sources of oil imports. The need to cope with oil price swings also necessitates action to be taken to decrease the oil dependency of the economy, perhaps via fiscal measures to lower consumption, and by developing alternatives to oil, such as natural gas and renewable energy.

Some of the solutions that may be employed to tackle future energy security problems include:

- Constructing stockpiles
- Diversification of energy sources
- Increasing fuel switching capacity
- Requiring restraint
- Development of renewable energy sources
- Energy efficiency
- Long-term development

All of these alternatives are realistic, but they will require time to execute. Also, due to resource restrictions, dependence on stocks would be sluggish in countries like India. Furthermore, neither the market nor the monitoring organizations are smart enough to foresee the supply situation in time to take essential measures. Inadequate storage capacity is another source of concern that must be addressed if India's energy stockpile is to be increased.

However, of all of these solutions, lowering demand via consistent energy-saving measures is the easiest and most achievable.

15.14.1 THE IMPORTANCE OF ENERGY CONSERVATION

Coal and other fossil fuels, which took three million years to produce, are on the verge of becoming depleted. We have depleted 60% of all resources in the previous two centuries. Energy efficiency measures are required for long-term development.

Energy efficiency and conservation are two different but related concepts. When the growth of energy consumption is reduced in physical terms, energy conservation is achieved. As a consequence, energy conservation might be the outcome of a variety of processes or advancements, such as increased production or technological advancement. Energy efficiency, on the other hand, is achieved when the energy intensity of a specific product, process, or area of production or consumption is reduced without compromising output, consumption, or comfort. Energy efficiency promotion contributes to energy conservation and is thus an important component of energy conservation policies.

Energy efficiency is often thought of as a resource, similar to coal, oil, or natural gas. It adds value to the economy by maintaining the resource base and lowering pollution. To illuminate a room, for example, replacing standard light bulbs with compact fluorescent lamps (CFLs) would require just one-fourth of the energy. Pollution levels fall by the same percentage.

Nature establishes some basic limits on the amount of energy that can be used efficiently, but most of our products and manufacturing processes are still far from reaching this theoretical limit. Simply put, energy efficiency means using less energy to accomplish the same task.

Although energy efficiency has been practiced since the 1973 oil crisis, it has now taken on even greater significance as the most cost-effective and dependable strategy for reducing global climate change. Because of this potential, there are high hopes for future CO_2 emissions to be controlled by even greater energy efficiency advances than in the past. The industrial sector consumes around 41% of global primary energy and emits almost the same amount of CO_2.

15.14.2 THE FUTURE ENERGY STRATEGY

The future energy plan may be divided into three categories: immediate, medium-term, and long-term. The following are the various components of these strategies:

- Prompt-response strategy:
- The tariff structure of various energy products should be rationalized.
- Getting the most out of existing assets.
- Production system efficiency and distribution losses, particularly those in conventional energy sources, are being addressed.
- Promoting research and development, as well as the transfer and use of technology and practices for ecologically sound energy systems, including new and renewable energy sources.

15.14.2.1 Strategy for the Medium Term

The medium-term strategy includes demand management through increased energy conservation, optimum fuel mix, structural changes in the economy, and an appropriate model mix in the transportation sector, i.e. greater reliance on rail for the movement of goods and passengers and a shift away from private to public modes for passenger transport; and changes in product design to reduce the material intensity of those products, recycling, and so on.

It is necessary to transition to less energy-intensive forms of transportation. This would include measures to enhance transportation infrastructure, such as the construction of new roads, roadways, car design, compressed natural gas (CNG) and synthetic fuel usage, and so forth. Similarly, smarter urban design will lower energy consumption in the transportation sector.

It is necessary to transition from non-renewable to renewable energy sources, such as solar and wind. Solar, wind, biomass, and other renewable energy sources are all available.

15.14.2.2 Long-Term Strategy

- The efficient development of energy resources.
- Coal, oil, and natural gas production that is efficient.
- Flaring of natural gas is being reduced.
- Infrastructure improvements in the energy sector.
- Constructing new refineries.
- Construction of a gas transmission and distribution network in cities.
- Increasing the efficiency of coal transport by rail.

- New coal and gas-fired power plants are being built.
- Increasing the effectiveness of energy use.
- Improving energy efficiency to meet national, socioeconomic, and environmental objectives.
- Energy efficiency and emission requirements are being promoted.
- Large-scale enterprises are implementing product labeling programs and adopting energy-efficient technology.
- Energy industry deregulation and privatization.
- Cross-subsidies on oil products and electricity tariffs should be reduced.
- Getting rid of the price ceiling on coal and making natural gas more competitive.
- Oil, coal, and electricity industries should be privatized to increase efficiency.
- To encourage foreign investment, legislation has been enacted.
- Streamlining the approval process to encourage private sector involvement in power generation, transmission, and distribution.

15.15   FEATURES OF THE ENERGY CONSERVATION ACT OF 2001

The Government of India has adopted the Energy Conservation Act of 2001 – Policy Framework in order to effectively overcome the barrier by bridging the gap between demand and supply, lowering environmental emissions via energy savings, and bridging the gap between demand and supply. The Act creates the necessary legal and administrative framework for launching an energy-saving campaign.

The Bureau of Energy Efficiency was founded on March 1, 2002, under the requirements of the Act, replacing the Ministry of Power's former Energy Management Center. The Bureau is in charge of carrying out policy programs and coordinating the execution of energy-saving measures.

The following are key components of the Energy Conservation Act:

- Labeling and standards
- Standards and labeling (S & L) have been highlighted as a critical activity for increasing energy efficiency. When the S & L program is fully implemented, customers will have access to only energy-efficient equipment and appliances.
- The following are the key provisions of the EC Act on standards and labeling:
- Minimum energy usage and performance criteria for notified equipment and appliances should be developed. Prohibit the manufacture, sale, and import of equipment that does not meet the required standards.
- Introduce a mandatory labeling scheme for notified equipment appliances so that customers can make educated decisions.
- Consumers should be informed about the advantages.
- Consumers who have been designated.

- The following are the major clauses of the EC Act concerning designated consumers:
- As designated customers, the government would inform energy-intensive companies and other facilities.
- The Act's Schedule included a list of authorized customers, which primarily included energy-intensive enterprises, railways, port trusts, the transportation sector, power plants, transmission and distribution companies, and commercial buildings or institutions.
- A qualified energy auditor will perform an energy audit for the specified customer.
- The designated customers must select or designate energy managers who have the relevant qualifications.
- The designated users must adhere to the central government's energy usage regulations and requirements.
- Energy managers must be certified, and energy auditing firms must be accredited. The following are the key activities outlined in the Act:
- Through the Certification and Accreditation program, a cadre of professionally qualified energy managers and auditors with expertise in policy analysis, project management, financing, and implementation of energy efficiency projects would be developed. The Bureau of Energy Efficiency (BEE) will develop training modules and administer a national test for the certification of energy managers and auditors.

15.15.1 BUILDING CODES FOR ENERGY CONSERVATION

The following are the main provisions of the European Commission Act on Energy Conservation Building Codes.

The BEE will draft Energy Conservation Building Codes (ECBC) guidelines.

For commercial buildings created after the regulations pertaining to energy conservation building codes have been declared, the owners would be notified by the different states to meet local climatic circumstances or other compelling considerations. Furthermore, these structures must have a connected load of 500 kW or a contract demand of 600 kVA or more, and they must be used for commercial purposes.

It is also mandatory to conduct an energy assessment of specified commercial building customers.

The BEE is a government agency that promotes energy efficiency.

The Bureau of Energy Efficiency's purpose is to institutionalize energy efficiency services, allow delivery mechanisms throughout the nation, and lead energy efficiency in all sectors of the economy. The primary goal is to reduce the Indian economy's energy intensity.

The Governing Council, which has 26 members, has broad supervision, direction, and control of the Bureau's business. The Council is headed by the Union Minister of Power and consists of members represented by Secretaries of various Ministries, the CEOs of technical agencies under the Ministries, members representing equipment and appliance manufacturers, industry, architects, and consumers, and five power regions representing the states. The ex-officcio member-secretary of the Council should be the Director General of the Bureau.

The BEE will be initially financed by the central government via budgetary allocations; however, it will become self-sufficient in five to seven years. It will be allowed to charge an appropriate price in order to carry out the responsibilities that have been allocated to it. In order to stimulate energy-efficient investment, the BEE will also employ the Central Energy Conservation Fund and other monies obtained from other sources for the creative financing of energy efficiency projects.

15.15.2   Bureau of Energy Efficiency's Role

BEE's responsibilities include developing standards and labels for appliances and equipment, compiling a list of designated consumers, defining certification and accreditation procedures, preparing building codes, administering the Central EC Fund, and conducting promotional activities in collaboration with federal and state agencies. Energy service companies (ESCOs) will be developed, the market for energy efficiency will be transformed, and awareness will be raised via measures such as a clearinghouse.

15.15.3   The Central and State Governments' Roles

The Act envisions the central and state governments playing the following roles.

Central – to announce rules and regulations under different parts of the Act, offer initial financial support to the BEE and EC Fund, and coordinate with various State Governments for notice, enforcement, fines, and adjudication.

State – to amend energy conservation building codes to reflect regional and local climates, to establish a state-level agency to coordinate, regulate, and enforce the Act's provisions, and to establish a State Energy Conservation Fund to promote energy efficiency.

15.15.3.1   Self-Regulation as a Means of Enforcement

Only two items would be required to be inspected under the EC Act. The following self-regulatory approach is recommended to be used for confirming regions that only need inspection of two items.

Accredited Energy Auditors may certify energy consumption norms and manufacturing process standards, which can be used to ensure effective energy efficiency in designated consumers.

Manufacturers' reported values will be tested in accredited laboratories for energy performance and standards by taking a sample from the market. Any manufacturer, consumer, or consumer association can question another manufacturer's values and bring it to BEE's attention. In contested circumstances, BEE might perceive challenge testing as a self-regulation technique.

15.15.3.2 Penalties and Judgment

The penalty for any infraction under the Act will be in monetary terms, i.e. for each offense, a fine of Rs.10,000 is imposed, with an additional fine of Rs.1,000 every day of non-compliance.

The first five years would be spent on promotion and building infrastructure in preparation for the Act's implementation. During this time, no penalties would be effective.

The state Electricity Regulatory Commission has been given the authority to adjudicate, and any of its members can be appointed as an adjudicating officer to conduct an investigation into the penalty imposed.

15.16 LIST OF ENERGY-INTENSIVE INDUSTRIES AND OTHER FACILITIES LISTED AS DESIGNATED CUSTOMERS

1. Aluminum
2. Fertilizers
3. Steel and iron
4. Cement
5. Paper and pulp
6. Chlor alkali
7. SugarTextiles
8. Chemicals
9. Railways
10. Trust for the port
11. Industries and services in the transportation sector
12. Petrochemicals, gas crackers, naphtha crackers, and petroleum refineries
13. Electricity transmission and distribution businesses, thermal power plants, and hydel power plants
14. Buildings or institutions that are used for business

15.17 FUTURE RECOMMENDATION FOR GREEN ENERGY

Renewable energy has never been a more hotly debated – or vital – topic than it is now. Many governments, industries, enterprises, and individuals worldwide have begun to recognize the need for more renewable energy. But how should clean energy be defined?

Solar power and wind energy are the two cleanest methods of energy generation, and both are growing in popularity in both residential and national grid applications. These renewable energy sources have a lot of promise for the future, not just in terms of the clean (zero-emissions) energy they provide, but also in terms of what they can do for local economies. Solar panels present a possible long-term solution open to everybody with a rooftop, but installation prices remain extraordinarily high. Hydropower, wind power, and solar power are expected to increase their capacity by 50% over the next four years.

15.18 POTENTIAL SECTOR OF DEVELOPMENT OF ENERGY FROM BIONANOTECHNOLOGY

In more recent times, scientists are getting interested in using biological materials to absorb and store solar power. This could lead to the development of bio-reactors which will produce more energy than current technologies.

Bionanotechnology is a field that has been used in many different industries including medicine, cosmetics, and even cooking! There are many possibilities with respect to applications of nanotechnology and this is just one of them.

15.19 CONCLUSION

India is one of the major producers of renewable power in the world. The country has an estimated solar capacity of 18 GW and wind power capacity of 31 GW. This has helped India to have a share of more than 5% in the global energy mix. With this, India has become a leading nation in developing green energy resources, infrastructures, economies, energy schemes, and sustainability (Renewable Energy Agency).

15.19.1 THE STORY CONTINUES BELOW

India faces a lot of challenges when it comes to renewable resources for power generation. These include:

- A lack of large-scale storage facilities for renewable energy generation
- A lack of necessary infrastructure such as grid transmission lines
- A lack of financial incentives to make the switchover from coal-based fuel sources to renewable sources

BIBLIOGRAPHY

Aggarwal, P. (2017). 2°C target, India's climate action plan and urban transport sector. *Travel Behaviour and Society* 6:110–116.

Bhattacharya, S., Giannakas, K., and Schoengold, K. (2017). Market and welfare effects of renewable portfolio standards 30 in United States electricity markets. *Energy Economics* 64:384–401. https://doi.org/10.1016/j.eneco.2017.03.011.

Blondeel, M., and Van de Graaf, T. (2018). Toward a global coal mining moratorium? A comparative analysis of coal mining policies in the USA, China, India and Australia. *Climatic Change* 150(1–2):89–101.

BP Statistical Review of World Energy (2003, June). https://www.bp.com/en/global/corporate/energy-economics/statistical-review-of-world-energy.html

Canadian Environmental Sustainability Indicators. (2017). *Global Greenhouse Gas Emissions.* http://www.ec.gc.ca/indicateurs-indicators/54C061B5-44F7-4A93-A3EC-5F8B253A7235/GlobalGHGEmissions_EN.pdf. Accessed 27 June 2017.

Central Electricity Authority (CEA). (2016). *National Electricity Plan, Volume 1, Generation.* Ministry of Power, GOI. http://www.cea.nic.in/reports/committee/nep/nep_dec.pdf. Accessed 31 January 2018.

Charles Rajesh Kumar, J., Mary Arunsi, B., Jenova, R., and Majid, M. A. (2019a). Sustainable waste management through waste to energy technologies in India—Opportunities and environmental impacts. *International Journal of Renewable Energy Research* 9(1):309–342.

Charles Rajesh Kumar, J., Vinod Kumar, D., and Majid, M. A. (2019b). Wind energy programme in India: Emerging energy alternatives for sustainable growth. *Energy & Environment* 30(7):1135–1189.

Kumar, S. (2016). CO2 emission reduction potential assessment using renewable energy in India. *Energy* 97:273–282.

Lapedes, D. N. (1976). *McGraw Hill Encyclopedia of Energy.* McGraw Hill Publishing.

Loftness, R. L. (1979). *Energy Handbook.* Von Nostrand Reinhold Company.

National Institution for Transforming India. (2015). *Government of India, Report of the Expert Group on 175 GW RE by 2022.* http://niti.gov.in/writereaddata/files/writeread-data/files/document_publication/report-175-GW-RE.pdf. Accessed 31 December 2016.

National Productivity Council. (2004). *Cleaner Production - Energy Efficiency Manual.* United Nation Publications.

Pappas, D. (2017). Energy and industrial growth in India: The next emissions superpower? *Energy Procedia* 105:3656–3662.

Paris Agreement. (2015). https://unfccc.int/sites/default/files/english_paris_agreement.pdf. Accessed 20 August 2017.

Sholapurkar, R. B., and Mahajan, Y. S. (2015). Review of wind energy development and policy in India. *Energy Technology & Policy* 2:122–132.

Thumann, A., and Mehta, D. P. (2008). *Handbook of Energy Engineering.* The Fairmont Press Inc.

US Department of Energy. (2002, March). *International Energy Outlook.* Energy Information Administration, Office of Integrated Analysis and Forecasting.

Von Zabeltitz, C. (1994). Effective use of renewable energies for greenhouse heating. *Renewable Energy* 5:479–485.

World Energy Council. (2013). *World Energy Scenarios Composing Energy Futures to 2050.* https://www.worldenergy.org/wp-content/uploads/2013/09/World-Energy-Scenarios_Composing-energy-futures-to-2050_Full-report. Accessed 01 January 2017.

www.eia.doe.gov www.epa.org www.bp.com/centres/energy

Index

For Product Safety Concerns and Information please contact our EU
representative GPSR@taylorandfrancis.com
Taylor & Francis Verlag GmbH, Kaufingerstraße 24, 80331 München, Germany

www.ingramcontent.com/pod-product-compliance
Lightning Source LLC
Chambersburg PA
CBHW060815220326
41598CB00022B/2619